# PRINCIPLES OF INSECT MORPHOLOGY

RECENT ADVANCES IN ENTOMOLOGY SERIES—2

# PRINCIPLES OF INSECT MORPHOLOGY

*Edited by*

**A. B. Saxena**

ANMOL PUBLICATIONS PVT LTD

NEW DELHI - 110002 (INDIA)

**ANMOL PUBLICATIONS PVT LTD**
4374/4B, Ansari Road, Daryaganj,
New Delhi - 110002

*Principles of Insect Morphology*


**First Edition, 1996**
Reprinted – 2000

**ISBN 81-7488-347-9**



PRINTED IN INDIA

---

Published by J. L. Kumar for Anmol Publications Pvt Ltd, New Delhi, and Printed at Tarun Offset Press, Delhi.

# Preface

This fascinating title has been carefully organised to meet the long-felt needs of increasingly large number of those who deals with different aspects of Entomology. It provides a balanced and integrated treatment of the entire field of entomology. The title is intelligible to the educated laymen but it deals with some complex ideas. It is an adequate text for all requirements in this area for most university students. This ground work with its many new and unfamiliar names is, at any time, dry, uninteresting and difficult, especially as the beginner does not at first see the necessity for it. We have endeavoured to make it interesting by introducing as much comparative morphology and phylogeny, as possible in the limited space at our disposal.

The primary aim throughout has been clarity so as to make the text selective and meaningful. We hope also that general Entomologists and students of Agricultural Entomology will find in this book valuable and recent informations about the subject and also allied branch of the subject. The book provides a nucleus around which the teacher can plan a successful course without much bearing on complementary and supplementary readings. The theme of the book is novel and for the first time the various chapters in different volumes have been organised reviewing the available literature. All the topics have been illustrated with line diagrams, graphs and tables wherever necessary.

We hope that the book will definitely prove to be a boon to teachers, students and research scholars.

**Author**

# Contents

# 1

# The Insect Body

The body of an insect is elongated and slender shaped. The morphological and anatomical features of each part of the body begins with the features seen on common adult insects, such as grasshoppers, and concludes with a brief discussion of some distinctive modifications of other insects. The application of anatomical terms to different insects requires the determination of *homology*. Recall that this is the structural likeness of anatomical parts due to a common evolutionary origin, i.e., a structure possessed by two or more organisms is homologous because the genes responsible were inherited from a common ancestor. Homologies are established in insects by matching the structures with regard to segmental position, location on the segment, time of embryonic development, size, shape, and the association of other structures whose homologies have been determined. The structures named in this chapter are usually recognizable by simple inspection, yet some insects possess uniquely modified parts whose homology is unknown. Another kind of homology, called *serial homology*, is applied to the same structures on different segments of an individual. For example, each body segment of an arthropod may have a pair of serially homologous, jointed appendages. In insects these appendages are modified into mouthparts, legs, certain types of gills, and external organs for reproduction. The serial homology of structures on different segments is sometimes not immediately apparent; careful study of embryology is then required. Considerable argument, for example, surrounds the nature of the antennae. Are they true appendages, like legs, or special sensory structures peculiar to the head? To date this question has not been resolved.

**Some Important Anatomic Terms**

In order to read the directions intelligently the student should know the meanings of the following terms. Terms of direction and position do not necessarily indicate that the structure to which they are applied lies near or nearer the part suggested by the term. They mean that the structure lies *in the direction* taken by a particle moving from the center of the body toward and continuing beyond, the part in question. Thus "cephalic' relates to the *cephalon* or head but the second segment of an antennae which projects anteriorly beyond the head is cephalic to (or cephalad of) the first segment, although it is farther away from the head.

*Anterior* refers to the forward or head end of the animal. In some insects the anterior end of the body is formed by the mouthparts, in others by the vertex, the mouthparts being directed ventrally or even posteriorly.

***Caudad,*** posteriorly; in the direction of the posterior or "tail" end. *Caudad of* = posterior to.

***Caudal,*** pertaining to the posterior or "tail" end.

***Cephalad,*** anteriorly; in the direction of the head.

***Cephalic,*** pertaining to the head; anterior.

***Distal,*** toward the distal end.

***Distal,*** refers to the free end of a part or appendage, or the end furthest away from the point of attachment.

***Dorsad,*** dorsally; toward the dorsum or back.

***Dorsal,*** pertaining to the dorsum or back.

***Dorsum,*** the entire back region or upper portion of an insect; may also be used for the back of a single segment.

***Ectad,*** toward the outer surface. Used to indicate the relative position of internal structures.

***Ectal,*** pertaining to the outer surface.

***Entad,*** toward the inside, i.e., the long axis.

***Ental,*** pertaining to the inside of the body.

***Laterad,*** toward the sides (right or left).

***Lateral,*** pertaining to the sides.

***Mesad***, toward the meson.

***Mesal***, pertaining to the meson.

***Meson***, the sagittal plane dividing the insect into right and left halves.

***Metamere***, a primary segment, which in the embryo usually contains a pair of coelomic sacs.

***Notum*** = tergum.

***Pleural region***, the lateral or ventrolateral parts of the body or of a segment, where the legs, when present, are found.

***Pleurite***, a sclerotized subdivision of the pleuron; small plates in the membranous pleural region.

***Pleuron***, the lateral sclerotized wall of a segment (usually thoracic), formed by the incorporation of a portion of the limb base into the pleural region.

***Segment***, one of the linear series of ring into which the body is divided. A body segment may be *primary*, in which case the term is synonymous with *metamere*, or it may be *secondary* having its definitive limits different from those of the embryonic segment. Used also for the divisions of an appendage such as the leg or antenna.

***Sternite***, a sclerotized subdivision of the sternum.

***Sternum***, the ventral sclerotized plate or plates of a body segment.

***Tergite***, a sclerotized subdivision of the tergum.

***Tergum***, the dorsal sclerotized plate or plates of a body segment.

***Venter***, the entire undersurface of the body; may be also used for the lower surface of a single segment. Not to be confused with *sternum* which refers to the sclerotized plates of the venter.

***Ventrad***, toward the venter.

***Ventral***, pertaining to the venter.

*Properties of the exoskeleton.* The body wall of an insect is called the *integument*. The outer layer is a noncellular *cuticle*, chemically composed of chitin and proteins that are secreted by the inner layer of *epidermal cells* (Fig. 1.1). In some areas the cuticle is chemically tanned, or sclerotized, by the epidermis to form rigid, platelike *sclerites*. Areas of the cuticle between sclerites remain tough and

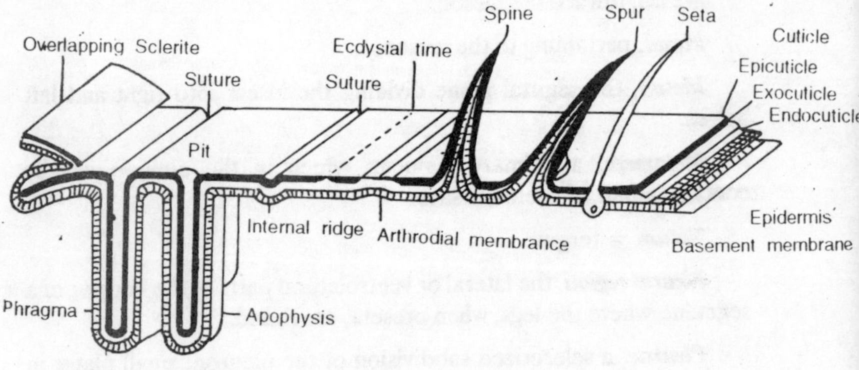


**Fig. 1.1. Diagram of integumental feature.**

flexible, creating membranous joints or articulations. The sclerites and membranes together provide an external support, or *exoskeleton*, for the insect body. Single cells of the epidermis may be modified into hairs, or *setae*. The seta-forming cell is the *trichogen*, and it is accompanied by a socketforming cell, or *tormogen*. Setae are greatly varied in both form and function. The common types are hairlike or *simple* setae, featherlike or *plumose* secae, and platelike *scales. Poison setae* are hollow and filled with toxic fluids for defense. Setae of diverse shapes are innervated by the peripheral nervous system and function as sensory organs called *sensilla*.

Many cells are involved in local outgrowths or ingrowths of the integument. Rigid outgrowths are commonly called *spines*. Some outgrowths, especially on the legs, are movable and called *spurs*. Rigid fingerlike ingrowths of the integument associated with muscle attachments are commonly called *apodemes*, or if larger and armlike, *apophyses*. The latter term is sometimes used for external protuberances, but in this text we restrict it to invaginations. Large apophyses in the head and thorax buttress the exoskeleton and provide areas for muscle attachment. These apophyses function as the *endoskeleton*. Platelike invaginations associated with the dorsal flight muscles are *phragmata*.

The integument may also be infolded linearly to produce internal ridges for strengthening the body wall or for muscle attachment. The internal ridge is usually traced externally by a line or groove. Virtually any line in an insect is loosely called a *suture*, regardless of whether it is a line of flexibility (e.g., an articulation) or a line marking an internal ridge, or phragma. Sutures are used as convenient boundaries to delimit areas of the exoskeleton. In the strict sense, a suture is a seam along which two plates have fused. The lines between plates of the vertebrate skull are properly called sutures. The cuticle of the insect integument, however, is a continuous layer; adjacent sclerites usually become unified in the course of evolution without leaving a trace.

## TAGMATA, SEGMENTS, SCLERITES AND SUTURES

### Tagmata

*Melanoplus sp. (grasshopper)*

The body of an insect is made up of three distinct regions or *tagmata*. The anterior tagma is the *head*, and it bears the eyes, the antennae, the jaws and other mouthparts. The middle tagma is the *thorax* which bears three pairs of legs and two pairs of wings. The posterior tagma, the *abdomen*, bears no appendages apart from the genital apparatus and a pair of minute processes *(cerci)* near the free end.

### Segments

The trunk is distinctly divided into a number of segments or *metameres*, separated from each other by intersegmental grooves. The three anterior segments, each with a pair of legs, constitute the thorax. The most anterior segment is the *prothorax*, the middle one the *mesothorax* and the posterior one the *metathorax*. The remaining segments constitute the abdomen. The outer covering of an insect is known as the *cuticle*. Note that the cuticle on the thoracic segments is for the most part hard with considerable rigidity. Such cuticle is said to be "sclerotized." That of the abdomen, on the other hand, is for the most part soft and flexible. It is said to be *"membranous"* or *"unsclerotized."*

Note that the lateral regions of each abdominal segment in *Corydalis*, except the last, are expanded to form large tubercles which bear a tapering filament or *stylus* laterally and *gills* ventrally. These tubercles are regarded by some authors as the vestiges of abdominal limbs. This is doubtful although they do occupy the same positions as would the limb bases if these were present. They divide the segment into four regions: (a) a *dorsal region*, bearing a spiracle on each of its lateral margins, just dorsal to the "limb base"; (b) a *ventral region* between the "limb bases" on the ventral side, (c) two *pleural regions* occupied almost completely by the "limb bases." The abdominal segments are distinct, but differ from those of *Corydalis* in that they are sclerotized. Each sclerotized abdominal plate is separated from that of the next segment by an unsclerotized intersegmental membrane or *conjunctiva*. The prothorax, which bears legs but not wings, is distinct and freely movable on the mesothorax. The mesothorax and metathorax, however, are closely united with no intersegmental membrane between them. This close association is always found in insects with functional wings. Because it bears the wings the united structure is sometimes referred to as the *pterothorax*. The head also consists of several segments but these are so completely united that their limits cannot be determined.

### Sclerites and Sutures

Examine the lateral sides of the wing-bearing region of the thorax and note that each consists of several hardened or *sclerotized* plates separated from each other by impressed lines or grooves. These grooves mark inflections or infoldings of the integument and their position is marked on the inner surface by a corresponding ridge. The grooves are known as *sutures* and the hardened plates which they define are *sclerites*. The term *suture* (literally, a seam) is used in entomology to designate any line which defines the limits of two united or closely adjacent sclerites, but the structures so designated may differ morphologically. Grooves such as those in the thorax of *Melanoplus* are distinguished as *sulci* (sing. *sulcus*). Occasionally sclerites are separated by a very narrow strip of unsclerotized cuticle having the appearance of a pale line. These may be called line sutures or *lineae*.

The cuticle of insects and all other arthropods contains a horny substance known as "chitin," and because of a mistaken idea that the hardening of the cuticle was associated with heavy deposition of chitin, the hard parts were said to be "chitinized." This terms which occurs widely in the early literature is now displaced by the more accurate term *"sclerotized."* Insects are very active animals and require a skeleton which is at once light and rigid. The hard exocuticle is relatively thin, light and—except in some forms such as the beetles—more or less flexible. Rigidity is provided by the development of internal struts and ridges at points where much muscular force is exerted, such as the articulations of the appendages. The ridges are formed by inflection or infolding of the integument, therefore they are marked externally by grooves or sulci. Consequently, most of the sutures mark the positions of functional strengthening ridges and have no relation to segmentation. Nevertheless, these sutures and the sclerites which they limit are fundamental features in the topographical anatomy of insects and much of external anatomy is concerned with their presence and absence. The more constant sutures and sclerites bear distinctive names.

## Evolution of the Body Regions

The organization of an insect's body is more readily understood if we recall that insects and other arthropods evolved from the segmented worms, or Annelida. Snodgrass (1935) proposed a sequence of steps leading to the insects that is instructive in this

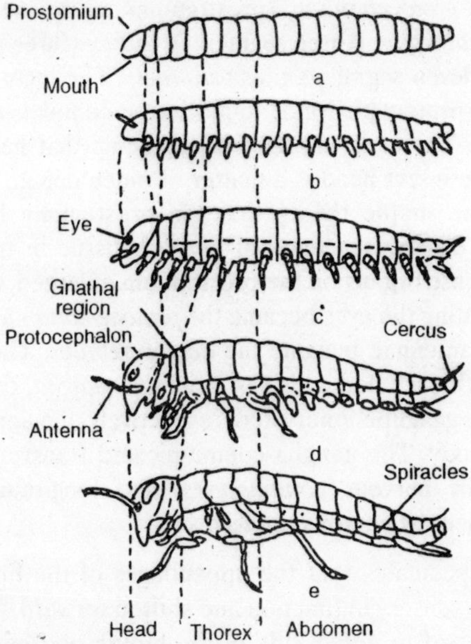


**Fig 1.2. Diagrams of the steps in the hypothetical origin of the insect body from a wormlike ancestor. See text for explanation.**

context. In the earliest stage the wormlike ancestor has a cylindrical body with the tubular intestine running nearly the full length. Between the mouth and anus the body is divided transversely into segments. In front of the mouth is an unsegmented part called the *prostomium*. The last part of the body encircles the anus and is the *periproct*. This is the endpiece and also is not a true segment. In a second hypothetical stage, the prostomium acquires a pair of light-sensitive receptors, or *eyes*, and a pair of sensory antennae. Each true body segment develops a pair of movable, ventrolateral appendages. In the third stage the ancestor is an arthropod with jointed legs and enlarged sensory apparatus. Evidently it has become terrestrial, because most of the segments have a pair of openings, called *spiracles*, leading to the internal air tubes, or *tracheae*, of the respiratory system. The appendages of the last true segment are called *cerci*. The features of the insects are foreshadowed in the fourth stage as regions of the body become specialized. The modification of segments into functional units is called *tagmosis*. At this stage, four such units, or *tagmata*, can be recognized: *protocephalon* (prostomium plus one to three segments), *gnathal region* (three segments), *thorax* (three segments,), and *abdomen* (eleven segments plus *periproct)*. The factual basis for the hypothetical protocephalon is found in the complex structure of the insect brain. The exact number of segments that have become fused to form the insect head is a matter of much debate. Snodgrass has defended the simple theory that the prostomium housed the primitive brain. This was a mass of nervous tissue in front of the mouth. As the sense organs on the prostomium enlarged, that part of the brain innervating the eyes became the *protocerebrum* and the part innervating the antennae became the *deutocerebrum*. The segments behind the mouth each had a pair of primitive *ganglia*, or masses of nerve cells. Each ganglion innervated, respectively, the appendage on its side of the body. The ganglia communicated transversely within each segment by nervous *commissures*, and longitudinally from segment to segment by nervous *connectives*.

Snodgrass speculates that the appendages of the first postoral segment became sensory in function and shifted forward. The ganglia also moved forward to either side of the mouth and became more closely associated with the brain. In the insects, the appendages atrophied and were lost in the course of evolution, yet the ganglia of the first segment still persist in insects as the third part of the brain, or

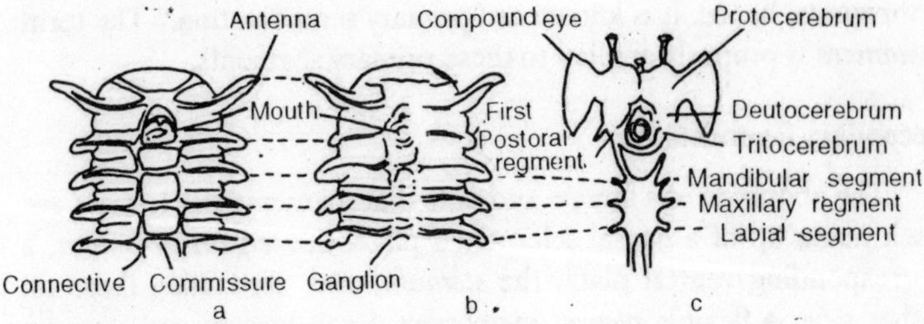


**Fig. 1.3.** Diagrams of the steps in the hypothetical origin of the insect brain and head segmentations. See text for explanation.

*tritocerebrum*. The tritocerebral commissure is evidence of its former segmental position behind the mouth. In Crustacea the appendages of this segment still persist as the second pair of antennae. The appendages of the next three segments are specialized for feeding. This is the gnathal region. The three segments following are modified for walking and became the thorax. The last of the four tagmata, the abdomen, contains no more than 11 true segments plus the periproct. Most of the visceral organs are located here. With the walking appendages concentrated in the thorax, the appendages of the abdomen became modified for other functions or atrophied. In the final stage, the body regions have been reduced to three by the tagmosis of the protocephalon and gnathal region to form the *head* tagma. The head combines the sense organs and brain with the gnathal appendages to form a center for sensing the environment and ingesting food. The thorax is the center for locomotion and houses the large muscles that attach to the leg bases, as well as others that act on the thoracic skeleton to move the wings. The abdomen is the center for the reproductive organs and intestines. By pumping movements, the abdomen also aids airflow in the tracheal system.

## Primary Segmentation

Each segment is separated from its neighbours by a simple annular groove. The principal longitudinal segmental muscles consist of continuous bands which are inserted on the internal surface of the intersegmental line. The muscle segmentation is therefore coincident with the body-wall segmentation. This is a primitive condition, essentially similar to that found in annelids and in the embryos of all arthropods, hence, it is known as "primary segmentation." The term *metamere* is properly applied to these primary segments.

**Secondary Segmentation**

The abdomen of a female and note that segments one to eight are each made up of a dorsal sclerotized plate, the *tergum* or *notum*, a corresponding ventral plate, the *sternum*, and, separating these on either side, a flexible *pleural membrane*. Each tergum and sternum overlaps the one behind and is separated from it by the flexible *conjuctiva* or intersegmental membrane. The abdomen may then be said to be made up of a number of segments, sclerotized dorsally and ventrally, and separated from each other by intersegmental membranes. This condition is typical of insects with a sclerotized cuticle. The limits of the sclerotized segments are not the same as those of the primitive segments; therefore, the former are known as *secondary segments*.

If we cut longitudinally through the pleural membranes and across the anterior end of the dorsum and remove the entire dorsal wall of the abdomen. Then observe it under the binocular dissecting microscope. The surface will be found to be partly covered with irregularly arranged, opaque white tissue. This is a portion of the fat body and should be removed carefully with fine-pointed forceps. The body wall will be left, consisting of the integument and the longitudinal muscles. The latter are semitransparent and may be seen more clearly if the specimen is immersed for a few seconds in methylene blue solution, or other suitable stain. Examine the longitudinal muscles nearest the mid-dorsal line of the anterior segments and note that they originate at the anterior edge of the segment, run posteriorly through the segment and across the intersegmental membrane to be inserted in the anterior edge of the segment behind. The lines along which the muscles are attached to the integument represent the *primitive intersegmental* lines such as are found in the caterpillar, but in the cricket and other sclerotized insects the intersegmental membranes lie in front of these lines. In fact, they are posterior bands of the primary segments.

This arrangement, by means of which the muscles can exert a definite pull on the sclerites, allows for a certain amount of specificity of movement as compared with the very generalized types of movement characteristic of worms and soft-bodied insects. The movements of the anterior segments of the abdomen in the cricket are, however, still relatively generalized, consisting of telescoping, bending and twisting, and there are no marked specializations in the areas to which the muscles are attached. The posterior segments, however, make more specific movements in relation to the movements of the ovipositor or of the copulatory organs of the male, and the anterior edges to which the muscles are attached are modified.

The anterior edges of the apparent three last segments (8, 9 and 10) and they are folded in to form strong sclerotized ridges for the attachment of the muscles. These ridges are known as the *antecostae* and the corresponding groove or sulcus on the external surface in the *antecostal suture*. In front of the antecosta a very narrow sclrotic rim may be observed. This is the *acrotergite*, which in some insects forms a distinct sclerite. It is particularly well developed in certain of the thoracic segments and attention will be drawn to it later. Note that the antecosta of the tenth segments is produced into two flattened processes or *apodemes* for the attachment of certain of the muscles of the cercus. This also is a common feature, especially in the thorax. The *sterna* of the abdominal segments have a similar structure to that of the terga. When an anterior sclerite is developed in front of the antecosta it is known as an *acrosternite*.

**Structure of a Typical Secondary Segment**

A typical segment of an adult arthropod is in general one of the secondary annular sections of the body defined by the lengthwise extent of its dorsal and ventral plates. A considerable part of the true segmental area, however, is formed by the posterior conjunctival membrane, which is usually infolded and more or less concealed within the posterior ends of the tergal and sternal plates. The wall of each segment, as we have seen, contains typically a dorsal sclerotic area, or tergum, and a ventral sclerotic area, or sternum, while the lateral or pleural walls may contain each one or more pleural sclerites.

*The Tergum.*

In its typical form the back plate of a secondary segment includes the sclerotization of the dorsum of the primary intersegmental area preceding. The primitive intersegmental fold, therefore, becomes a submarginal ridge near the anterior edge of the inner surface of the tergum. This ridge is the *antecosta* of the tergum. The corresponding external groove, or transverse line of inflection forming the antecosta, is the *antecostal suture*. The narrow precostal lip of the tergum in the *acrotergite*. The postcostal tergal sclerotization usually forms a simple plate, which is subject to modifications in various ways, especially in the wing-bearing segments of pterygote insects, but it may be broken up into smaller sclerites, as in many holometabolous larvae. The dorsal longitudinal muscles, primarily attached on the intersegmental folds usually retain their attachments on the antecostae in the secondary segments, though some o" all of their fibers may migrate to the precostal or postcostal areas of the tergum.

Tergal plates are present in the adult stages of nearly all arthropods. Their principal variation in size with relation to the shape of the segment is in a transverse direction. They may be limited to the median part of the back, or their lateral areas may be extended downward a varying distance in the lateral walls of the dorsum, and the lateral margins may project as free folds either horizontally extended or ventrally deflected over the sides of the segment, sometimes concealing the bases of the legs. Frequently laterodorsal sclerites are distinct from a principal median tergite. In the thoracic segments of wined insects the typical structure of the dorsal plates is generally obscured by a modification in the intersegmental sclerotization, correlated with the development of the wings as efficient organs of flight. The intertergal parts of the conjunctive between the mesothorax and metathorax and between the metathorax and first abdominal segment are almost obliterated by an anterior extension of the acrotergites of the matathoracic and first abdominal terga (Fig. 1.4B). Furthermore a secondary membranous suture may appear in the tergal region just behind the antecosta of each of these segments. There is thus formed posterior to each wing-bearing tergal plate a *postnotal plate* bearing the intersegmental attachments of the dorsal muscles. Since the dorsal muscles are greatly enlarged in the thorax, the antecostae develop large lobes, the *phragmatc*, for their accommodation. The postnotal plates are usually regarded as belonging to the segment preceding in each case, but it is clear that they are intersegmental structures analogous to the intersternites of the venter between the prothorax and mesothorax and between the mesothorax and metathorax, which are usually more closely associated with the preceding sterna.

## *The Sternum*

Sternal plates are not so constant a feature of the arthropod skeleton as are the tergal plates. They may be present or absent within the same major group, and, where present, they are often highly variable in both shape and size between closely related forms, and even in different body regions of the same species.

The ventral plate of a segment usually have the same structure as the tergum, as in the abdomen of insects, where the sternum generally is an inverted replica of the tergum. In this case each sternum bears anteriorly a submarginal antecosta on its inner surface, on which the ventral longitudinal muscles are usually attached, and presents a narrow precostal lip, or *acrosternite*, corresponding to the acrotergite of the dorsal plate. In the thorax of many insects, however, the intersegmental sclerotizations of the venter form small independent plates, or intersternites, as they do in the whole length of the body in most of the chilopods. The intersternites of the thorax of insects,

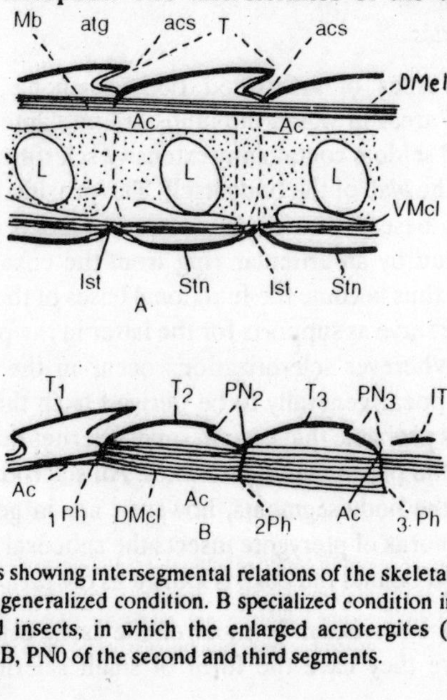


**Fig 1.4.** Diagrams showing intersegmental relations of the skeletal plates in secondary segmentation. A generalized condition. B specialized condition in the dorsum of the thorax of winged insects, in which the enlarged acrotergites (A *atg*) become the postnotal plates (B, PN0 of the second and third segments.

which may occur between the prothorax and mesothorax and between the latter and the metathorax, are known as *spinasterna* because each usually bears an internal spine-like process giving attachment to some of the ventral muscles. The spinasterna are often united with the segmental sterna preceding them, but never with those following. A definitive thoracic sternum, therefore, never has a true antecostal. Most of the ventral muscles of the thorax are attached (probably secondarily) on apodemal processes of the segmental plates; a few fibers, however, usually retain the primitive connections with the median processes, or *spinae*, of the intersegmental spinasterna.

The definitive sterna of insects are usually composite plates, each comprising a median region representing the area of the primitive segmental sternum and lateral parts derived from the limb bases. In the thorax the adjoined lateral elements are the ventral ares of the subcoxal parts of the leg bases; in the abdomen the sternal plates appear to contain, in most cases, the entire basal parts of the otherwise suppressed limbs. The term "sternum" is usually applied to the principal ventral plate of a segment regardless of the real or theoretical composition of the lattter.

### *The Pleural Sclerites*

The podial areas of arthropod body segments, that is, the so-called pleural areas in which the limbs are implanted, are usually membranous and seldom contain any extensive sclerotization that can be attributed to the wall of the body itself. The proximal, or subcoxal, parts of the limb bases, however, are often expanded in the pleural wall and separated by an articular ring from the coxal parts of the limbs. The coxae thus become the functional bases of the appendages, and the subcoxae serve as supports for the latter in the podial areas of the body wall. Wherever sclerotizations occur in the podial areas, therefore, they appear generally to be derived from the bases of the limbs, though it is probable that certain small sclerites may belong also to the region of the peripodial membranes. All sclerotizations of the podial areas of the body segments, however, are in general termed *pelurites*. In the thorax of pterygote insects the subcoxal sclerotization above, before, and behind the coxa is known as the *pleuron*.

Subcoxal pleurites occur in most of the body segments of the Chilopoda, where they have the form of small sclerites of various shapes more or less closely associated with the bases of the coxae. Similar sclerites are present in the pleural areas of the thorax in apterygote Hexapoda, but here they frequently appear as two crescentic arches over the bases of the coxae. In adult pterygote insects the thoracic pleural plates are much enlarged, especially in the wing-bearing segments, where they form supports for the wings as well as giving articulation to the coxae of the legs. In the decapod Crustacea the inner walls of the gill chambers are evidently of subcoxal origin, but in most of the Crustacea and in the Arachnida there is little evidence of the presence of pleurites derived from the limb bases. In the abdominal segments of insects the limb bases are sometimes represented by distinct plates occupying the pleural areas between the terga and sterna, as in certain Thysanura and in many larval forms; in the genital segments they are retained as the basal plates of the gonopods. In general, however, the abdominal pleurites appear to be fused with the primitive sterna in continuous plates, which are the definitive sterna.

**Intersegmental Relations**

The primary intersegmental grooves, we have seen, are the functional segmental limits only soft-bodied arthropods or in forms with but a weak or partial sclerotization of the integument. In all arthropods with well-developed body-wall plates, the definitive segmentation is a secondary one; but the limits of the secondary segments differ according to the relations of the sclerotization in the primary intersegmental regions to that of the segmental regions before and behind them.

In the insect abdomen, where both the dorsal and the ventral primary intersegmental areas of sclerotization are continuous with the segmental plates following, a typical secondary segmentation prevail, and the functional intersegmental rings are the membranous posterior parts of the primary segments. The primitive intersegmental fold (*Isg*) forms an antecosta on both the tergum and the sternum, and the precostal lip forms an acrotergite of the dorsal plate and an acrosternite of the ventral plate. Wherever, there is a difference between the dorsum and the venter in the sclerotization of the intersegmental region, however, the intersegmental relations are less simple. If the dorsal plates retain the form typical of secondary segmentation, while the ventral sclerotization takes the form of independent sternites and intersternites, as often occurs in the thorax, the ventral half of each intersegmental membrane will embrace the intersternite and will include a part of the two adjoining primary segmental regions. Again, if the intersternite is united with the segmental sternum preceding, while the intertergal sclerotization remains continuous wit the tergum following, the conjunctival membrane will cross obliquely on the side of the body from the posterior part of the anterior segment to the anterior part of the posterior segment, and the ventral postcostal lip becomes a *posternite* of the anterior segment. Finally, if both the dorsal and ventral intersegmental sclerotizations are united or closely associated with the segmental plates preceding, the functional intersegmental membrane becomes the anterior part of the posterior segment and the post costal lip forms a *posttergite* dorsally and a poststernite ventrally.

## Tagmosis

In all adult arthropods some of the segments are more or less united into groups forming distinct trunk sections, or *tagmata*. The number of tagmata, and the number of segments in each tagma vary in different arthropods. The most constant tagmosis of the trunk is that which differentiates, in the embryo, the protocephalic head region from the primitive body. The definitive head of most mandibulate arthropods,however, contains a second tagma, which is that of the gnathal segments. In the Hexapoda a third embryonic tagma becomes the thorax, or second tagma of the adult, which is usually composed of three segments, though in most Hymenoptera it contains four. A fourth embryonic tagma becomes the adult hexapod abdomen, which has at most 12 segments, including the periproct. Tagmosis is more

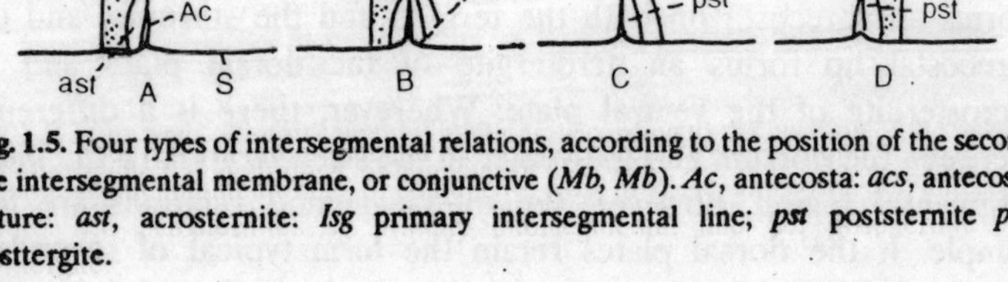


**Fig. 1.5.** Four types of intersegmental relations, according to the position of the second are intersegmental membrane, or conjunctive (*Mb, Mb*). *Ac*, antecosta: *acs*, antecosta suture: *ast*, acrosternite: *Isg* primary intersegmental line; *pst* poststernite *ptg*, posttergite.

variable in the Crustacea; in the Chilopoda and Diplopoda it results only in the formation of a head, including the gnathal segments, and a body; in the Chelicerata again it is variable, but the principal division of the trunk is into a "cephalothorax" and an "abdomen."

## The Head

The anterior most part of the body is the head or cephalic region and its differentiation is called the *cephalization*. It bears a pair of large *compound* eyes on the dorsolateral regions, the many jointed *antennae* anterior to the eyes, and the *mouthparts* on the ventral edge of the head. A head like that of the locust, having the mouthparts directed ventrally, is said to be *"hypognathous."* When, as in most beetles, the mouthparts are directed anteriorly the head is said to be *"prognathous."* In relatively few insects, such as many of the Hemiptera, the head is reflexed ventrally to such an extent that the mouthparts are directed posteriorly. Such a head is said to be *"opisthognathous."* Note that the antennae and mouthparts are movably attached to a strongly sclerotized "box" which forms the skeleton of the head and is comparable to (though in no sense homologous with) the skull or cranium of the vertebrates, and the same term, *"cranium"*, used for it.

The most ventral sclerite on the anterior part of head is a bilobate flap known as the *labrum*, it is movably hinged to the cranium by a membranous *clypeolabral suture*. The labrum is functionally a sort of upper lip and regarded as one of the mouthparts, but morphologically it is probably an integral part of the head wall and not an appendage. The *clypeus* lies dorsal to the labrum. It is a rectangular, transverse sclerite bounded dorsally by a sulcus, the *epistomal* or *frontoclypeal suture*. The median area dorsal to the clypeus is known as the *frons* and the lateral areas ventral to the compound eyes as the *genae*. The top of the head, dorsal to the frons and genae is the *vertex*. The right and left halves of the vertex together with the corresponding genae constitute the *parietals*. Usage differs as to the precise limits of these regions, as will be shown later.

In some species there is a pale Y-shaped line in the vertex. This is known as the *ecdysial suture*, and is the line along which the cuticle of the nymph splits at ecdysis. It is usually absent or weakly developed in adult insects. The stem of the suture lies in the mid-dorsal line of the top of the head and is known as the *coronal suture*. The arms, which diverge and extend toward the eyes, form the V-shaped *frontal suture*. Pull the head gently away from the thorax and note that these two tagmata are connected by the membranous *cervix* (*cervical membrane* or *neck*) which, because of its flexibility, allows free movement of the head.

The posterior *occipital region* the cranium consists of an incomplete ring of sclerotized tissue forming lateral and dorsal walls surrounding a large central opening known as the *foramen magnum* or *occipital foramen*. The foramen is closed ventrally, not by the walls of the cranium, but by the base of the labium or lower lip. Most of the easily visible portion of the occipital region consists of a somewhat horseshoe-shaped sclerite. The dorsal portion of this sclerite is very narrow and is known as the *occiput*. The two ventrolateral portions expand to form triangular sclerites known as *postgenae*. The three regions are not separated by sutures. The large *compound eyes* have already been observed. In addition three simple eyes or *ocelli*, one in the middle of the face between the bases of the antennae, and one in the vertical ridge mesal to each compound eye, just anterior to the upper edge of the latter are present.

The *mouthparts* are mandibulate, i.e., adapted for chewing. They consist of two lips and two pairs of jaws. The *labrum* or upper lip has already been observed. Remove this and find a pair of strongly sclerotized unsegmented *mandibles* immediately beneath them. Posterior to the mandibles is a pair of segmented jaws, the *maxillae*, each bearing a 5-segmented, antenna-like *palpus*. The *labium* or lower lip is also segmented and bears a pair of 3-segmented *palpi*. There is a blunt, tongue-like *hypopharynx* lying between the bases of the jaws. The hypopharynx, like the labrum, is formed from part of the original head wall. The mandibles, maxillae and labium are segmental appendages serially homologous with the legs.

## The Thorax

Next to head is the thorax which is made up of the three segments. The most anterior segment is known as the *prothorax*, the middle one as the *mesothorax* and the hindmost as the *metathorax*.

The prothorax is quite large, more or less independent of the other segments, and is movable on them. The large shield which covers the dorsal and lateral sides of the prothorax is the *pronotum*. In the grasshoppers this is greatly enlarged and not only does it expand backward, covering part or all of the two succeeding terga, but it also extends laterally, almost completely covering the two pleura or lateral walls of the segment. The pleura are greatly reduced, visible externally only as minute triangular sclerites ventral to the anteroventral edges of the tergum and just in front of the bases of the forelegs. This condition is specialized and not typical.

The *prosternum* which lies ventrally between the forelegs and not that it consists of an anterior *basisternite* extending laterally in front of the legs to fuse with the pleura, and a posterior *sternellum* separated from the basisternite by a suture. The basisternite bears a distinct spine, and the sternellum a longitudinal pit. The lateral parts of the intersegmental membrane a pair of *mesothoracic spiracles* having the form of a small vertical slit between two minute sclerites.

When the prothorax is removed the mesothorax and metathorax are exposed. Observe that these two are closely united and not easily movable on each other. Because they form a structural wing-bearing unit these two segments are often referred to collectively as the *pterothorax*. The structure is essentially the same in both segments. Only the larger and more obvious sclerites will be referred to at this time. Note that the terga, pleura and sterna are all conspicuously developed. Spread the wings laterally at right angles to the body and note that they are attached to the thorax by a flexible *articular membrane* between the terga and pleura.

In the *mesonotum* the following sclerites have been observed. (1) The *presscutum* is a narrow anterior sclerite with expanded lateral areas. These lateral areas are evident as triangular plates at the anterolateral angles of the tergum, immediately anterior to the wing bases. This sclerite is usually narrow and inconspicuous, but in some insects it is of considerable width. (2) The *scutum* lies posterior to the prescutum and is, in this insect, as in most others, the largest and most prominent sclerite of the tergum. (3) The *scutellum* is the posterior transverse sclerite overlapping the metatergum. The small prominence in the middle of the hinder portion of the tergum is also part of the scutellum.

There are three vertical sutures which divide the *pleura* of the pterothorax into a series of four sclerites. The middle suture, which passes just behind the second pair of legs is intersegmental, separating the anterior *mesopleuron* from the posterior *metapleuron*. The two pleura have essentially the same structure and either may be studied. Each is composed principally of two plates, the *epistrnum* in front and the *epimeron* behind. The former, in some species, is transversely divided into two secondary sclerites. The suture separating the epoisternum from the epimeron is the *pleural suture*. The *metathoracic spiracle* lies in the intersegmental area, dorsal to the base of the middle leg. The *sternum* in the mesothorax and metathorax consists apparently of a single large bilobate scleite in each segment. In the mesosternum, however, you will find a pair of sutures which partially separate the two lobes from the anterior transverse plate. The latter is the *basisternite* while the lobes form the *sternellum*. In the inner angles between the lobes and the hind margin of the basisternite, look for a pair of small pits. These are the *furcal pits* which mark the point of invagination of a pair of internal skeletal rods known as the *furcal arms* or *sternal apophyses*. MIdway between the furcal pits is a *spinal pit*, marking the position of the *spina*, a third internal rod. The position of the furcal pits typically marks the division between basisternite and sternellum.

Whenever, the nature of the articulation of legs with the thorax is studied, it is seen that it pivots against the thorax at the ventral end of the pleural suture. This suture marks he position of a strong inflection or internal ridge known as the *pleural ridge*, and a ventral extension of this ridge provides an articulatory condyle for the leg. Identify the following leg segments: (1) the *coxa*, a short, stout, basal segment: (2) the *trochanter*, a very small segment, movable on the coxa and closely united with; (3) the *femur*, the largest segment of the leg; (4) the *tibia*, a long, slender segments; (5) the *tarsus*, made up of three subsegments; (6) the *pretarsus*, consisting chiefly of a pair of strong curved *claws,* and between them a somewhat concave disc, the *arolium*.

**The Abdomen**

The male and female differ in respect to structures at the posterior end of the abdomen. The female can be distinguished by the presence of the ovipositor, consisting of two pairs of short, hard processes with sharp curved ends. There is a clear cut distinction between the arrangement of the segmental *terga* and *sterna* the abdomen male and female. In female there is a narrow *pleural membrane* which is infleced in such a way that the tergum overlaps the sternum. Note also that each segment overlaps the one behind, and by bending the abdomen dorsally and ventrally, find the *intersegmental membranes* which allow for free movements of this region.

There are the eleven abdominal terga. The first bears the *sound-receiving* or *tympanal organs*, two lateral light-coloured, somewhat bean-shaped taut membranes. The *first abdominal spiracles* are to be found at the anterior border of the tympanal membranes. The next seven segments also bear a pair of spiracles each. Look for these near the outer margins of the terga, just above the pleural membranes.

The first eight terga are essentially alike. The ninth and tenth differ in being much narrower, in the absence of spiracles, which are never found behind the eighth segment in insects, and in being partially fused with each other. The eleventh tergum has the form of a convex, diamond-shaped plate. It is known as the *epiproct* because of its position above the anus. It partly overlies laterally a pair of triangular sclerites, the *paraprocts* which are also part of the eleventh segment. Lateral to the proximal border of the epiproct there is a pair of small tapering processes. These are the *cerci* which represent appendages of the eleventh segment, serially homologous with the legs and mouthparts. Find the structures of the eleventh segment enumerated above; with the point of a needle raise the epiproct dorsally and note its relations to the paraprocts. Find the *anus* which opens mesally between the bases of the epiproct and paraprocts.

On the ventral side you will find only eight *sterna*, those of segments 9, 10 and 11 being reduced, displaced or absent. The first sternum is separated from its tergum by the leg cavity. The seven others are closely associated with their terga. The eighth sternum which is somewhat elongated is called the *subgential* plate because it lies ventral to the genital opening and the base of the ovipositor. On its posterior border it bears a small up-curved spine, the egg guide, which passes between the ventral plates of the ovipositor. Note the *dorsal* and *ventral valves* of the *ovipositor*, two divergent pairs of stout, curved, pointed appendages near the posterior end of the abdomen.

Whenever, the abdomen of a male insect is observed it is seen that it resembles with that of the female except for the parts of the venter posterior to the seventh sternum. The eight sternum is similar to those anterior to it. The ninth sternum, sometimes called the *hypandrium* because it lies beneath the male genitalia, is enlarged and divided transversely into two parts so that the venter appears to have ten visible sterna. The posterior portion of the ninth sternum probably represents the fused bases of the appendages which this segment originally bore.

# 2

# The Head

Head is the anterior most part of the body bearing the major sense organs and the mouthparts. Proper differentiation of head from the rest of the body occurs in the group Arthropods. It is called the *cephalization*. Considerable controversy still surrounds the problem of segmentation of the insect head, especially concerning the number and nature of segments anterior to the mouth. At various times it has been argued that there are from three to seven segments in the insect head although it is generally agreed that there are three pastoral (gnathal) segments. The *Mandibular, Maxillary* and *Labial*. It is not feasible to discuss here the many theories concerning the segmental composition of the insect head, and the reader is referred to the work of Rempel (1975) for a recent review of the subject. The major points of contention appear to have been (1) whether a preantennal segment occurs between the acron and the antennal segment and what appendages are associated with such a segment, and (2) whether the antennae are segmental appendages or merely outgrowths of the acron. Recent embryological studies (for example, those of Rempel and Church, 1971) have demonstrated convincingly that the head comprises the *Acron* (which is non-segmental and homologous with the annelid prostomium) three preoral segments, and the three gnathal segments. The first preoral segment is the *Labral* segment whose appendages fuse to form the labrum. The second preoral (*Antennal*) segment bears the antennae, which are therefore true segmental appendages. The appendages of the third preoral (*Intercalary*) segment appear briefly during embryogenesis, then are lost.

## GENERAL CONSIDERATIONS

### Orientation

As stated already that the head is said to be *"hypognathous"* when the mouthparts are borne on its ventral margin and are therefore directed ventrally. This is the primitive orientation and in the following pages, unless otherwise stated, reference to position or direction of head structures relates to a head of this type. In many insects the head is orientated in such a way that the mouthparts are directed anteriorly. Such a head is said to be *"prognathous."* In a few insects the mouthparts are directed posteriorly on the ventral side and the head is then said to be *"opisthognathous."*

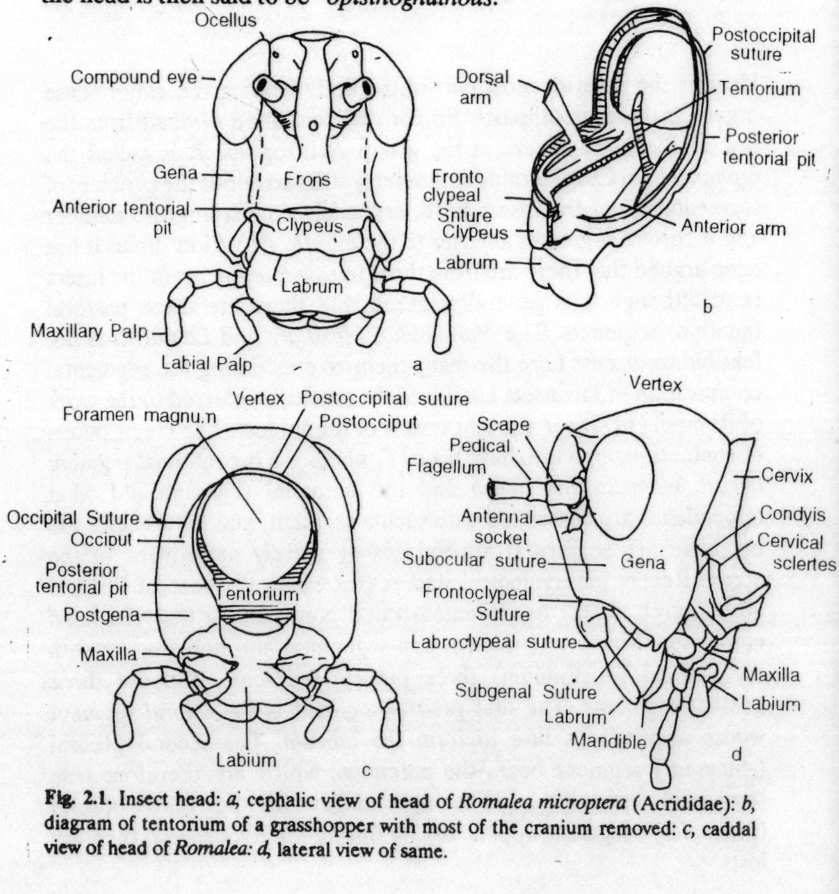


**Fig. 2.1.** Insect head: *a*, cephalic view of head of *Romalea microptera* (Acrididae): *b*, diagram of tentorium of a grasshopper with most of the cranium removed: *c*, caddal view of head of *Romalea: d*, lateral view of same.

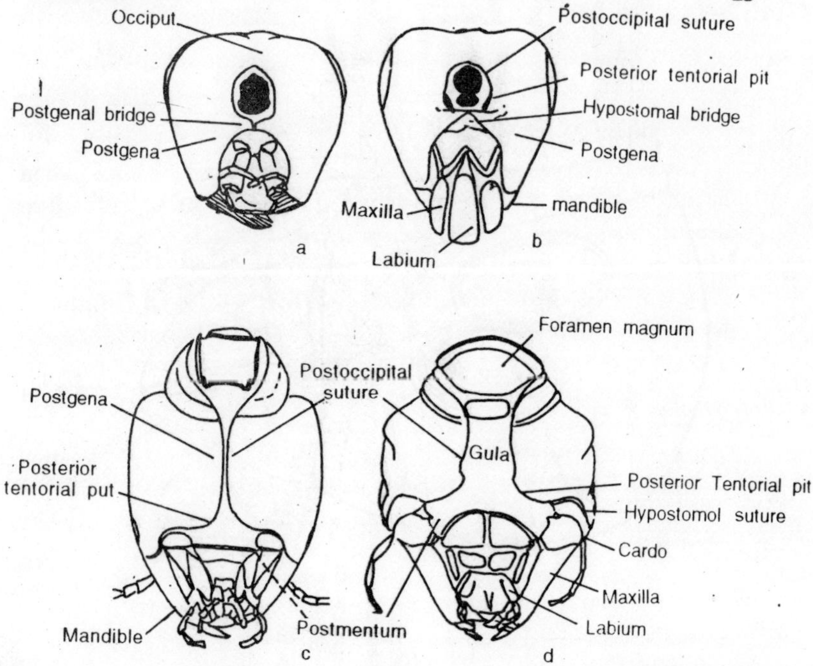


**Fig. 2.2.** Modifications of the posteroventral region of the head: *a*, postgenal bridge of a wasp, Vespa spp. (Vespidae): *b*, hypostomal bridge of the honeybee, *Apis mellifera* (Apidae): *c*. narrow gula of a beetle, *Staphylinus spp.* (Staphylinidae): *d*, broad gula of the dobsonfly larva *Cordydalus spp.* (Corydalidae).

## The Generalized Pterygote Cranium

The cranium is the sclerotized skeleton of the head to which the antennae and mouthparts are attached. Typically and *occipital suture*, which runs from a point immediately in front of the posterior articulation of the mandible along the sides and top of the cranium to the corresponding point on the other side, divides the head into an anterior *facial* or *frontoparietal region* and a posterior *occipital region*. Ferris believes this suture to be the intersegmental line between the mandibular segment and an oculoantennal segment but there is no confirmation of this from embryological studies. Its close association with the mandibular articulation suggests that the inflection which it marks maybe a functional strengthening ridge. Nevertheless it is convenient to study the two regions separately in order to trace consecutively the changes which take place in each.

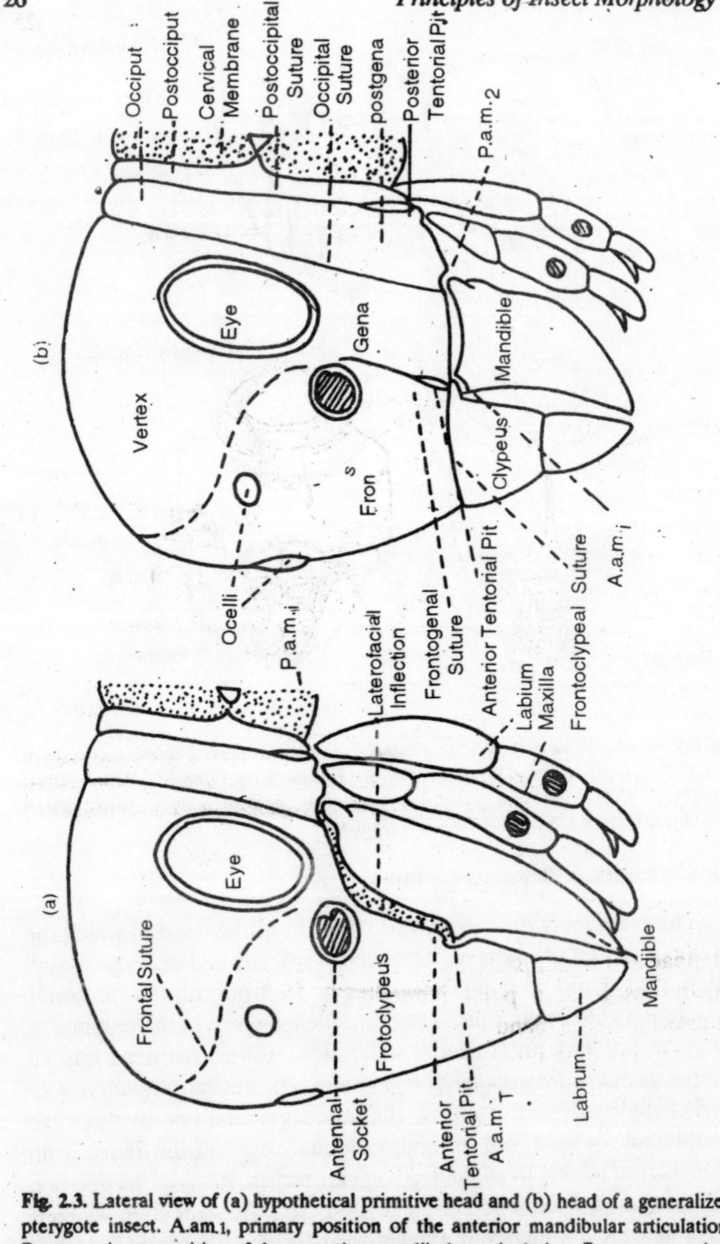


**Fig. 2.3.** Lateral view of (a) hypothetical primitive head and (b) head of a generalized pterygote insect. A.am.1, primary position of the anterior mandibular articulation: P.a.m.1, primary position of the posterior mandibular articulation; P.a.m.2, secondary position of the posterior mandibular articulation.

## The Frontoparietal Region

This region includes the large bulging front, sides and top of the head. It bears a pair of compound eyes laterally and a pair of antennae mesoventral to the eyes. Usually there are one median and two lateral ocelli between the compound eyes.

The frontoparietal region consists essentially of a single large sclerite divided ventrally by two vertical *frontogenal sutures* into a median *frontoclypeus* and two lateral *genae*. The ventral edges of the genae lie at the level of the mouth in the generalized head but the frontoclypeus project ventrally beyond the level of the mouth and is usually divided transversely by the *frontoclypeal suture* into a dorsal or postoral *frons* and a ventral or preoral *clypeus*. The frontogenal sutures mark the position of internal strengthening ridges, the *frontogenal inflections* which extend vertically from the lateral edges of the clypeus to the outer edge of the antennal sockets. Their ventral ends form the anterior articulations of the mandibles ($A.a.m._1$). The anterior tentorial arms join the frontogenal inflections near their ventral extremities at approximately the level of the mouth and are marked externally by the *anterior tentorial pits*. The ventral edges of the genae may be turned inward to form a strengthening ridge, the *subgenal inflection*. Sometimes the inflection is submarginal and marked externally by a *subgenal suture* which cuts off a narrow sclerite, the *pleurostoma*, from the ventral region of the gena. The frontoclypeal inflection and the subgenal inflections are at the same horizontal level. They meet and fuse with the frontogenal inflection near the point of attachment of the tentorial arms. These ridges and struts are therefore continuous and converge on the anterior mandibular articulation thus reinforcing the cranial wall against the force exerted by contracting muscles, especially the adductor muscles of the strong crushing mandibles.

The dorsal undivided region of the face is the *vertex* and each lateral half of the vertex together with the gena of that side is known as a *parietal*. In immature insects and some generalized adult insects there is a Y-shaped suture in the head known as the *ecdystat suture*. The stem runs in the middle line of the top or dorsal side of the cranium and at some point between the eyes it forks, the two branches running in a lateroventral direction. The stem is the *coronal suture*, and the arms the *frontal suture*. The latter, primitively, probably terminated at a point between the eyes and antennae, near the dorsal end of the frontogenal sutures, but in modern insects the course followed varies greatly. The ecdysial suture is the line of weakness along which the cuticle splits at ecdysis. The pigmented exocuticle is not developed along this line, therefore it appears as a pale line against the darker sclerite through which it passes. It is not formed by inflection and is not associated with an internal ridge. *A mid-cranial inflection*, marked externally by a sulcus, is found in the face of many insects. It extends for various distances down the face, and in immature insects, the coronal and sometimes part of the frontal suture are carried inward with the inflection. The mid-cranial suture, which is a sulcus, should not be confused with the coronal which is a line suture.

This interpretation of the cranial structure differs from that commonly found in the literature and is based on the recognition of the nature of the ecdysial and frontogenal sutures. The primitive arthropod head, including that of some apterygote insects, resembles the head shown diagrammatically in. There is at first a single mandibular articulation dorsal and posterior to the level of the mouth. The face or front of the head is elongated and extends ventrally for beyond the posterior portion of the head with which the mandible articulates. The mandible is attached by membrane to the lateral edge of the face between the articulation and the mouth level and this edge is inflected for strength (laterofacial inflection). In lepismatids and pterygotes the mandible made a second articulation with the edge of the face at the mouth level and in pterygotes the region ventral to the eye and posterior to it descended ventrally until the two mandibular articulations were at the same horizontal level. Thus, a pair of lateral regions, the *genae*, were added to the face and the lines where these united with the edges of the original face are marked externally by the *frontogenal sutures*. A transverse *fronto- clypeal inflection* between the anterior mandibular articulation divides the frontoclypeus into a dorsal *frons*, lying between the frontogenal sutures, and a ventral *clypeus* which is free laterally.

In the literature the frons and genae are defined, not by the position of the frontogenal sutures, but by the so-called epicranial suture. Usually this is the ecdysial suture but other sutures have been confused with it. Because of this confusion and because of the very different positions taken by the frontal suture the *"frons"* and *"genae"*

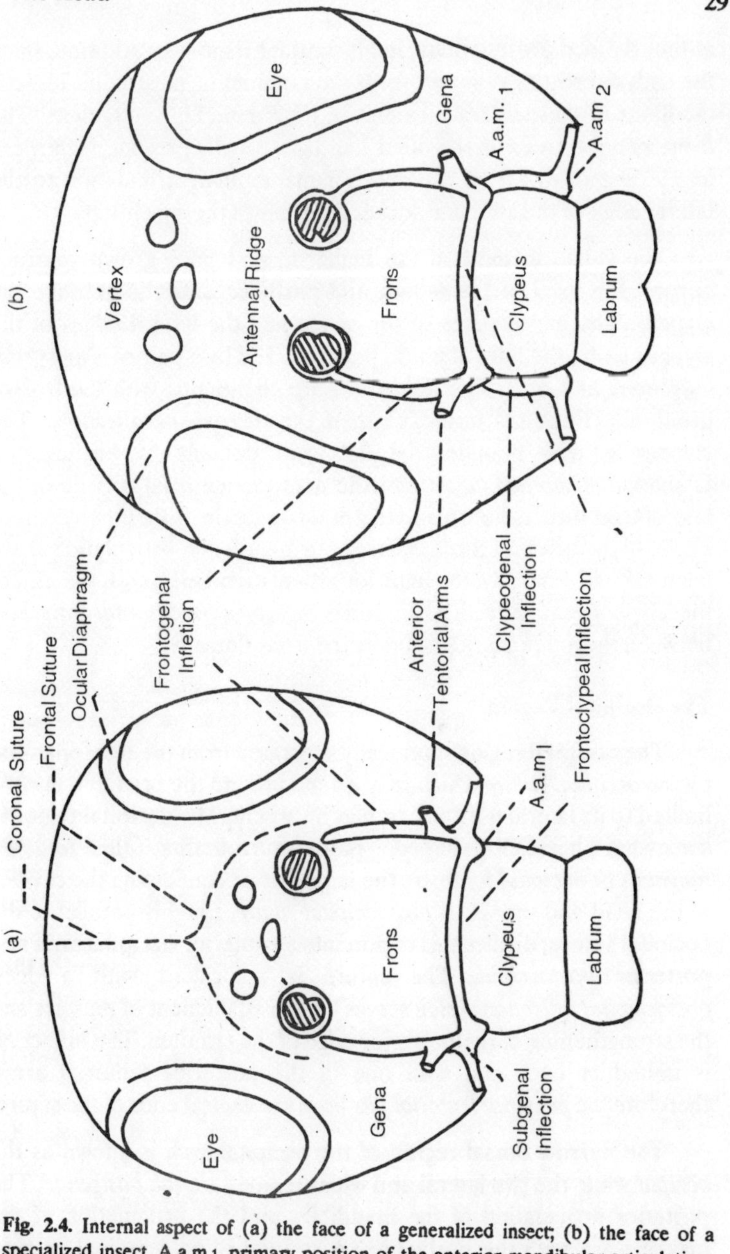


**Fig. 2.4.** Internal aspect of (a) the face of a generalized insect; (b) the face of a specialized insect. A.a.m.1, primary position of the anterior mandibular articulation; A.a.m.2, secondary position of the anterior mandibular articulation.

as thus defined are indefinite and inconstant regions In addition, since the ecdysial suture is absent from the cranium of most adult insects, the limits of these sclerites cannot be determined in such adults. The frontogenal sutures on the other hand are usually present. If they are lost, a line drawn from the anterior mandibular articulation to the lateral edge of the antennal socket determines their position.

The genae in most of the higher insects have grown ventrally beyond the level of the mouth and partly or entirely embrace the clypeus. The mesal edges of the genae and the lateral edges of the clypeus unite, the lines of union being marked by a pair of *clypeogenal inflections* and *sulci*. The inflections are continuous with the fronto-genal inflections and form with them the *laterofacial inflections*. The clypeus is now bounded laterally and dorsally by an inverted U-shaped *epistomal suture* and the anterior tentorial pits lie in the laterofacial sutures approximately at the junction of their two components. In the heads of these insects both mandibular articulations have been carried ventrally, the anterior articulation being with the end of the clypeogenal inflection. In some insects a *transfrontal inflection* between the antennal ridges limits the frons dorsally.

**The Occipital Region**

The occipital region is typically separated from the frontoparietal by the *occipital suture* which may extend around the cranium, may be limited to its lateral portions, or may be absent. The occipital region is somewhat horseshoe-shaped, partly surrounding the *foramen magnum* or *occipital foramen*, the large orifice connecting the cavities of the head and thorax. A *postoccipital suture*, roughly parallel to the occipital suture, divides this region into an anterior occipital arch and posterior *postocciput*. The suture is associated with a deep *postoccipital inflection* which serves for the attachment of muscles and the strengthening of the posterior edge of the cranium. The inflection is united at each end with one of the posterior tentorial arms, therefore the *posterior tentorial pits* lie in the ventral ends of the suture.

The narrow dorsal region of the occipital arch is known as the *occiput* while the two lateral and wider regions are the *postgenae*. The posterior articulation of the mandibles and the articulation of the maxillae are with the postgenae. A strengthening marginal or sub-marginal thickening or inflection of the postgena is known as the *hypostomal inflection*. If submarginal it is marked externally by a *hypostomal suture* running from the mandibular articulation to the posterior tentorial pit, and a narrow *hypostoma* is cut off from the edge of the postgena. The inflection takes part in the articulation of the mandible. The postocciput is usually a very narrow sclerite, to the free margin of which the neck membrane are attached. It bears a pair of small lateral processes, the *occipital condyles*, which articulate with the lateral cervical sclerites in the neck membrane and serve as pivots for the movements of the head. The cranium is open posteriorly and ventrally, the mouthparts being suspended from the ventral margin. The entire free margin is strengthened by a continuous series of ridges, the postoccipital, hypostomal, subgenal and frontoclypeal inflections. There are many variations in head structure, including differences in the proportions of the various regions, the suppression of some sutures and the appearance of new sutures. In higher insects

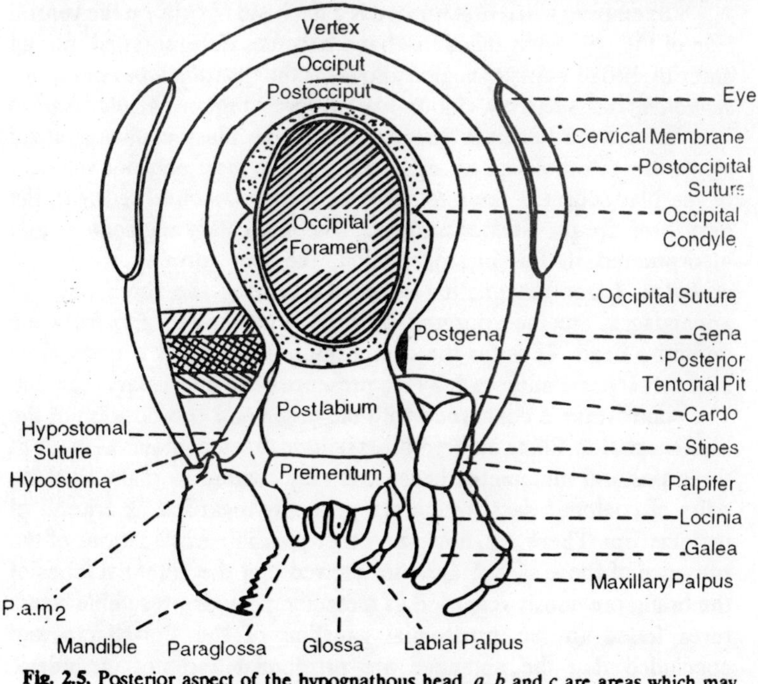


**Fig. 2.5.** Posterior aspect of the hypognathous head. *a, b* and *c* are areas which may undergo rotation and elongation in the prognathous head.

the occipital foramen is entirely surrounded by sclerites and in progna- thous insects there is a ventral wall to the cranium. Examples of some of the commoner modifications will be found in the insects selected for study.

**Metameric Constitution of the Head**

The head is a compound structure and includes several metameres which originally were postoral in the annelid-like ancestors as well, probably as the prostomium of these ancestors. There is, however, no general agreement as to the number of these metameres nor as to the extent of the contribution, if any, made by the prostomium. Some authors have tried to identify the transverse sutures of the head with intersegmental lines but most authors agree that, with the possible exception of the postoccipital suture, they are secondary, resulting from the formation of functional inflections in the cranium.

The embryo when first formed is a flat band of cells on the ventral side of the egg. Soon this germ band becomes differentiated into an anterior, broad bilobate region known as the *cephalic lobes* or *blastocephalon*, and a narrow, ribbon-like posterior region, the *blastocorm*. The latter soon becomes segmented but the blastocephalon never shows any external sign of segmentation. The most anterior segment of the blastocorm is the *intercalary* or *second antennal segment*, the next three are the *gnathal segments*. The intercalary segment is early incorporated in the blastocephalon, the two forming the *protocephalon*. Later the gnathal segments, bearing the three pairs of appendages, are incorporated with the protocephal onto form the definitive head. Thus far the composition of the head is quite clear except that some authors deny the presence of an intercalary segment. The controversy is concerned with the origin and composition of the blastocephalon. Some embryolotists believe that it contains as many as three postoral metameres because it may contain as many as three pairs of coelomic sacs, which are generally regarded as criteria of metamerism. There are, however, other plausible explanations of the presence of these sacs. Hanström showed that the antennal lobes of the brain, previously regarded as metameric ganglia, resemble structures found in the prostomial ganglion of the Polychaeta and concluded that the antennae are prostomial and not metameric appendages. If this is so, the antennal region of the head, and parts anterior to it, all of which are derived from the blastocephalon, must bee the homologue of the prostomium of the Polychaeta. Butt, however, has recently flamined to have found evidence that the labrum is the united appendages of the intercalary segment.

Thus at one extreme it is claimed that the head is composed of seven postoral metameres, with or without contributions from the prostomium; at the other, that it consists of the prostomium and three or four metameres. While it is still desirable to keep an open mind, the evidence for the prostomial origin of the blastocephalon appears to be quite strong.

### The Frontoparietal or Facial Region

Whenever, the relation of the head to the trunk is studied its typical hypognathous orientation has been noted. The *cervical* or *neck membrane* which connects the cranium to the prothoracic skeleton is also present. Note the small cervical sclerites in the membrane. The two most prominent, and which are most constantly present in insects, are the *lateral cervical sclerites* which articulate anteriorly with two small occipital condyles on the posterior margin of the cranium, and posteriorly with the episternum of the prothorax. These sclerites from a pivot for the movements of the head.

It has already been observed that the head consists of an anterior bulging frontoparietal region, and a smaller, flat occipital region to which the neck membrane is attached. On the frontoparietal region find the long multiarticulate *antennae*, the two large *compound eyes* and the three small *ocelli* or simple eyes. Remove the antennae completely to expose their sockets. Note that there is a suture surrounding the compound eyes which cuts off a very narrow *ocular sclerite*. A similar *antennal sclerite* surrounds the antennal sockets.

The *labrum*, which is the flap-like sclerite movably hinged to the ventral edge of the face forms the upper lip. It has the heavily sclerotized mandibles beneath.

There is an *occipital suture* which runs around the cranium, terminating on each side just anterior to the posterior mandibular articulation. The suture runs along a ridge at the back of the head and separates the frontoparietal and occipital regions.

In the frontoparietal region there is the *clypeus* and the median quadragular sclerite. The labrum is hinged to the ventral margin of this sclerite. The membranous suture between the clypeus and labrum is the *clypeolabral suture* and the sulcus which bounds the clypeus dorsally is the *frontoclypeal suture*. Note that the clypeus is partially divided by two transverse grooves into a dorsal *postclypeus* and a ventral *anteclypeus*. The dorsolateral angles of the clypeus project laterally and take part in the articulation of the mandibles.

The *ecdysial suture* may not be visible in the untreated head of the adult but will be seen later in the head treated with potassium hydroxide. If it is not visible examine the head of a nymph. It consists of a pale line running along the middle of the dorsal region of the head to a point between the eyes, where it forks into two divergent branches which run obliquely to the margins of the lateral ocelli. The median line is the *coronal suture*, the branches constitute the *frontal suture*. Note that the coronal line in the nymph is continued along the mid-dorsal line of the thorax. That part of the face which lies dorsal to the frontoclypeal suture is divided into three ventral lobes by two short vertical sutures which extend from the extremities of the frontoclypeal suture, just above the mandibular articulations, to the lateral margin of the antennal sockets. These are the *frontogenal sutures*. The median lobe is the *frons* and the lateral lobes the *genae*. The large undivided dorsal region is the *vertex*. The term *parietal* is sometimes used for one-half of the vertex with the gena of the same side. When a coronal suture is present it separates the two parietals. The ventral edges of the genae are bounded by *subgenal sutures*, continuous with and in the same horizontal plane, as the frontoclypeal suture. Between the subgenal suture and the base of the mandible there is a small triangular sclerite known as the *pleurostoma*.

The prominent sclerotic framework within the head is the *tentorium*, which will be studied later. The *anterior arms* of the tentorium arise by a broad base from the ridge which marks internally the position of the frontoclypeal and subgenal sutures. Turn to the outer surface and note that the position of the arm is marked by a deep cleft which extends through part of both sutures. This is the *anterior tentorial invagination* or *pit*. Add it to your drawing of the external features.

From the margin of the eye socket the thin flange projects obliquely inward. This is the *ocular diaphragm* whose position is marked externally by the ocular suture. Note that the antennal sockets are also surrounded by a narrow internal *antennal ridge*.

The *frontoclypeal* and *subgenal inflections* form a continuous internal ridge marked externally by the corresponding sutures. The *frontogenal inflection* joins this ridge at the junction of the two other inflections and runs dorsally to the lateral portion of the antennal ridge with which it merges. On the ventral side of this system of ridges, directly opposite the frontogenal ridge, find a small rounded protuberance which is the *anterior condyle* of the mandibular articulation. It should perhaps be regarded as the ventral extremity of the frontogenal inflection.

### *Anisolabis Maritima*

The head is prognathous and the areas which are dorsal and anterior in the cricket are both dorsal in the earwig. The terms "anterior" and "posterior" as used here will therefore have the same morphological significance as the terms "ventral" and "dorsal" as used in describing the head of the cricket. In examining this and the succeeding heads look for the structures described in the cricket as well as the special features to which your attention is drawn.

The *clypeus* is divided into a sclerotized *postclypeus* and a membranous *anteclypeus*. In the mid-dorsal line of the vertex you will find a sulcus in the position occupied by the coronal suture in the cricket. This is the *mid-cranial suture* which marks the position of an inflection which has carried the coronal suture inward so that it is not evident externally. Its extreme tip can be seen emerging from the sulcus and forking to form the *frontal suture*. The latter curves laterally and terminates near the mesal margins of the eyes in close proximity to the posterior ends of the *frontogenal sutures* which run anteriorly, around the lateral margin of the antennal sockets, and terminate at the lateral ends of the frontoclypeal suture just behind the anterior mandibular articulation. They can be seen more clearly from the lateral view of the head. The *anterior tentorial pit* is a long fissure in the frontogenal suture extending from its anterior end parallel to the antennal socket. Superficially the frontal and frontogenal sutures appear to be continuous and are commonly thus described in the literature under the name of "frontal suture." The fact that one is a line suture and the other a sulcus distinguishes between them.

It is seen that the genal region is partly membranous and there is no subgenal sulcus or pleurostoma. The membranous area has been called by Ferris the *subocular sinus*.

The mantid head presents two new features. There is a pair of sulci on the vertex, mesal to the eyes, which extend a short distance down the face. These are the *temporal sutures* and they occur in some other insects. The *frontogenal sutures* extend, as in the cricket, from the anterior mandibular articulations to the lateral edges of the antennal sockets. *A transfrontal suture* extends between the antennal sockets. Since the frontogenal inflections merge with the antennal ridges and the transfrontal inflection extends between the antennal ridges, this inflection may be regarded as forming the dorsal boundary of the frons. A similar inflection is found in some other insects. The *anterior tentorial* pit is, as in the earwig, an elongated slit extending through most of the length of the frontogenal suture.

### *Leptinotarsa (Larva)*

In Leptinotarsa larve the *ecdysial suture* consists of a short coronal suture and a *frontal suture* and their arms diverge widely and terminate at the dorsal edges of the antennal sockets. The *mid-cranial sulcus* extends from the end of the coronal suture to the frontoclypeal suture. It is marked internally by a deep inflection which terminates in the frontoclypeal inflection. It is seen that in other larvae the antennae are borne well forward on the genal region and the eyes are represented by stemmata which also are usually well forward.

### *Leptinotarsa (Adult)*

The ecdysial suture is absent but the *mid-cranial suture* persists though it is not as distinct as in the larva. It meets the *frontoclypeal suture* which arches dorsally in such a way that the relation of the two sutures resembles superficially that of the coronal and frontal suture. This combination in the heads of some beetles has been confused with the ecdysial suture and described as the "epicranial suture."

There is a pair of prominent *mandibular condyles* at the lateral ends of the frontoclypeal inflection. A very slender thread-like process arises from the ridge opposite the condyle. This is the greatly reduced *anterior tentorial arm*. The thickness and rigidity of the cuticle reduces the need for internal scaffolding in most beetles, and in many species the frontoclypeal inflection is the only prominent ridge. The ventral edge of the gena is inflected, the *subgenal inflection* being continuous with the frontoclypeal. The ventral position of the eyes and antennae suggests that in this insect the vertex has expanded ventrally at the expense of the frons and genae. This may explain the absence of the frontogenal inflection.

### *Apis Mellifera (Worker)*

The *clypeus* is the large median plate ventral to the antennae. Its sides are not free as in the more generalized insects but are joined to the mesal edges of the genae. The lines of demarcation between the clypeus and the genae are the vertical components of an inverted U-shaped sulcus, the *epistomal suture*. Toward the dorsal end of this suture find the *anterior tentorial pits*. They do not lie entirely in the suture, their dorsal ends being somewhat lateral to it. From these dorsal ends a pair of very short sutures run to the lateral margins of the antennal sockets. These are the greatly reduced *frontogenal sutures*. The explanation of this structure is as follows.

In a generalized insect like the cricket the genae have grown ventrally to the level of the mouth, which is also approximately the level of the tentorial pits and the frontoclypeal suture. In the honey bee this ventral growth has continued and the lateral edges of the clypeus have fused with the mesal edges of the genae. The sutures thus formed between the sclerites, i.e., the vertical components of the epistomal suture, are the *clypeogenal sutures*. The clypeogenal suture is continuous with the frontogenal suture, the two constituting the *laterofacial suture*. The tentorial pit lies at the junction of the two sutures. The horizontal component of the epistomal suture is the *frontoclypeal suture* in its normal position near the level of the mouth.

The anterior articulation of the mandible has shifted to the ventral end of the clypeogenal suture.

It is seen that the very short *frontogenal inflection* unites with the lateral side of the antennal ridge as in the cricket, and that the *tentorial arm* is attached at the junction of the frontogenal and *clypeogenal inflections*. On the mesal side of each laterofacial ridge, ventral to the base of the tentorial arm, there is a small tubercle which is probably the ventral extremity of the frontogenal inflection, i.e., the original *mandibular condyle* corresponding to that in the cricket.

***Protoparce Quinquemaculata (Larva)***

The most conspicuous feature of the face is the ***mid-cranial sulcus*** which forks at its ventral end to form an inverted V, the two arms enclosing a median triangular sclerite. The midcranial inflection has carried the coronal suture and part of the frontal suture inward but the *frontal suture* emerges from it dorsal to the fork and can be seen as a pale line lateral and more or less parallel to the V-shaped sulcus. The *tentorial pits* lie near the middle of each arm of the *V* but may not be very evident. Their positions maybe checked later on the inner surface.

The triangular sclerite and the V-shaped sulcus have been variously interpreted. For a long time the sulcus was believed to be the frontal suture and the sclerite the frons. Snodgrass interpreted the sulcus as the epistomal suture and the sclerite as the clypeus. Hinton believes the sutures to be secondary, cutting through the fronto-clypeus, and the sclerite to be a median portion of the frontoclypeus. Judging by the position of the frontal ganglion and the anterior tentorial pits in relation to that of the morphological mouth, and by the origins of the cibarial and anterior pharyngeal dilators, it appears that the region of the sclerite dorsal to the tentorial pits is the frons and that ventral to the pits the clypeus. The entire sclerite is therefore the *frontoclypeus*, and the arms of the V are *laterofacial sutures*, the dorsal ends of which have been drawn together as a result of the very deep mid-cranial inflection.

The frontoclypeal suture is not developed but that the clypeal region is traversed by a *transclypeal suture*. The clypeal region dorsal to the suture is the *postclypeus*, that ventral to the suture the *anteclypeus*. The latter consists of a narrow sclerite followed by a membranous area to which the labrum is attached.

## The Occipital Region

*Acheta assimilis*

If we remove the head with the neck membrane attached to it. We find the labium and maxillae on the posterior side of the head, and the large *foramen magnum* or *occipital foramen* which is enclosed dorsally and laterally by the cranial wall and ventrally by the base of the labium. The neck membrane is attached to the dorsal and lateral edges of the cranium and to the proximal margin of the labium. The sclerotic cranial wall is not completed ventrally, the sclerite bounding the foramen being more or less horseshoe-shaped. This is the primitive condition. The remnants of the neck membrane may now be removed.

There is a strong transverse bar near the ventral edge of the foramen. It is known as the *tentorial bridge* and is formed by the union of the two posterior tentorial arms. It is, of curse, an internal structure and does not close the foramen externally.

There is a deep inflection of the cranial wall, close to and parallel with the margin of the foramen. This is the *postoccipital inflection*, marked externally by the postoccipital suture, which divides the occipital region into an anterior *occipital arch* and a posterior *postocciput*. The neck membrane is attached to the margin of the postocciput.

In *Acheta* the postocciput is reduced, forming little more than the posterior wall of the inflection. Divide it by cutting through the middle, and bend the right half back, i.e., away from the occipital arch. Its relation to the remainder of the cranium will be more evident and the postoccipital suture will be fully exposed on this side. Near its ventral ends the postocciput bears a pair of small triangular processes, the *occipital condyles*, which articulate with the little cervical sclerites. The junctions of the tentorial bridge with the postoccipital inflection are marked externally by the *posterior tentorial pits* which lie at the ventral ends of the postoccipital suture. This relation to the suture appears to be constant in spite of the change which take place in this region. Find the pits and note that the proximal angles of the labium meet the cranial wall at the level of the pits.

The *occipital arch* lies between the occipital and postoccipital sutures and forms the major portion of the occipital region. It is a single sclerite consisting of a narrow dorsal area, the *occiput*, and two wider triangular areas, the *postgenae*. The *articulation of the maxilla* is with the postgena just ventral to the tentorial pit. The *posterior articulation of the mandible* is with the ventral edge of the postgena. The *hypostomal sulcus* which extends from the mandibular articulation to the posterior tentorial pit and cuts off a narrow marginal sclerite, the *hypostoma*, from the postegena. In *Acheta* the hypostoma is little more than a marginal thickening but in many insects it is a more distinct sclerite. The hypostomal sulcus marks the position of an internal hypostomal ridge, the lower end of which provides the socket for the mandibular condyle. It will be observed that the postoccipital, hypostomal, subgenal and frontoclypeal inflections from what is, in effect, a continuous strengthening ridge to the free margin of the cranium.

*Protoparc quinquemaculata*

In many insects, although the hypognathous condition is retained, the occipital foramen is closed ventrally and the labium is separated from the neck membrane by sclerotic tissue. The closure is effected by the mesal growth of the hypostomae. In lepidopterou larvae the ventral portion of the head is complete but the posterodorsal region is reduced. This reduction is slight in *Protoparce*, but in some larvae it is carried so far that the entire vertex proximal to the fork of the ecdysial suture is lost. The occiput is probably always lost.

To the ventral edges of postgenal the mandibles are articulated, and note that they are not separated from the genae by an occipital suture. The foramen is closed ventrally by two triangular sclerites whose bases are attached to the mesal edges of the postgenae and whose apices meet in the middle line. The *tentorial bridge*, the slender bar which stretches across the cranial cavity dorsal to the triangular sclerites. From the position of the bridge find the *tentorial pits*. There are sulci which bound the triangular sclerites laterally run from the posterior mandibular articulation to the tentorial pits. They are therefore the *hypostomal sulci* and the sclerites are the *hypostomae* which have extended mesally to form a *hypostomal bridge* beneath the foramen.

The tentorial bridge lies between the ventral ends of two deep inflections. These are the lateral portions of the *postoccipital inflection*. The dorsal portions are lost as the result of the reduction of the cranium in this region, and the postocciput has disappeared as an external sclerite.

Dorsal to the postoccipital inflections there is a marginal inflection of the cranium continuous with the mid-cranial inflection. It appears superficially to be continuous with the postoccipital inflections but careful inspection will show that it is not. It is more probably a secondary strengthening ridge but may be the dorsal portion of the occipital inflection.

*Apis mellifera (worker)*

In the hypognathus head of *Apis* there is a *postoccipital suture* which cuts off narrow *postocciput* dorsal and lateral to the foramen. The slit- like *tentorial pits* in the suture mark its ventral limits an those of the postocciput. The base of the proboscis fits into a U-shaped depression, the *fossa of the proboscis*. Pull the proboscis ventrally to expose the fossa and note that the lateral walls of the latter with which the maxillae articulated are sclerotized. These sclerites are the inflected *hypostomae*. The *hypostomal sutures* between them and the postgenae run from the mandibular articulation to the tentorial pits. The two hypostomae meet and fuse beneath the foramen forming thc *hypostomal bridge*.

The postgenae send short triangular processes mesally into the subforaminal bridge. Only the middle portion of the bridge, therefore, is formed from the hypostoma; in some other Hymenoptera the postgenal processes apparently meet to complete the bridge externally, and the hypostoma portion is invaginated into the head cavity.

**Ventral Region of the Prognathous Head**

The cranium of the hypognathous head has no ventral wall. It terminates ventrally in the *peristome,* which is formed by the clypeal and lateral cranial borders surrounding the mouth, to which the mouthparts are attached. In typical prognathous insects, such as the Coleoptera, the occipital foramen retains its vertical orientation but the head is rotated dorsally so that the primitively cephalic aspect is now dorsal and the peristome opens anteriorly instead of ventrally. This change is accompanied by the elongation and dorsal rotation of the postgenal and, to some extent, of the postoccipital region, and by the development of a ventral wall to the cranium, between the postgenae, which closes the foramen ventrally. Judging by the structure of the ventral wall of the prognathous head it appears that the rotation and elongation may have been restricted to different regions of the postgena in different insects. In some, the region ventral to the posterior tentorial pits seems to have been involved, in others the region of the pits, and in still others the region dorsal to the pits. In some orthopteroid insects the area involved is ventral to the pits which retain their primitive position. The base of the labium retains its relationship to the pits and to the cervical membrane. No gula is formed but the postlabium is greatly elongated. When the region of the pits is involved the pits become greatly elongated in the gular sutures. The labium is carried forward, its basal angles being adjacent to the ventral ends of the pits. The area between the labium and the neck is sclerotized to form the gula. When the change occurs dorsal to the pits, the pits and the labium are carried forward and again a gula is formed.

In the head there is a *postoccipital suture* and the *tentorial pits* having essentially the same relations to the cranium as they have in *Acheta*. The opening into the head is vertical. Actually the foramen is somewhat oblique, but this is compensated by a large ventral cervical sclerite adjacent to the base of the labium. Those portions of the postgenae which, in *Acheta*, lie ventral to the tentorial pits are, in *Anisolabis*, greatly elongated and lie in the horizontal instead of the vertical plane, so that the mouthparts are directed anteriorly. The two postgenae now form a partial ventral wall to the head but are still widely separated and the foramen is incomplete ventrally. As in *Acheta*, the labium is suspended from the neck membrane at the level of the tentorial pits and the maxillae also retain their primitive position, articulated with the hypostoma just anterior to the pits. As a result of the elongation of the postegenae, both the mentum and the stipes are also elongated. The posterior articulation of the mandible is still which the ventral (now anterior) margin of the postgena and is therefore far removed from that of the labium and maxilla.

The *hypostomal suture* which extends from the posterior articulation of the mandible to the tentorial pit and cuts off a narrow *hypostoma* from the mesal edge of the postgena. The elongation of the postgenae had already begun in the cricket, but there is no rotation of the head. Apart from the greater length of postgenae, labium and maxillae, and the rotation of the head, the structure of the head in the earwig is essentially the same as in the cricket.

### *Phyllophaga anxia*

In the head of *Phyllophaga*, the occipital foramen is vertical and is completely closed ventrally. The ventral wall of the cranium consists of three sclerites, a median *gula* and two lateral *postgenae*, the three separated by a pair of *gular sutures*. Find the *postoccipital suture* and the *postocciput*, visible as a very narrow plate bordering the foramen laterally and dorsally. Note that the postoccipital and the gular sutures are continuous. Internally the gular sutures are marked by high ridges. The tentorial bridge formed by the posterior tentorial arms arises from the anterior ends of these ridges and retains its primitive attachments at the lateroventral margins of the foramen. Because of this structure, it is assumed that the elongation and rotation of the postegenae involved the region of the tentorial pit and that the ventral ends of the postocciput therefore accompanied the elongating postgenae. The gular sutures therefore would be elongations of the postoccipital suture and the gular ridges elongations of the postoccipital inflections. Since the labium is attached at the ventral ends of the postocciput it has been carried forward and is now attached to the ventral edge of the gula.

On this theory, the elongation of the narrow postocciput should be acompanied by an elongation of the neck membrane stretched between its mesal edges, and the gula is formed either by sclerotization of this neck membrane or by the mesal growth and fusion of the elongated postoccipital regions.

There is, however, another possible explanation of the origin of the gula. Refer to your drawing of *Anisolabis* and note the narrow hypostomae along the mesal eedges of the postgenae. If these grew mesally and united, a gula would be formed similar to that of *Phyllophaga*. The gular sutures would then be the hypostomal sutures and not the postocciptal, and the gula would be equivalent to the hypostomal bridge in such insects as the Hymenoptera and Lepidoptera. Since, however, the posterior tentorial pits often lie far forward in the gular sutures it would have tobe assumed that the tentorial arms have migrated anteriorly into the hypostomal sutures. Some authors maintain that the gula is a basal region of the labium fused with the cranial wall. This seems unlikely and the first of these theories of the gula is the one most strongly supported. The *submentum*, the basal sclerite of the labium, is attached to the anterior edge of the gula, far removed from the neck. A distinct suture makes the boundary between the two sclerites.

*Dytiscus*

The ventral region of the cranium has a structure essentially similar to that of *Phyllophaga*. Note the following differences : There are two deep *postoccipital inflections* at the lateral margins of the foramen, and these are definitely continuous with the gular ridge. The *tentorial bridge* lies farther forward. There may be a shallow groove or depression between the gula and the submentum but the two are not separated by a distinct suture. When the two sclerites are thus united they are often referred to as the *gulamentum*. In all cases a transverse line through the ventral ends of the posterior tentorial pits may be regarded as the boundary between the gula and the submentum.

**Suppression of the Gula**

It will be remembered that in the bees and wasps there is a tendency for the postgenae to grow mesally and obliterate the hypostomal bridge. A similar tendency is found in the prognathous head. The gula in the two specimens studied is a broad sclerites but there is a tendency for it to contract in width and finally disappear as the postgenae grow mesally and meet in the middle line. Successive stages may be seen in the following heads :

*Harpalus*. The gula here is still quite distinct but much narrower. There is a *mid-gular suture* which may mark an invagination resulting from its lateral compression.

*Necrophorus*. The postgenae have met in the proximal region of the head and have forced out the gula which is reduced to a narrow triangular sclerite between the distal ends of the postgenae.

*Cantharis or Chauliognathus*. The mesal edges of the postgenae have united completely and the gula is obsolete.

**The Tentorium**

The tentorium consists of a set of strengthening struts within the cranium. It is not a true endoskeletons as it is formed by two pairs of integumental invaginations known as the *tentorial arms*. It serves not only for strengthening the cranial wall against the pull of the mandibular adductors and other muscles but also for the origin of certain muscles and for the support of the pharyngeal region. The *anterior arms*, as we have seen, originate just dorsal to the anterior

(a)

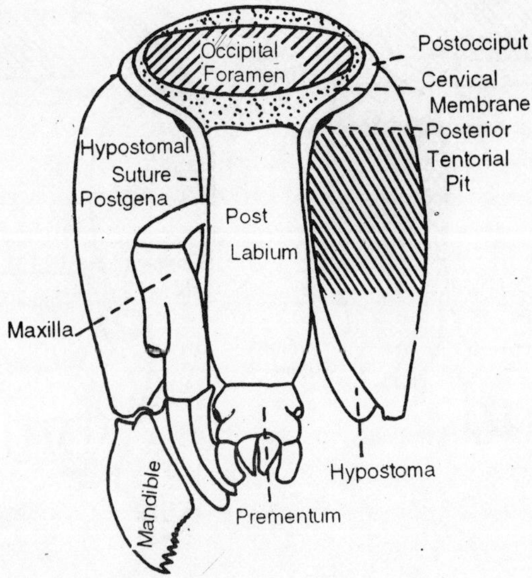


**Fig. 2.6.** Prognathous Heads. (a) Rotation and elongation have taken place distal to the tentorial pits. No true gula is formed, but the postabium is greatly elongated and the other mouthparts carried forward.

articulation of the mandibles while the *posterior arms* arise at the ventral ends of the postoccipital inflections, their points of origin being marked by the *tentorial pits*. In many insects the anterior arms give rise to a third pair of process, the *dorsal arms*, which are attached to the inner wall of the face near the antennal sockets. Since these are not independent invaginations their attachment to the cranium is not indicated externally by pits. The tentorium undergoes many modifications and in some insects it is greatly reduced. It has been examined in part in several of the heads already studied. At this time two relatively primitive examples will be studied in detail.

*Protoparce quinquemaculata (larva)*

The *tentorial bridge* is a slender transverse bar across the cranial cavity near the ventral edge of the foramen. Cut away the postgenal areas, the maxillae and the labium, leaving the bridge intact. The tentorial bridge is formed by the fusion of the *posterior tentorial arms*, the origins of which are marked by the *posterior tentorial pits*. The

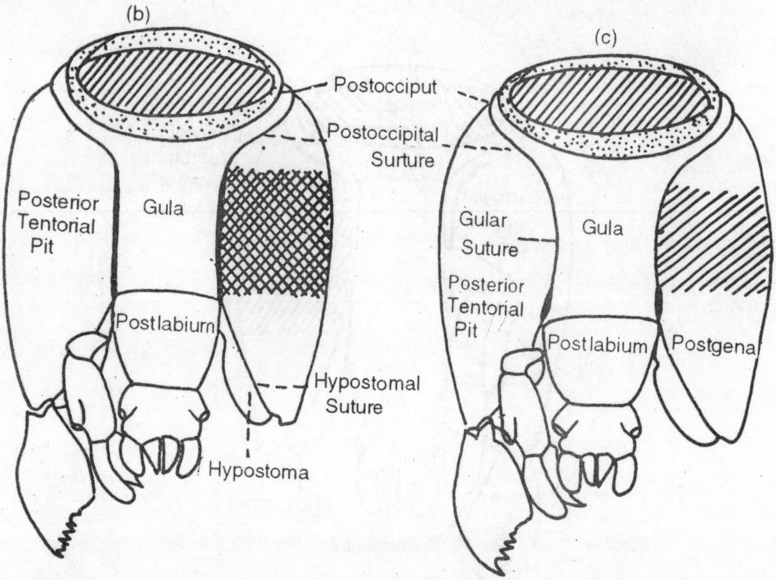


**Fig. 2.6. (contd.)** (b) Rotation and elongation have taken place in the region of the pits. These are greatly elongated in the gular sutures, the labium is carried forward and a true gula is formed.
(c) Rotation and elongation have taken place proximal to the pits. The pits and the labium are carried forward and a true gula is formed proximal to them.

*anterior tentorial arms*, originating near the middle of the *laterofacial inflections*, have been previously identified. Note that they unite with the tentorial bridge a short distance from the ends of the latter, so that the combined structure is shaped like the Greek letter $\pi$.

*Acheta assimilis*

The *posterior arms* unite to form the stout, vertically flattened *tentorial bridge*. The *anterior arms* converge posteriorly and fuse to form a horizontally flattened triangular plate, known as the *body of the tentorium*. This is turn unites with the dorsal edge of the tentorial bridge. From the dorsal side of the anterior arms another pair of processes, the *dorsal arms*, are given off. They extend anterodorsally to meet the cranial wall in the lateral regions of the antennal ridges with which they unite.

## Compound Eyes

Pterygote insects typically bear a pair of compound eyes on the lateral regions of the head. Each eye is made up of a number of elements known as *ommatidia* whose cuticular portions form the six-sided *facets,* or *corneal lenses*, seen on the surface of the eye. The shape, size and position of the eyes are subject to considerable variations. The number of facets also varies from a very few in some insects to many thousands in others. Insects with strong vision have many-facetted eyes. A few insects have lost their eyes and are totally blind. The ocular foramen may be surrounded by a narrow, usually thickened or heavily sclerotized, *ocular sclerite*. This in turn is sometimes surrounded by an *ocular suture* which marks the position of an oblique inflection which slopes inward forming a marginal shelf beneath the cuticular portion of the eye. This is the *ocular diaphragm* and the extent of its development varies widely in different insects.

### *Acheta assimilis*

Boil in potassium hydroxide that portion of the head which bears the compound eye. Remove all soft parts completely, including the pigmented material on the inside of the eye. Examine the outer surface of the eye and note the ocular sclerites which is little more than a slight thickening of the margin of the ocular foramen. On the inner surface find the *ocular diaphragm* which surrounds the *ocular foramen*, the opening through which the sensory structures of the eye pass to the facets. Examine a portion of the cuticle of the eye under the high power and note the *facets*, each of which forms part of the lens system of an ommatidium or single element of the compound eye.

## Variations in Eyes

A few of the more marked variations in the external features of the eyes will be seen in the following examples :

*Aeschna sp. (dragonfly)*. Eyes are greatly enlarged with many facets. The range of vision is increased by the great convexity of the eye which enables slight to strike it from many directions.

*Camponotus sp. (ant) worker caste*. The eyes are greatly reduced in size with relatively few facets.

*Apis mellifera (honeybee) drone and worker.* Note the sexual differences in the eye.

*Gyrinus sp. (whirligig beetle).* Each eye is completely divided so that there are apparently two pairs, one dorsal the other ventral.

**Antennae**

All insects except the Protura, which are doubtfully included among the insects, have a single pair of antennae which are sensory in function. These were probably borne primitively in the dorsal region of the frons, ventromesad of the compound eyes, but their definitive positions vary widely in modern insects. The antennae is set in a *foramen* or *socket* of the cranium and an *articulatory membrane* connects the base of the antenna with the edge of the socket, thus permitting free movements of the antennae. The rim of the socket may be thickened or more heavily sclerotized. This thickening sometimes has the appearance of a distinct sclerite and is often referred to as the *antennal sclerite*. It is associated with the internal reinforcing *antennal ridge* surrounding the foramen seen earlier in the examination of the internal aspect of the face. The rim of the socket may bear an articulatory process, the *antennifer*, which makes contact with the base of the antenna, forming a pivot against which the latter works. The basal segment of the antenna is known as the *scape*, the second segment as the *pedicil*, and the remaining portion as the *floagellum* or *clavola*. Muscles are inserted on the scape, pedicil and the base of the flagellum. The individual segments of the flagellum are not provided with muscles except in the Diplura and Collembola in which Imms has shown that all segments are provided with muscles. It would seem therefore that in the higher insects the antenae are really three-segmented, the flagellum being a single subdivided segment.

The antennae. many have a great variety of forms which have greater taxonomic than morphological significance. The following are some of the terms descriptive of the form of the antennae :

*Filiform*. Slender, thread-like, the segments nearly uniform in width.

*Setaceous*. Resembling a tapering bristle. The escape is fairly stout and the remaining segments progressively smaller in diameter.

*Moniliform*. Segment globose or ovoid so that the antenna has the appearance of a string of beads.

*Serrate*. The segments are triangular and produced slightly on one side giving the appearance of the teeth of a saw. When the segments are thus produced on both sides the antenna is said to be *biserrate*.

*Pectinate*. Segments with longer processes like the teeth of a comb. If such processes occur on both sides the antenna is *bipectinate*.

*Plumose*. The antenna bears numerous long slender processes and thus resembles a feather.

*Clavate*. Distal segments become gradually broader so that the antenna resembles a club.

*Capitate*. Distal segments broaden suddenly so that the antenna has a pronounced knob or head at the end.

*Lamellate*. Distal segments as in the capitate type but produced on one side into flat plates arranged like the leaves of a book.

*Aristate*. Three short segments, the distal one bearing a lateral bristlelike structure known as the "arista."

*Geniculate*. Bent abruptly at an angle resembling a bent knee.

*Acheta assimilis*

The base of antenna is called the *antennal socket*. There is *articulatory membrane* by which the base of the antenna is attached to the rim of the socket. The basal segment of antenna or *scape* is distinctly larger than the other segments. It is dorsoventrally compressed and its lateral edges about against the rim of the socket so that the horizontal movement of the scape is limited. Vertical movement, however, is quite free. Check the movements by manipulating the antenna with a needle. A small *pivotal sclerite* in the articulatory membrane is broadly united with the lateral edge of the base of the scape. It tapers to a point which pivots against the rim of the socket.

The second segment or *pedicil* has a vertical docondylic articulation with the scape and moves in the horizontal plane. The *flagellum* consists of an indefinite number of short, cylindrical rings, each slightly recessed into the one behind and united with it by a narrow membrane, so that the flagellum is perfectly flexible.

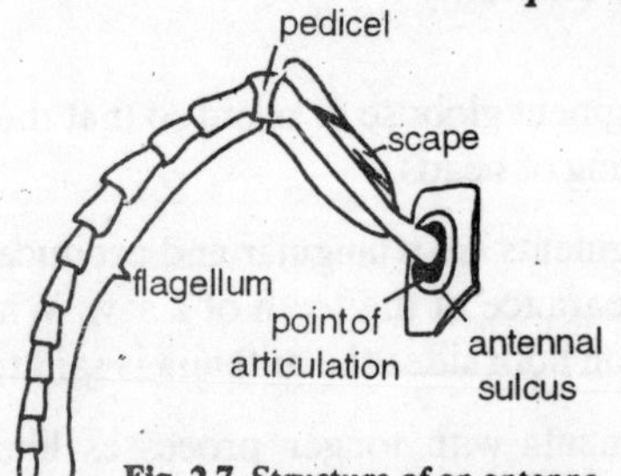


**Fig. 2.7. Structure of an antenna.**

*Apis mellifera*

In Apis the antennal socket bearing the scape here can move in any direction. On the lateral side find a somewhat depressed triangular process of the rim of the socket which articulates with the base of the scape. This is known as the *antennifer* and is the functional equivalent of the pivotal sclerite found in *Acheta*.

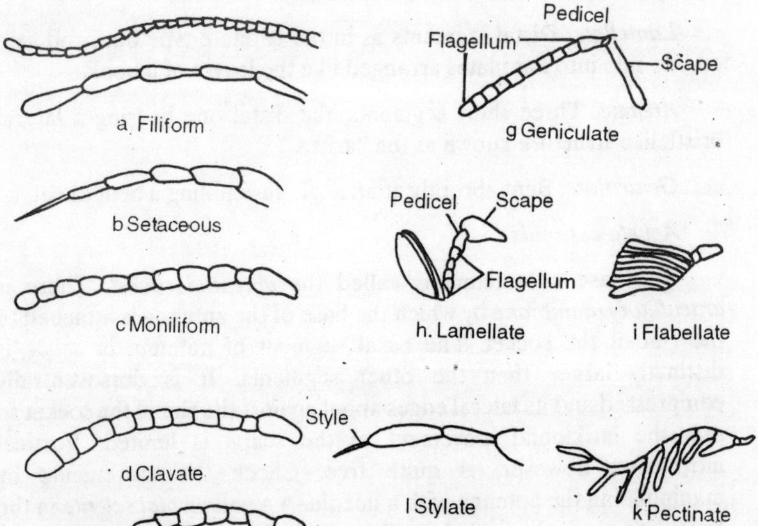


**Fig. 2.8.** Types of insect antennae. (a through g, j).

## MOUTHPARTS

The mouthparts consist of the labrum, a pair of mandibles, a pair of maxillae, the labium and the hypopharynx. The latter structure is not, of course, an appendage in the strict sense, but simply part of the ventral surface of the head. In Collembola, Protura, and Diplura the mouthparts are enclosed within a cavity formed by the ventrolateral extension of the genae which fuse in the midline (the *Entognathous* condition). In Thysanura and Pterygota and mouthparts project freely from the head capsule, a condition described as *Ectognathous.* The form of the mouthparts is extremely varied, and it is appropriate to describe first their structure in the more primitive chewing condition.

### Typical Chewing Mouthparts

It a typical chewing insect the *Labrum* is a broadly flattened plate hinged to the clypeus. Its ventral (inner) surface is usually membranous and forms the lobelike epipharynx, which bears chemosensilla.

The *Mandible* is a heavily sclerotized, rather compact structure having usually a dicondylic articulation with the subgena. Its functional area varies according to the diet of the insect. In herbivorous forms there are both cutting edges and grinding sur- faces on the mandible. In carnivorous species the mandible possesses sharply pointed "teeth" for cutting and tearing. In apterygotes (except the Lepismatidae) and larval Ephemeroptera the mandible has a single articulation with the cranium and, as a result, much greater freedom of movement. Of all the mouthparts the *Maxilla* retains most closely the structure of the primitive insect an limb. The basal segment is divided by a transverse line of flexure into two subsegments, a proximal *Cardo* and a distal *Stipes.* The cardo carries the single condyle with which the maxilla articulates with the head. Both the cardo and stipes are, however, attached on their entire inner surface to the membranous head pleuron. The stipes bears an inner *Lacinia* and outer *Galea,* and a *Maxillary Palp*. This basic structure is found in both apterygotes and the majority of chewing pterygotes, although in some forms reduction or loss of the lacinia, galea, or palp occurs.

The *Labium* is formed by the medial fusion of the primitive appendages of the postmaxillary segment, together with, in its basal

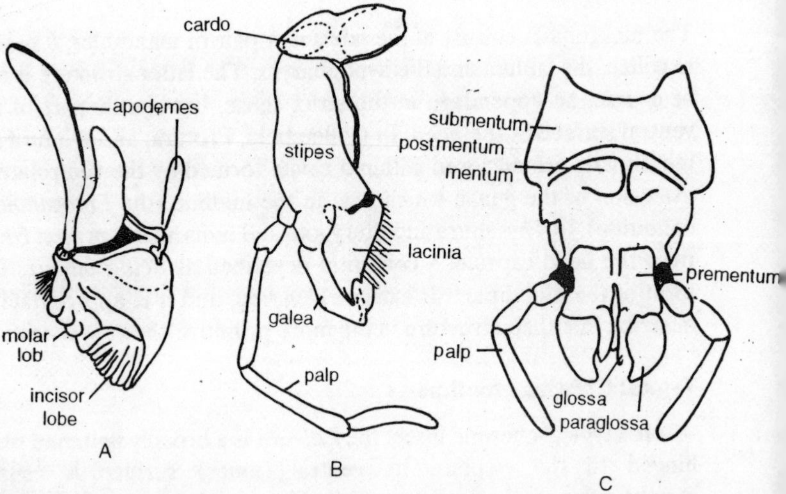


**Fig. 2.9.** Structure of mandible (A) maxilla (B), and labium (C) of a typical chewing insect.

region, a small part of the sternum of that segment. The labium is divided into two primary regions, a proximal *Postmentum* corresponding to the maxillary cardines plus the sternal component, and a distal *Prementum* homologous with the maxillary stipites. The postmentum is usually subdivided into *Submentum* and *Mentum* regions. The prementum bears a pair of inner a *Glossae* and a pair of outer *Paraglossae*, homologous with the maxillary laciniae and galeae, respectively, and a pair of *Labial Palps*. When the glossae and paraglossae are fused they form a single structure termed the *Ligula.*

Arising as a median, mainly membranous, lobe from the floor of the head capsule and projecting ventrally into the preoral cavity is the *Hypopharynx*. It is frequently fused to the labium. In some insects (Thysanura, Ephemeroptera and Dermaptera) a pair of lobes, the *Superlinguae*, which arise embryonically in the mandibular segment, become associated with the hypopharynx. The hypopharynx divides the preoral cavity into anterior and posterior spaces. The upper parts of which are the *Cibarium* (leading to the mouth) and *Salivarium* (into which the salivary duct opens), respectively.

**Mouthpart Modifications**

The typical chewing mouthparts described above can be found with minor modifications in Odonata the orthopteroid orders. Neuroptera and Coleoptera (with the exceptions mentioned below). Mecoptera, primitive Hymenoptera, and larval Ephemeroptera. Trichoptera, and Lepidoptera. However, the basic arrangement may undergo great modification associated with specialized feeding habits (especially the uptake of liquid food) or other, non-trophic functions. Suctorial mouthparts are found in members of the hemipteroid orders, and adult Siphonaptera. Diptera, higher Hymenoptera and Lepidoptera. The mouthparts arc reduced or absent in non-feeding or endoparasitic forms.

The structure of the mouthparts provides information on an insect's diet and feeding habits, and is also of assistance in taxonomic studies. Some of the more important modifications for the uptake of liquid food are described below. It will be noted that all sucking insects have two features in common. Some components of their mouthparts are modified into tubular structures, and a sucking pump is developed for drawing the food into the mouth.

**Coleoptera and Neuroptera**

In certain species of Coleoptera and Neuroptera the mouthparts of the larvae are modified for grasping, injecting, and sucking. In the beetle *Dytisus,* for example, the laterally placed mandibles are long, curved structures with a groove having confluent edges on their inner surface. The labrum and labium are closely apposed so that the cibarium is cut off from the exterior. When prey is grasped digestive juices from the midgut are forced along the mandibular grooves and into the body. After external digestion liquefied material is sucked back into the cibarium. In *Dytiscus* the suctorial pump is constructed from the cibarium, the pharynx, and their dilator muscles.

**Hymenoptera**

In adult Hymenoptera a range of specialization of mouthparts can be seen. In primitive forms, such as sawflies, the mandible is a typical biting structure, and the maxillae and labium, though united, still exhibit their component parts. In the advanced forms, such as bees, the mandibles become flattened and are used for grasping and

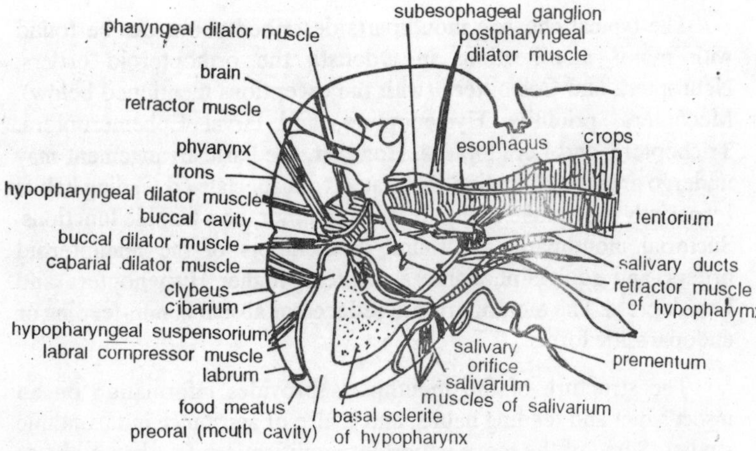


**Fig. 2.10.** Simplified sectional diagram through the insect head showing the general arrangement of the parts.

molding materials rather than biting and cutting. The maxillolabial complex is elongate, and the glossae form a long flexible "tongue," a sucking tube capable of retraction and protraction. The lacinia are lost and the maxillary palps reduced. But the galeae are much enlarged, flattened structures, which in short-tongued bees are used to cut holes in the flower corolla to gain access to the nectary. When the food is easily accessible, the glossae, labial palps, and the galeae form a composite tube up which the liquid is drawn. When the food is confined in a narrow cavity such as a nectary, only the glossae are used to obtain it. The sucking mechanism of the Hymenoptera includes the pharynx, buccal cavity, and cibarium and their dilator muscles.

### Lepidoptera

In the adults of only one family of Lepidoptera, the Micropterygidae, are functional biting mouthparts retained. In all other groups the mouthparts are considerable modified in conjunction with the insect's diet of nectar. The mandibles are usually completely lost, the labrum is reduced to a narrow transverse sclerite, and the labium is a small flap (though its palps remain quite large). The long, suctorial proboscis is formed from the interlocking galeae.

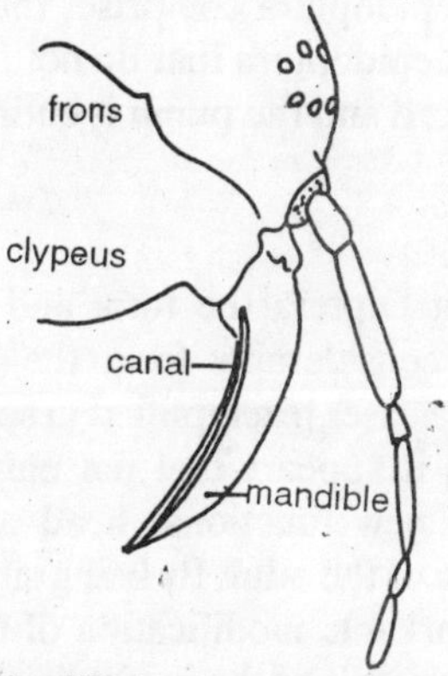


**Fig. 2.11.** Left mandible of Dytiscus larva, seen dorsally, showing the canal on its inner side.

Whose outer walls comprise a succession of narrow sclerotized arcs alternating with thin membranous areas: presumably this arrangement facilitates coiling. Extension of the proboscis is brought about by a local increase in blood pressure.

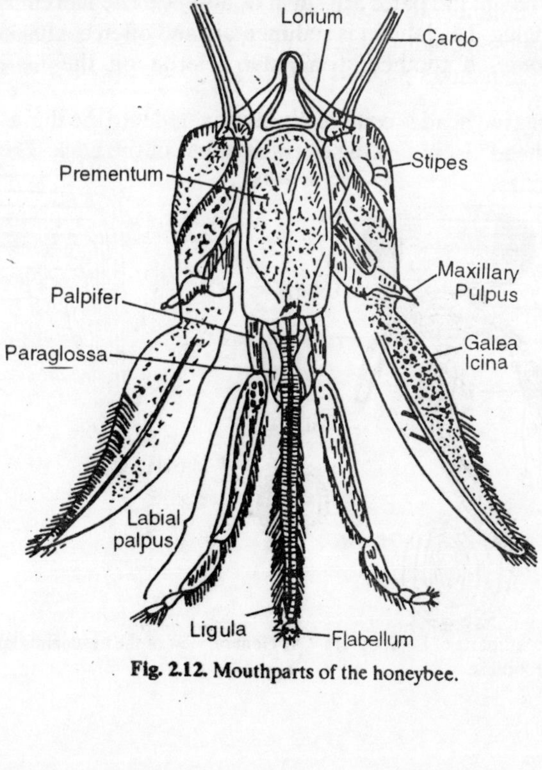


**Fig. 2.12.** Mouthparts of the honeybee.

The sucking pump of Lepidoptera comprises the same elements as that of Hymenoptera. In Lepidoptera that do not feed as adults all mouthparts are greatly reduced and the pump is entirely absent.

## Diptera

In both larval and adult Diptera the form and function of the mouthparts have diverged considerably from the typical chewing condition. Indeed, in extreme cases [exemplified in some of the larvae (maggots) of Cyclorrhapha] it appears that not only a new feeding mechanism but an entirely new functional head and mouth have evolved, the true mouthparts of the adult fly being suppressed during the larval period. This remarkable modification of the head and its appendages is of course, the result of the insect living entirely within its food.

*Larva*. In many of the more primitive Orthorrhapha the head is retracted into the thorax and enclosed within a sheath formed from the neck membrane. The mandibles and maxillae possess the typical biting structure (though the palps are small or absent). The labrum is large and overhanging. The labium is rudimentary and often confused with the *Hypostoma*, a toothed, triangular sclerite on the neck membrane.

In maggots the true head is completely invaginated into the thorax and the conical "head" is, in fact, a sclerotized fold of the neck. The

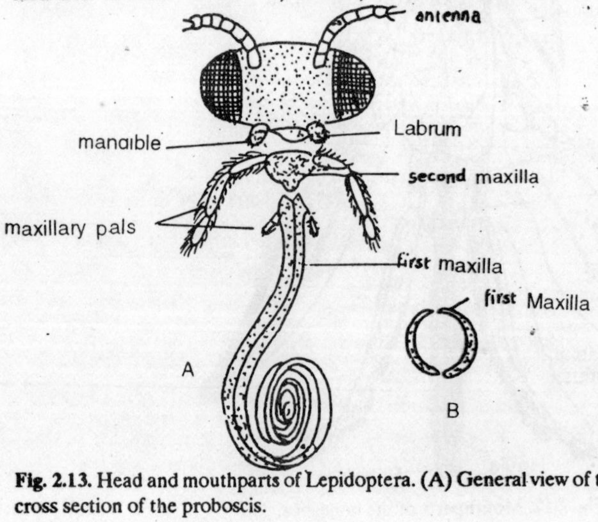


**Fig. 2.13.** Head and mouthparts of Lepidoptera. (A) General view of the head, and (B) cross section of the proboscis.

functional "mouth" is the inner end of the preoral cavity, the *Atrium* from which a pair of sclerotized hooks protrude. The cibarium is transformed into a massive sucking pump, and the true mouth is the posterior exit from the pump lumen.

*Adult*. No adult Diptera have typical biting mouthparts, although, of course, many are said to "bite" when they pierce the skin. The mouthparts can be divided functionally into those that only suck and

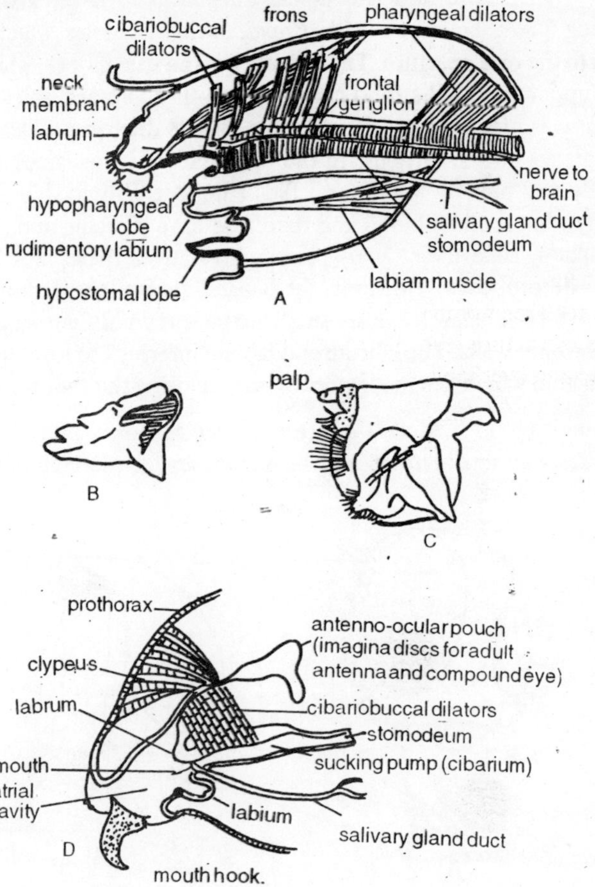


**Fig. 2.14.** Head and mouthparts of larval Diptera. (A) Diagrammatic section through the retracted head of *Tipula*, (B) right mandible of *Tipula* (C) left maxilia of *Tipula*, and (D) diagrammatic section through the anterior end of a magoot.

those that first pierce an then suck. In the latter the piercing structure may be the mandibles, the labium, or the hypopharynx.

In the Diptera which merely suck or "sponge" up their food (for example, the housefly and blowfly) the mandibles have disappeared and the elongate feeding tube, the *Proboscis,* is a composite structure that includes the labrum, hypopharynx, and labium. The proboscis is divisible into a basal *Rostrum* bearing the maxillary palps, a median flexible *Haustellum*, and two apical *Labella*. The latter are broad sponging pads, equipped with *Pseudotrachea*E along which food passes to the oral aperture. The latter is not the true mouth, which lies at the upper end of the food canal. As in other Diptera, the sucking apparatus is formed from the cibarium and its dilator muscles which are inserted on the clypeus. In the same family (Muscidae) ass the housefly and blowfly are several flies (the tsetse fly, stable fly, and horn fly) that feed on blood and therefore have piercing and sucking mouthparts. Like their "non- piercing" relatives these flies have a composite proboscis. However, the haustellum is elongate and rigid, and the distal labellar lobes are small but bear rows of *Prestomal Teeth* on their inner walls. The labrum and labium interlock to form the food canal within which lies the hypopharynx enclosing the salivary duct.

Other blood-feeding flies (for example, horse and deerflies, blackflies, and mosquitoes) use the mandibles for piercing the host's

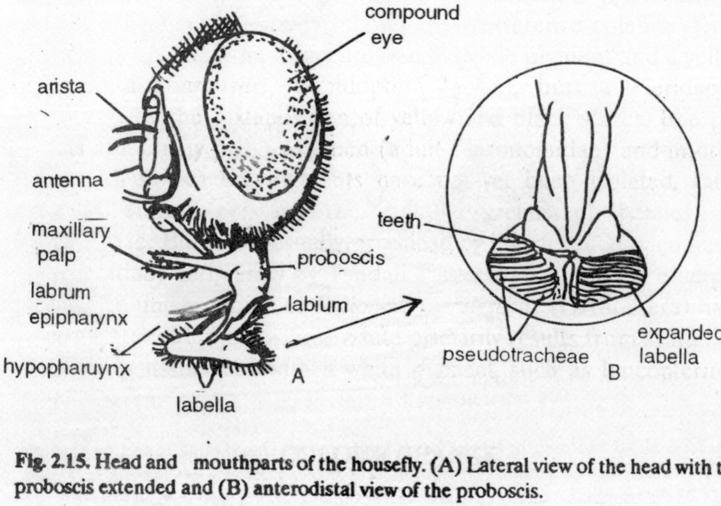


**Fig. 2.15. Head and mouthparts of the housefly. (A) Lateral view of the head with the proboscis extended and (B) anterodistal view of the proboscis.**

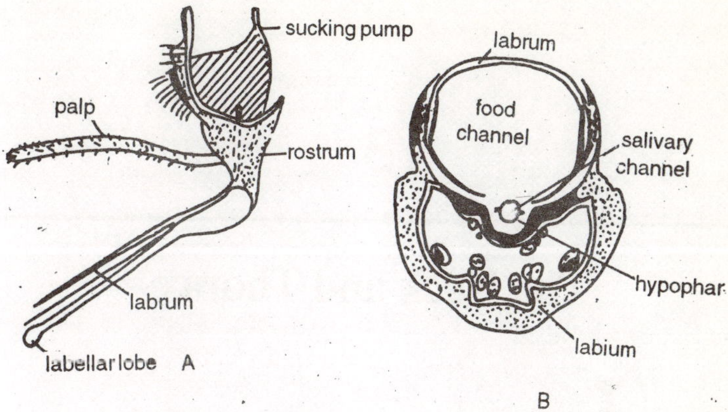


**Fig. 2.16.** Mouthparts of the testse fly. (A) Lateral vieew and (B) cross section of the proboscis.

skin. The mouthparts of the horsefly Tabanus may be taken as an example. The labrum is dagger-shaped but flexible and blunt at the tip. On its inner side in a groove closed posteriorly by the mandibles to form the food canal. The mandibles are long and sharply pointed. The maxillae retain most of the components of the typical biting form (except the laciniae) but the galeae are long bladelike structures. The hypopharynx is a styletlike structure and contains the salivary duct. The labium is a large, thick appendage with a deep anterior groove into which the other mouthparts normally fit. Distally it bears two large labellar lobes. Blood flows along the pseudo-tracheae to the tip of the food canal.

**Hemiptera**

The major contributor to the hempteran proboscis is the labium, a flexible segmented structure with a deep groove on its anterior surface. Within this groove are found the piercing organs, the *Mandibular* and *Maxillary Bristles*. The two maxillary bristles are interlocked within the labial groove and form the food and salivary canals. Because of the great enlargement of the clypeal region of the head associated with the opisthognathous condition, the cibarial sucking pump is entirely within the head.

# 3

# The Neck and Thorax

As stated already that the body of an insect is divided into three parts : Head, Thorax and Abdomen. The thorax is the locomotory center of the insect. Typically each of its three segments (pro-, meso-, and metathorax) bears a pair of legs, and in the adult stage of the Pterygota the meso- and metathoracic segments each have a pair of wings. Between the head and thorax is the membranous neck *(Ceervix)*.

## THE NECK

Study of its embryonic development shows that the neck contains both labial and prothoracic components and therefore the primary

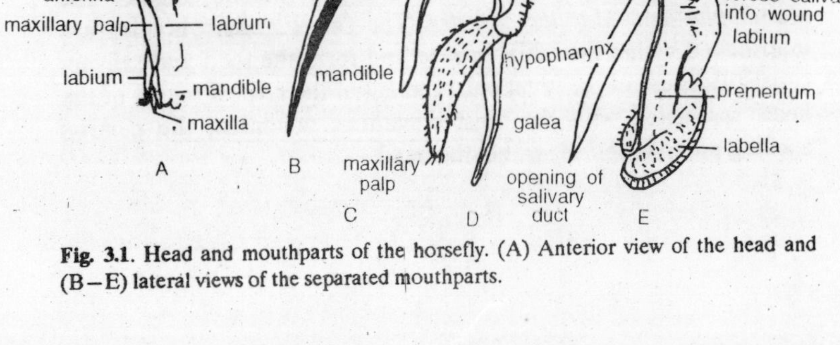


**Fig. 3.1.** Head and mouthparts of the horsefly. (A) Anterior view of the head and (B–E) lateral views of the separated mouthparts.

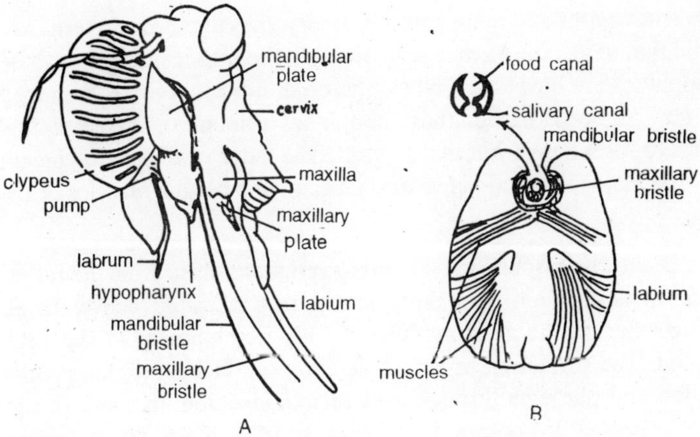


**Fig. 3.2.** Head and mouthparts of Hemiptera, (A) Head with the mouthparts separated and (B) cross section of the proboscis.

intersegmental line must be within the neck membrane. The muscles that control head movement arise on the postoccipital ridge and are attached to the antecosta of the prothoracic segment. Thus, they must include the fibers of two segments. This modification of the basic segmental structure was apparently necessary to provide sufficient freedom of head movement. Usually supporting the head and articulating it with the prothorax are the cervical sclerites, a pair of which occurs on each side of the neck. Either one or both sclerites maybe absent. When only the occurs it is often fused with the prothorax. Occasionally additional dorsal and ventral sclerites are found.

## STRUCTURE OF THE THORAX

In the evolution of the typical insectan body plan there have been two phases associated with the development of the thorax as the locomotory center: in the first the walking legs became restricted to the three thoracic segments, and in the second atriculated wings were formed on the meso- and metathoracic terga. Accompanying each of these developments were major changes in the basic structure of the secondary segments of the thoracic region. These changes were primarily to strengthen the region for increased muscular power. In the apterygotes and many juvenile apterygotes the thoracic terga are little that unites the *Sternal Apophyses*. In the higher pterygotes these apophyses are borne on a median internal ride and form a Y-shaped *Furca.* As noted earlier these apophyses combine with the pleural arms to form a rigid internal support. The latter provides attachment for the major longitudinal ventral muscles and certain muscles of the leg.

It must be emphasized that many variations occur from the rather general description of a thoracic segment provided above. In all insects the prothorax is modified by the development of the neck region. The pronotum especially is different, lacking the antecostal region and phragma through neck membranization. In some groups (for example, Orthoptera. Hemiptera, and Coleotera) the pronotum is greatly enlarged; in others it is reduced to a narrow band between the head and mesothorax. In those orders whose members have a single pair of functional wings, the tergal plates of the segment from which the wings are absent are usually reduced in size.

**The Tergum**

The *Pronotum* (the tergum of the prothorax) is typically a single sclerite, but may be subdivided. Any sutures or sclerites present have no relationship with those of the mesonotum and metanotum. A peculiarity of the pronotum is the loss of the antecosta and acrotergite because the anterior region is unsclerotized and forms part of the neck membrane. The structure of the *mesonotum* and *metanotum* is

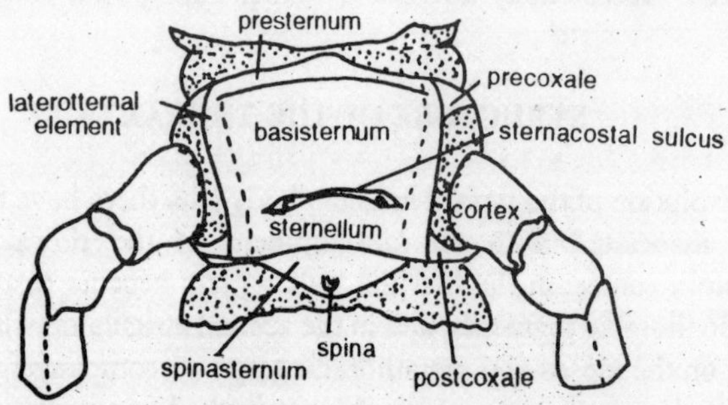


**Fig. 3.3. Ventral view of a generalized thoracic sternum.**

correlated with the development of wings. In immature insects and adult wingless insects they resemble the abdominal terga, but with the development of functional wings in the adult they undergo marked changes.

The principal flight movements are brought about, not by muscles inserted in the wing base, but by the bending and relaxation of the terga through the action of segmental muscles. To make these movements of the terga more definite and specific, the mesonotum, metanotum and the acrotergite with the antecosta of the first abdominal tergum are all consolidated, and the intersegmental membranes are eliminated or reduced to a minimum. In order to strengthen the terga, to afford attachment to muscles and to give more precise direction to the curvature of the terga during flight, a number of internal ridges or inflections have developed. These are marked externally by sulci which divide the tergum into several sclerites.

The most constant of these ridges are the *antecosta* and the *scuto-scutellar inflection. A prescutal inflection* also is usually present but is not always strongly developed. The sulci associated with these inflections divide the tergum into four primary sclerites. The *acrotergite* lies in front of the *antecostal suture* and when it remains attached to its own segment it is usually very narrow and inconspicuous. Behind the acrotergite is the *prescutum* which also is typically very narrow though in some insects it forms a conspicuous sclerite. The *prescutal suture* which marks its posterior limit is often poorly developed, but the prescutum can be recognized, by its lateral portions, the *prealar bridges*, which extend in front of the wing bases and usually meet the dorsal edges of the pleura. The third sclerite or *scutum* is always prominent and forms normally the most extensive region of the notum. Its lateral edges are parallel to the base of the wing and bear *anterior* and *posterior notal wing processes* which assist in the articulation of the wing. The *scutellum* is typically triangular with the apex directed anteriorly. Its posterior edge is continuous with the axillary cords which form the posterior margin of the base of the wings.

These four sclerites are the typical components of the tergum of a winged insect, but the prescutum is often incompletely marked off from the scutum and the acrotergite may be reduced to a narrow rim forming the anterior fold of the antecosta. In many insects the scutum and scutellum are subdivided by additional sutures.

The dorsal longitudinal muscles by their contraction cause an upward curvature of the tergum, which in turn brings about the downstroke of the wings. Thee muscles are usually inserted on the *phragmata* or expanded plates of the antecostae of the mesonotum and metanotum and those of the metanotum and the first abdominal tergum. When the anterior wings are the principal or the sole flight organs as in the Hymenoptera or the Diptera, the antecosta and acrotergite of the metathorax become closely attached to, and form an integral sclerite of, the mesothorax. The acrotergite is well developed and now forms the *postscutellum* of the mesothorax. This arrangement increases the ability of the contracting muscles to bring about a more precise curvature of the tergum, resulting of course in more definite wing movement. In such an insect the acrotergite may become detached from the metanotum behind the antecosta but more frequently it retains its attachment to the metanotum thus bringing about a consolidation of the entire pterothorax. When a postscutellum is present in the mesothorax there is of course no acrotergite or antecosta in the metathorax. In a like manner the acrotergite of the first abdominal tergum forms a postscutellum to the metanotum, the original intersegmental membrane between the two segments being eliminated or greatly reduced.

*Paragnetina, Acroneuria or related perlid stonefly*

The *pronotum* it consists of a large sclerotized plate covering the entire dorsal area of the prothorax. Apart from the division into right and left halves by a median longitudinal sulcus, there is no division into sclerites. There is an anterior submarginal groove but this must not be confused with the antecosta. Retain the specimen for further study.

**Romalea Mircroptera**

In this insect the *pronotum* is not confined to the dorsum but grows laterally over the pleural regions also. As a result the pleuron is greatly reduced. It also grows posteriorly over most of the mesonotum and metanotum. A series of transverse sulci divide it into five sclerites. Divisions of the pronotum such as found in *Romalea* and *Paragnetina* are peculiar to the species and have no relationship to the sclerites of the mesonotum and metanotum.

**Acheta assimilis (Nymph)**

Both the *mesonotum* and *metanotum* differ in no essential detail of structure from the terga of the abdominal segments. Each consists of a very narrow acrotergite and a large undivided postcostal plate. It will be seen, from the studies which immediately follow, that the mesonotum and metanotum of winged insects are subdivided into a number of sclerites, the principal ones of which are sufficiently constant to be recognized and named. The appearance of these sclerites with their dividing inflections and sulci is coincident ontogenetically and phylogenetically with the development of wings. They are therefore absent in primitively and secondarily wingless insects and in the larvae and nymphs of winged species.

**Paragnetina Media (Perlidae)**

Certain areas in the mesonotum of the prothorax are very lightly pigmented and might be mistaken for membrane, but testing them with a needle will show that they are hard and sclerotized. The *acrotergite* forms a narrow anterior sclerite. It is most evident in the media region where it follows the curve of the sclerite behind it, but it continues laterally in front of the prealar bridges mentioned below. The deep groove that separates it from the sclerite behind it is the *antecostal suture*.

The *prescutum* lies behind the acrotergite and consists of a median, somewhat diamond-shaped region and two lateral extensions known as the *prealar bridges* which extend to the episterna, from which they are separated by narrow membranes. This pronounced development of the prescutum is rare in insects. It is more usually a narrow, often imperfectly defined sclerite, which can be recognized most readily by the prealar bridges. The sulcus which separates the prescutum from the sclerite behind it is the *prescutal suture*. Manipulate the stumps of the wings and note that their posterior margins are continued mesally as thickened *axillary cords*. These in turn are continuous with the free posterior edge of the mesonotum which forms a transverse fold with a deep inflection behind it. The fold forms the posterior edge of the sclerite known as the *scutellum*, and its relation to the axillary cords is characteristic of all winged insects. The suture begins one each side at the junction of the axillary cord and the edge of the scutellum, runs straight forward to the posterior sides of two rounded shoulder-like areas, turns antero-mesally along the inner edges of these two areas and comes to a point in the middle line. This is the *scuto-scutellar suture* which separates the scutum from the scutellum.

The *scutum* is that sclerite which lies anterior and lateral to the scutoscutellar suture. It consists of the two bulging areas mentioned earlier, a narrow median bridge connecting them and two areas posterior to the bulges, one on either side of the scutellum. On the lateral side, near the anterior edge of the bulge, find a small triangular sclerite which projects laterally toward the anterior region of the wing base. This is the *anterior notal wing process*. A second lateral process, the *posterior notal wing process* is given off near the posterior end of the notum. This is long and narrow and projects obliquely forward toward the posterior region of the wing base. The *scutellum* consists of two sclerites separated by a U-shaped *scutellar suture*. The anterior sclerite may be called the *scutellar shield*, the other the *posterior scutellar plate*.

The median longitudinal sutures found in the prescutum and scutellum are characteristic of this insect and are not of general occurrence. Behind the scutellum there is a transverse sclerite which forms the posterior wall of the deep declivity between the mesonotum and metanotum. This is the *postscutellum* which is morphologically the acrotergite of the metanotum but functionally a part of the mesonotum. Note that it extends laterally behind the wings bases to form the *postalar bridges* which unite in part with the epimera. The *metanotum* is essentially similar to the mesonotum in size and structure except that its acrotergite is enlarged and consolidated with the mesonotum. The *postscutellum* of the metanotum is the acrotergite of the first abdominal segment. If the posterior portion of the abdomen is cut lengthwise through the pleural regions of the thorax and the remaining abdominal segments. Two sets of muscles exposed are visible. Laterally, there are several dorsoventral muscles cut across. Some of these are leg muscles, others are responsible for the upstroke of the wings in flight. Between these are the longitudinal muscles which bring about the downstroke of the wings. Note that the longitudinal muscles of the meso- and metathorax are attached anteriorly and posteriorly to inward-projecting cuticular plates or ridges.

The longitudinal muscles of the mesothorax are attached in front to the anterior region of the prescutum and to two flattened plates which grow inward from the antecosta. These plates are the *first phragmata*. Posteriorly the muscles are attached to the *second phragmata*, ingrowths from the antecosta between the postscutellum (metathoracic acrotergite) and the prescutum of the metathorax. The corresponding muscles of the metathorax are attached to the second phragmata and the strongly developed antecosta of the first abdominal segment, which here lies at the posterior edge of the metathoracic postscutellum. In other insects there may be a third pair of phragmata borne by this antecosta.

After removing the heavy inner longitudinal muscles, other muscles are exposed. Some of them are attached to a median longitudinal ridge in the scutellum and others to a ridge between the median region of the prescutum and the prealar bridge. The reason for the corresponding sulci in these two sclerites is now clear; they mark the position of inflections for the attachment of muscles.

**Acheta Assimilis**

Both pairs of wings in the stonefly are used actively in flight and the two terga are equally well developed. The acrotergite of the metanotum is well developed and forms the meso-postscutellum, but it has not lost its connection with the metanotum and the two terga are consolidated. In the Orthoptera, on the other hand, the forewings or tegmina are modified and the hindwings are the principal organs of flight; therefore, the metanotum is larger and more complex than the mesonotum. This is true of the field cricket although its hindwings are atrophied and are no longer functional in flight.

Anteriorly the metanotum might be likened to the human torso, the median portion representing the base of the neck and the rounded lateral portions the shoulders. To see the anterior structures it may be necessary to remove most of the abdomen and pin the specimen with the anterior end upward. At the anterior edge of the "neck" find the deep *antecostal sulcus* associated with an internal ridge, the *antecosta*, which bears two small lateral processes or *phragmata*. The *acrotergite* is very narrow, forming little more than the anterior lamina of the antecosta. It continues laterally as a narrow anterior border to the prealar bridge.

A second pronounced sulcus borders the anterior edge of each "shoulder" and joins the antecosta at the lateral edge of the "neck." These are lateral portions of the *prescutal suture* which is not continuous across the median region of the tergum. Between the prescutal suture and the lateral portions of the antecostal suture there is a forked sclerite on each side, the two constituting the *prescutum*. The anterior branch of the fork is the *prealar bridge*, which, in close association with the lateral portion of the acrotergite, extends laterally in front of the wing base. The posterior branch is triangular in form and rests on the "shoulder" of the scutum like an epaulet. In the membrane between the two branches there is a small rounded protuberance, sparsely clothed with hairs. This is the *tegula*, which in some other insects, such as the Lepidoptera, is more conspicuously developed.

The *scuto-scutellar suture* forms an incomplete arch on the notum. The *scutum* lies anterior and lateral to this suture and bears the *anterior* and *posterior notal wing processes* on its lateral edges. The anterior process is just behind the prescutum and is closely associated with a fairly large elongated sclerite which lies in the membrane between the notum and the wing. The posterior process is near the hinder end and is in contact with a smaller sclerite in the membrane. Thee sclerites form part of the articulatory mechanism of the wings and will be studied in detail later. The posterior edge of the scutellum can be identified by the fact that, as in the stonefly, it is infolded and is continuous laterally with the axillary cords of the wings. Note that a *scutellar suture* divides it into an anterior *scutellar shield* and a *posterior scutellar sclerite*. The post-scutellum is the narrow sclerite behind the scutellum. Note that it still retains its connection with the first abdominal tergum of which it is the acrotergite.

The structure of mesonotum, apart from the absence of a postscutellum and its much smaller size is essentially similar to that of the metanotum. The male usually shows a stronger development of the scuto-scutellar ridge. This is probably correlated with the fact that the anterior wings of the female are non-functional, while those of the male make specific movements in producing the chirp.

**The Pleuron**

The thoracic segments of all adult insects and the more active larvae or nymphs have definite and usually extensive sclerotizations in the pleural areas. These sclerotizations are formed from the dorsal or supracoxal arch of the subcoxa, the basal segment of the primitive leg, which has become incorporated in the body wall to form the *pleuron*. The fundamental structure of the pleuron is the same in the three segments, but the propleuron is usually much smaller than the pteropleura, and those structures which, in the dorsal region of the latter, are associated with the wing bases are absent in the propleura.

The *pleeral sulcus* divides the pleuron into an anterior *episternum* and a posterior *epimeron*. The sulcus is typically vertical or somewhat oblique but in extreme cases may be horizontal or may follow a tortuous course. It marks the position of a strong inflection, the *pleural ridge*. The ridge terminates ventrally in an articulatory process, the *pleural coxal process*, which takes part in the dorsal articulation of the coxa. Dorsally the pleural ridge terminates in the *pleural wing process* which forms a pivot for certain movements of the wing. Dorsal to the coxal process the pleural ridge gives off a strong apodeme, the *pleural apophysis*, which projects inward above the coxal cavity where it is closely attached to the sternal apophysis.

In the pterothorax the pleuron is well-developed dorsally and it may be in intimate contact or even sclerotically continuous with the tergum by means of the *prealar bridge* which connects the prescutum and episternum in front of the wing-base, and the *postalar bridge* uniting the postscutellum and epimeron behind. Similarly, contact or union is usually established between the episternum and sternum by means of a *precoxal bridge*, and sometimes between the epimeron and sternum by a *postcoxal bridge*. In the articulatory membrane between the episternum and coxa there is frequently a triangular sclerite, the *trochantin*. In lower insects it may be quite conspicuous and in some insects is articulated to the coxa. In higher insects it tends to atrophy. Small sclerites, known as *epipleurites*, usually become detached from the dorsal region of the episternum and epimeron in adult winged insects. They lie in the membrane between the pleuron and the wing base and serve for insertion of some of the wing muscles. Epipleurites anterior to the pleural wing process are known as *basalar plates*, those posterior to the process as *subalar plates*.

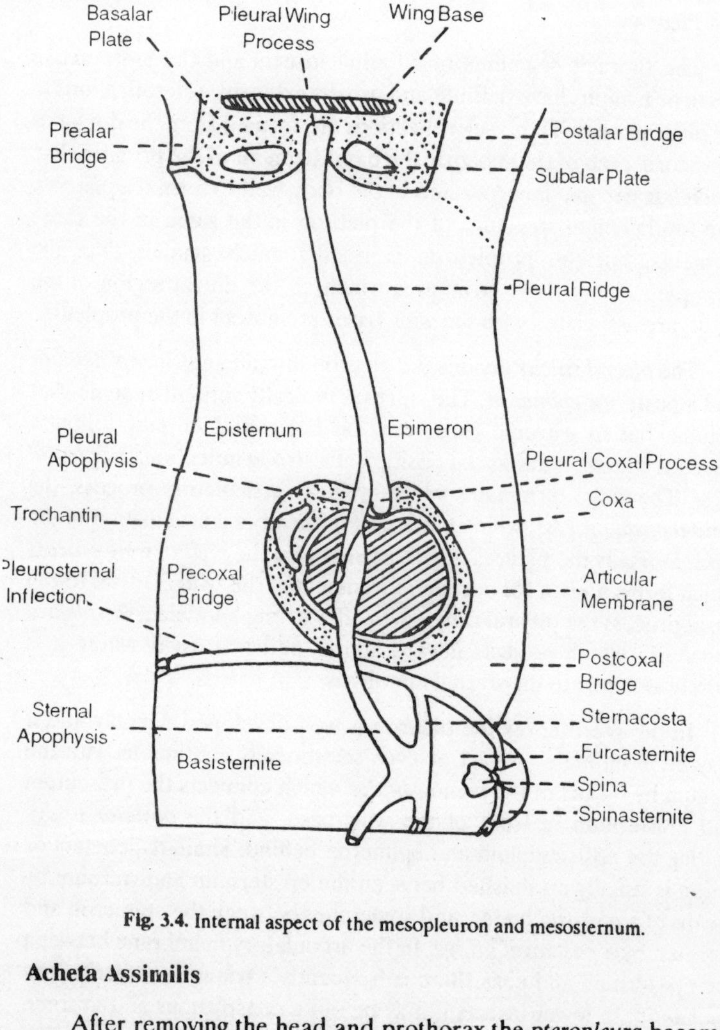


**Fig. 3.4. Internal aspect of the mesopleuron and mesosternum.**

## Acheta Assimilis

After removing the head and prothorax the *pteropleura* becomes visible. The structure of both the pleura is fundamen- tally the same. The mesopleuron overlaps the metapleuron but the two are connected by membrane and are not consolidated. Examine the mesopleuron and note that it consists largely of two obliquely elongated sclerites separated by a deep sulcus. The anterior sclerite is the *episternum*, the posterior the *epimeron* and the sulcus in the *pleural suture*. Note that dorsally the suture is continued to the base of the wing and ventrally to the proximal edge of the coxa. This is due to the fact that the suture marks the position of a strong internal *pleural ridge* which is produced dorsally to form a pivot for the movement of the wings and ventrally to form a condyle for the articulation of the leg. The processes are known respectively as the *pleural wing process* and the *pleurd coxal process*.

Dorsal to the episternum and anterior to the pleural wing process find a small sclerite known as the *basalare* or *basalare plate*. A similar but smaller and less conspicuous sclerite, the *subalare* or *subalar plate* lies dorsal to the epimeron. These sclerites are known collectively as *epipleurites* and are detached fragments of the episternum and epimeron which serve for the insection of certain wing muscles.

Partly overlapped by the ventral edge of the episternum, and lying between it and the sternum, there is a small triangular plate, the *precoxale*. It is separated by membrane from each of the adjacent sclerites. In the articular membrane in front of the coxa, and partly overlapped by the episternum and precoxale, there is a small elongated sclerite, the *trochantin* which articulates with the coxa by its ventral extremity. There is a large *metathoracic spiracle* in the intersegmental membrane ventral and posterior to the epimeron.

**Romalea Microptera**

When we observe the external surface we find that, unlike the condition found in *Acheta*, all of the sclerites of the pteropleuron, except the epipleurites and trochantin, are consolidated, i.e., they are continuous sclerotically, being separated merely by sulci. Furthermore the episternum is consolidated with the sternum. The pteropleuron consists chiefly of four nearly vertical sclerites separated by three sulci. The middle sulcus is the intersegmental suture, as can be determined by reference to its position in relation to the legs and the intersegmental line between the meso- and metathorax. Each segmental pleuron is divided by a pleural suture into an anterior *episternum* and posterior *epimeron*, as in the cricket, but the episternum is divided transversely into a dorsal *anepisternite* and a ventral *katepisternite*. A narrow sclerite, the *pre-episternite*, is cut off from the anterior edge of the katepisternite of the mesopleuron. It and its fellow on the opposite side are continuous with a similar sclerite (the presternite) cut off from the sternum. The three form a U-shaped "collar" known as the *prepectus*.

The katepisternite of the metapleuron extends ventrally in front of the coxal cavity forming a *precoxal bridge* which unites with the sternum, the two being separated by the *pleurosternal suture*. In the mesopleuron there is a *precoxale* interposed between the katepisternite and the sternum. The *epimeron* in the locust is an undivided sclerite which does not meet the sternum ventrally, but in many insects it is divided into anepimerite and katepimerite and may extend behind the coxal cavity, to the sternum forming a postcoxal bridge.

In each segment the *pleural wing process, pleural coxal process,* two *basalar plates,* one *subalar plate* and the *trochantin* are present. There are also the two prominent *spiracles,* one in the intersegmental membrane in front of the mesepisternum, the other in a small triangular plate at the posteroventral angle of the mesepimeron. These belong primitively to the meso- thorax and metathorax respectively. The sclerites surrounding the articulatory membrane of the leg have their edges strengthened by marginal and submarginal, inflections. In the mesothorax these ridges completely surround the leg, the coxal cavity being closed behind by the metepisternum. In the metathorax the cavity is open behind. As a result of these inflections a narrow sclerite is cut off in both segments anterior and ventral to the coxal cavity. This sclerite is referred to the literature as the *pericoxal ring* or *sclerite.*

On the internal face of the pleura each sulcus is represented by a well-defined inflection of ridge. The *pleural ridge*, corresponding to the pleural sulcus, is particularly prominent. Find its dorsal prolongation,s the *wing process*, and its ventral prolongation, the *coxal process,* which forms a strong condyle for the articulation of the coxa. Dorsal to the coxal process the pleural ridge gives off a strong spur-like process, the *pleural apophysis*. This extends inward above the leg opening and is intimately connected with the *sternal apophysis*, an ingrowth from the sternum. The two apophyses form a strengthening arch above the leg opening, from which certain of the leg muscles originate.

**Romalea Microptera**

The *propleuron* in the Orthoptera is greatly reduced and largely overgrown by the pronotum, only the ventral edge of the episternum being visible externally. The structure, however, is fundamentally the same as in the other pleura. Ventral to the anterior half of the edge of the pronotum, in front of and dorsal to the leg, find a triangular plate, which is continuous mesally with the basisternite, the spine-bearing sclerite of the sternum. This plate is all that is visible of the *episternum*, but the position and outline of the pleuron can be determined usually by indications on the pronotum. In front of the third sulcus or groove of the pronotum, a triangular area, slightly darker than the neighbouring parts and distinctly punctate, marks the position of the *episternum*. Behind the sulcus a very narrow horizontal, marginal, punctate area marks the position of the *epimeron*.

Whenever, the internal surface of prothorax is examined we find the following structures. The *episternum* is the anterior triangular piece, the outline of which you saw from the exterior. The *epimeron* is a very narrow sclerite posterior to the episternum and adhering closely to the lower margin of the pronotum. At the junction of the two there is a well developed *pleural ridge* starting near the apex of the triangular episternum and continuing to the leg socket where it is produced into the *pleural coxal process* which articulates with the coxa of the foreleg. There is no pleural apophysis distinct from the ridge, but the latter is very deep and the sternal apophysis is bound to its edge.

**The Sternum**

The sternum is the ventral sclerotized region of the segment. Its morphology is still somewhat obscure and the homologies in the higher insects difficult to interpret. It seems, basically, to have been formed from four distinct sclerotizations as follows. (1) A large segmental ventrosternite sclerotized in the venter of the segment. This may be termed the *mediosternite*. (2) A small intersegmental ventrosternite posterior to the mediosternite of the prothorax and mesothorax. This bears an internal process known as the *spina* and is therefore named the *spinasternite*. The anterior spinasternite may or may not unite with the prothoracic mediosternite but the posterior

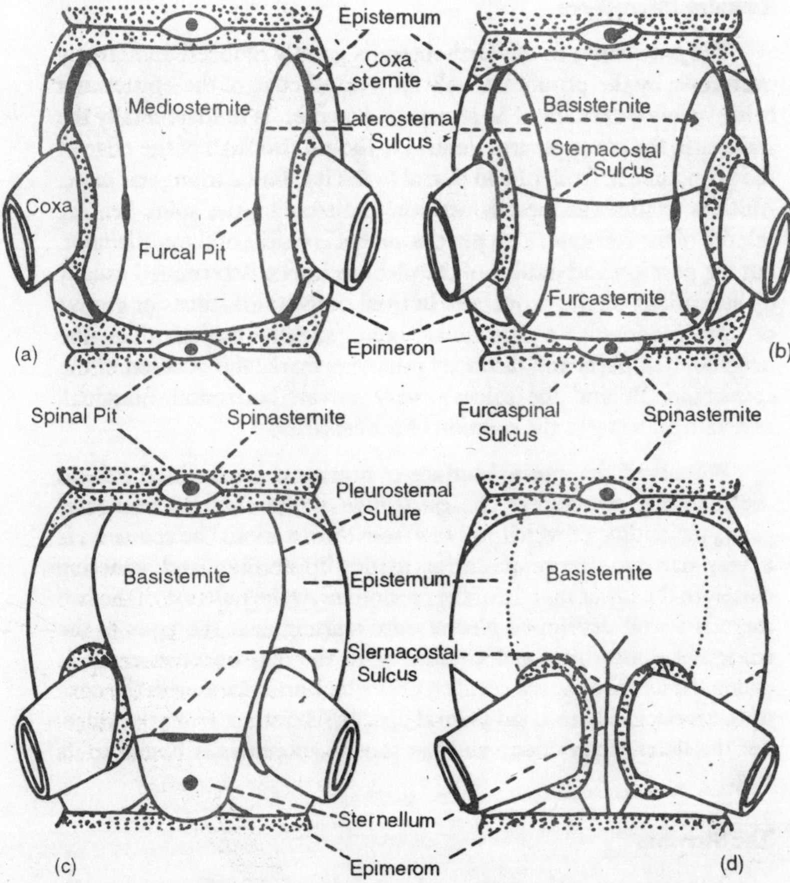


**Fig. 3.5.** Some stages in the evolution of the mesosternum. (a) Union of mediosternite and coxosternite. (b) Development of sternacosta. (c) Loss of laterosternal and furcaspinal sutures. (d) Posterior and ventral migration of coxae resulting in reduction and infolding of the sternellum.

one always unites with the mesothoracic mediosternite. (3) Two *coxosternites* derived from the infracoxal arcs of the subcoxae. These lie lateral to the mediosternite with which they later unite to form the *eusternum*. A pair of *laterosternal sulci* are formed at the lines of union and within each sulcus is a slit-like *furcal pit* which marks the invagination of the *sternal apophyses* two internal processes which lend support to the region of leg attachment. An internal strengthening ridge, the *sternacosta*, develops between the anterior edges of the bases of the apophyses and continues laterallv across the coxosternites. Thus, externally the eusternum is divided transversely by the *sternacostal sulcus* into an anterior *basisternite* and a posterior *furcasternite*. The laterosternal sulci disappear most insects. The legs in most insects lie toward the posterior end of the segment, therefore, the apophyses also are posterior in position. This results in a reduction of the furcasternite, leaving the basisternite as the most conspicuous sternal sclerite.

In the higher insects the leg bases are closely approximated and the portion of the sternum lying between them probably consists only of the coxosternites, the mediosternite being reduced and inflected along the middle line. In such insects there is usually an extensive sternal plate anterior to the legs. Whether this also consists only of the coxosternites or whether the mediosternite contributes to its structure is not definitely known. In these insects the episternum and basisternite are usually completely fused, as are the epimeron and furcasternite; consequently there are not dividing lines between the pleuron and sternum. Ferris contends that the entire ventral region in higher insects is derived from the subcoxae. He regards all structures derived from these as the pleuron, and would limit the sternum to the plates here called ventrosternites. He claims therefore that, at least in the higher insects, there are no sternal sclerites.

**Paragnetina, Acroneuria or Other Perlid Stonefly**

The distinct sternal plates are developed in each segment and that, unlike all other segmental plates, the anterior edges of the mesosternum and metasternum overlap and plates anterior to them. The metasternum also overlaps the first abdominal sternum. The component sclerites of the sterna are distinct and their arrangement is such as to present an almost schematic representation of the probable early stages in the development of the sternum.

**Prosternum**

Note that this is composed largely of three sclerites which together form the *eusternum*. The two somewhat crescentic lateral sclerites, adjacent to the *coxal cavities*, are *coxosternites* derived from the infracoxal arc of the subcoxa. The median sclerite or *mediosternite* is a ventrosternite sclerotized in the venter of the segment. The mediosternite and coxosternites are separated by the two *laterosternal sulci*. Find the elongated *furcal pit*, marking the position of the sternal apophysis or furcal arm, in the posterior half of each sulcus. Behind the mediosternite there is a small oval or lenticular intersegmental ventrosternite with a single invagination or pit. This is the *spinasternite* and the invagination is the *spinal pit*. Note that the spinasternite is separated by membrane from the eusternum.

### Metasternum

Note that the mediosternite is divided into *premediosternite* and *postmediosternite* by the transverse *sternacostal suture* which passes across the anterior ends of the furcal pits and continues across the coxosternites. The sulcus thus divides the entire eusternum into an anterior *basisternite* and posterior *furcasternite*, each of which is a composite plate made up respectively of the anterior and posterior regions of the mediosternite and coxosternites. The laterosternal sulci terminate some distance behind the anterior edge of the basisternite, therefore the mediosternite and coxosternites are completely fused anteriorly. This perhaps marks the beginning of the disappearance of the laterosternal sulci which are rarely present. In most insects the mediosternite and coxosternites are indistinguishably united. Note that there is no spinasternite in this segment.

### Mesosternum

The structure here is essentially like that of the metasternum, except for the presence of a *spinasternite* which is united with the eusternum. The united furcasternite and spinasternite constitute the *sternellum*. There is a median sulcus in the sternellum but this is characteristic of the species and of no special morphological significance.

### Romalea Microptera

The morphology of the sternum in the higher insects is difficult to interpret owing to the disappearance of some of the primary sutures described in *Paragnetina*, the appearance of secondary sutures marking the position of internal strengthening ridges and the reduction or loss of certain of the sclerites. This difficulty appears even in a relatively generalized insect such as *Romalea*.

**Prosternum**

All the primary sutures and sclerites are apparently recognizable in the prosternum. A very narrow marginal *presternite* is cut off from the anterior end. Immediately behind this is the *basisternite* which, as found earlier, is continuous, in front of the leg, with the episternum. The basisternite bears a large median spine which should be removed to expose the structures behind it more fully. A large *furcal pit* will be found on each side of the spine. A curved suture, not always very distinct, passes through each pit and cuts off a narrow crescentic sclerite lateral to the leg. This suture is almost certainly the *laterosternal suture* and the sclerite the *coxo-sternite*, but there is a possibility that the suture marks a secondary strengthening ridge in the basisternite. The student should be prepared to find these uncertainties in interpretation, which can be solved only by intensive comparative studies, and should not regard interpretations given here, or elsewhere, as necessarily final. Extending between the pits, and passing behind the spine,you will find the curved *sternacostal suture* which separates the basisternite from the transverse *furcasternite*. Identify the median *spinal pit* in the posterior sclerite of the sternum and note that a transverse suture separates the rather large *spinasternite*, in which the pit lies, from the furcasternite anterior to it. The acridids are among the few insects in which the spinasternite of the prothorax is fused with the eusternum.

**Mesosternum**

The mesosternum has a distinct *presternite*, conti- nuous with the preepisternite. Apart from this the sternum apparently consist of a single sclerite produced posteriorly into two lateral lobes. Find the two *furcal pits* and the *spinal pit* at the posterior border of the median region of the sclerite. Note that these pits connected by a short curved suture which continues a short distance lateral to the pits and partially separates the lateral lobes from the anterior portion of the sclerite. From its relation to the furcal pits this suture may be interpreted as the *sternacostal suture*. The lobes would, therefore, be the lateral portions of the *furcasternite*, and the body of the sclerite would be the *basisternite*. The spinasternite and the median portion of the furcasternite have disappeared and the three apophyses all arise from the internal ridge marked by the sternacostal suture.

### Acheta Assimilis

Remove the ventral half of the thorax of two specimens by cutting longitudinally through the pleura dorsal to the coxae. Boil one specimen in potassium hydrate for a study of the internal features.

### Posternum

Note that the prosternal area is not completely sclerotized but contains a considerable amount of membrane. This condition is common in the prothorax of insects and is probably the result of secondary descleŕotization. Find the sclerites whose descrip- tion follows. The *premediosternite* is the prominent, transverse, anterior sclerite which is continuous laterally with the edge of the episternum. Behind this sclerite and separated from it by the short, curved *sternacostal suture*, are two posteriorly divergent sclerites. These are lateral fragments of the *postmediosternite*. Examine the internal surface and note that the *sternal apophyses*. arise from their posterolateral edges. Lateral to the mediosternite are two crescentic lightly-sclerotized *coxosternites*, very similar in position and form to those of *Paragnetina*. They are semimembranous and not always easily distinguishable. In the membrane behind the mediosternite there is a small median shield- shaped *spinasternite*. On its inner face it bears the *spina*, a short process with a flattened palm-shaped ental surface.

### Mesosternum

Note that from the external view the sternum apparently consists of a single four-sides sclerite, the *basisternite*. Examination of the internal surface, however, will show that behind this there is a narrow sclerite bent entally and bearing the paired *sternal apophyses* as well as the *spina*. A common external invagination serves the three processes since the bases of the apophyses are juxtaposed and the spina arises from their dorsal wall at the point of junction. This juxtaposition of the apophyses to form a forked structure arising from a common base is a very common condition in higher insects and is the basis for the term *furca* generally used for them. The posterior sclerite bearing the three processes is obviously the sternellum composed of the greatly reduced *furcasternite* and *spinasternite*. Examine the inner surface and note the close relationship between the sternal and pleural apophyses.

## THE THORAX OF A SPECIALIZED INSECT

### Tipulidae (Crane Flies) (Diptera)

The preceding studies of the thorax have been concerned with comparatively generalized insects and have been designed to illustrate the fundamental structure of the thorax. In the higher insects the structure becomes more complicated and often difficult to interpret. The thoracic sclerites many be greatly distorted in position as well as in proportions. There may be additional subdivisions of the sclerites, and sclerites originally separate may become fused and their limits obliterated. It is impracticable here to study examples from each of

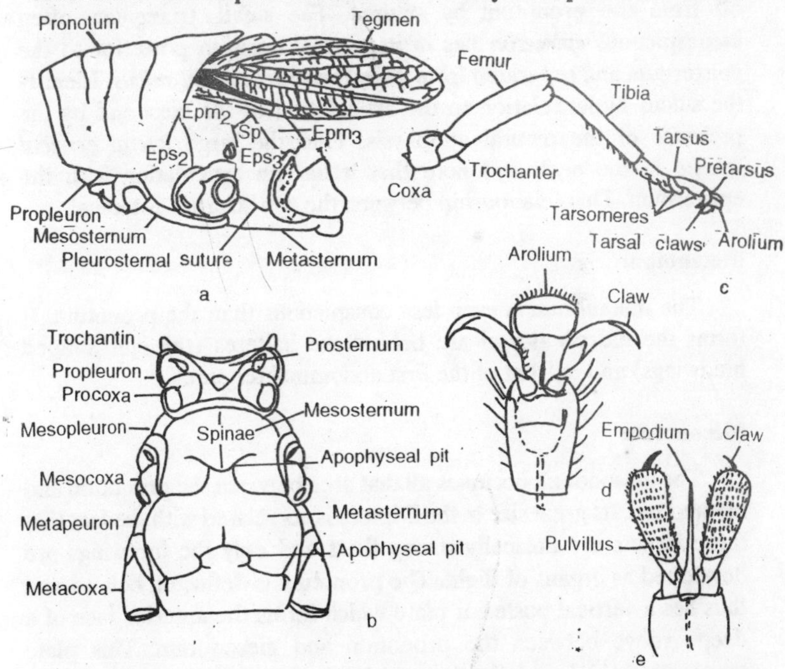


**Fig. 3.6.** Insect thorax and legs: *a* lateral view of thorax of *Romalea microptera* (Acrididae): *b*, ventral view of same: *c*, foreleg of same: *d*, pretarsus of a roach, *Periplaneta spp.* (Blttidae): *e*, pretarsus of an asilid fly. ventral view Epm epimeron Eps, episternum : sp, spiracle.

the higher groups, but the thorax of a crane fly will be studied in detail and its structure interpreted in the light of the principles already studied. Any of the common large crane flies may be used, as the differences in the structure of the thorax in different species are relatively slight. In some species the thorax is variegated and there maybe a sharp dividing line between two differently coloured regions of the same sclerite. Do not confuse such lines with suture. The whole specimens should be supplemented by sagittal and frontal sections boiled in potassium hydroxide.

### Pronotum and Propleuron

The *pronotum* and *proepisterna* together form a narrow arch or "collar" behind the neck. In some species the arch is a continuous sclerite, in others the episterna are more or less completely marked off from the pronotum by sutures. The small, triangular, often inconspicuous *epimeron* lies in its normal position posterior to the episternum and separated from its by the short *pleural sulcus*. Identify the sulcus by its relation to the articulation of the coxa and by the presence of the pleural apophysis. Find the large *lateral cervical sclerite* in the neck and note that it has an articulation with the episternum. This relationship between the two sclerites is typical.

### Metanotum

The metanotum is even less conspicuous than the pronotum. It forms the narrow dorsal arc behind the halteres (the club-shaped hindwings) and in front of the first abdominal tergum.

### Mesonotum

The mesonotum occupies all that area between the pronotum and metanotum. Its great size in the Diptera is correlated with the fact that these insects are normally strong fliers and only the forewings are developed as organs of flight. The pronotum is deflected behind and thus has a vertical posterior plate which forms the anterior face of a deep groove between the pronotum and mesonotum. This plate terminates in a short inflection which bears a small median *phragma*. The inflection is evidently the *antecosta*, but there is no acrotergite distinct from the posterior plate of the pronotum. The phragma is much smaller than would be expected in such a strong-flying insect. It no longer serves for the insertion of the longitudinal wing muscles but only for the origin of a pair of small neck muscles. The thick mass of longitudinal flight muscles arises secondarily along the entire length of the anterior sclerite of the scutum. The posterior face of the dorsal groove between the pronotum and mesonotum is formed by the anterior edge of the scutum. Forming the floor of the groove between the antecostal sulcus and the scutum there is a very narrow, usually unpigmented or lightly pigmented sclerite which is the median portion of the *prescutum*. Lateral to the groove the prescutum forks. The anterior branch runs ventrally adjacent to the pronotum. The posterior branch lies along the anterior lateral border of the scutum and terminates just in front of the wing base. The area between the two branches is desclerotized, the amount of desclerotization varying indifferent species. The *mesothoracic spiracle* lies in this area. Note that though greatly modified the prescutum forms an alar bridge, in front of the wing base, between the notum and episternum.

The *scutum* is partially or completely divided into a number of secondary sclerites. A pair of sulci, orginating near the middle of the lateral margins, in front of the wing bases, run posteromesally. They converge and in some species meet each other in the middle line, forming a *transscutal sulcus* which divides the scutum into an anterior and a posterior sclerite. A somewhat similar pair of sulci arise laterally in the anterior sclerite and run posteromesally. These do not meet but, in some species, continue posteriorly until they join the transscutal sure, thus dividing the anterior sclerite into a median and two lateral sclerites. These sulci seem to correspond to the sutures known as *notaulices*. The posterior sclerite of the scutum is frequently divided into right and left halves by a *median sulcus*.

The median shield-shaped *scutellum* is easily recognized behind the scutum. Note that, as in the more primitive insects studied, the posterior border of the scutellum is continued laterally into the axillary cord of the posterior margin of the wing base. Posterior to the scutellum lies the *postscutellum* which is extraordinarily well-developed and is divided by longitudinal sutures into dorsal and lateral sclerites. The lateral sclerites are often divided by vertical sutures into anterior and posterior portions. They occupy most of the space between the bases of the wings and halteres, and the lateral portions form what is in effect a *postalar bridge* uniting the tergum and the epimeron. After treating the specimen with potassium hydroxide and find the well-developed *phragmata* at the antecosta between postscutellum and metanotum. Examine a sagittal section of the thorax and find the prominent longitudinal muscles which have their origin on the inner surface of the median region of the anterior sclerite of the scutum and are inserted on the phragma of the postscutellum, except for one band which is inserted on the posterior edge of the postscutellum itself.

**Mesopleuron**

The mesopleuron occupies its normal position; the sclerites and the *pleural suture* are well-developed and easily recognized. The *episternum* is partially divided by an oblique suture into a dorsal *naepisternum* and ventral *katerpisternum*. The latter is indistinguishably united with the sternum and there is no pleurosternal suture. The epimeron also is divided into a large *anepimeron* and very small *katepimeron*. Ventral to the katepimeron, and lying between the bases of the middle and hind legs, there is a somewhat triangular sclerite, the *meron*. This is really a detached portion of the coxa or basal segment of the leg. In some species the katepimeron and meron may be partially or completely fused. Find the *pleural coxal* and *pleural wing processes,* the *subalar plate* and the *basalar plates,* the *pleural ridge* and the *pleural apophysis*.

**Metanotum and Metapleuron**

Examine the metanotum again and note that it consists of a single narrow sclerite forming the posterior border of the thoracic dorsum. In correlation with the degeneration of the hindwings the metanotum is reduced and not subdivided into sclerites, but its acrotergite forms the prominent and well-developed postscutellum of the mesothorax. The remaining portion of the metanotum is rolled into a narrow cylinder as can be seen by observing the cut edge of a sagittal section. Laterally the metanotum is indistinguishably fused with the *epimeron*, a narrow undivided sclerite which forms the posterior border of the lateral side of the thorax. In front of the epimeron, and separated from it by the *pleural sulcus*, is the small *episternum* divided by a distinct sulcus into *anepisternum* and *katepisternum*. The anepisternum is deeply cleft behind, and the *metathoracic spiracle* lies in the cleft. The *halteres* are borne dorsal to the pleural sulcus.

**Sternum**

The sternum is greatly modified in comparison with that of the more generalized insects studied. It is, however, possible to interpret its structure on the basis of the theory of the sternum outlined earlier. Examine both natural and cleaned (potassium hydroxide) specimens so that both external and internal elements can be seen, and verify the following description.

The *prosternum* consists of two median sclerites. The long narrow anterior sclerite is infolded so that only the narrow margins show externally on each side of the median groove. The comparatively wide spaces between this sclerite and the coxae are completely membranous. The posterior sclerite is triangular, its apex directed posteriorly. A sulcus running along its anterior margin separate it from the anterior plate and the *sternal apophyses* are borne at its anterior angles associated with this sulcus. It would appear that in this segment the coxosternites are completely desclerotized, the transverse sulcus in the *sternacostal sulcus*, the anterior infolded sclerite is the *premediosternite* and the posterior triangular plate the *postmediosternite*. There is no evidence of a spinasternite.

The mesothoracic coxae are posterior and ventral in their location and almost touch along the mid-ventral line. As a consequence the *mesosternum* consists of a broad anterior plate in front of the coxae and a very narrow posterior plate which lies chiefly between the coxae but has narrow lateral extensions which form the anterior margins of the coxal cavities. A transverse sulcus separates the two plates and examination of the internal surface will show that the sternal apophyses are associated with it. It is, therefore, the *sternacostal sulcus*, the large anterior plate is the *basisternite* and the posterior plate lying between the coxae is the *furcasternite*. There is no pleurosternal suture, consequently the basisternite is united with the katepisternum to form a continuous sclerite known to dipterologists as the "sternopleura." Presumably the basisternite includes both medisternite and coxosternite elements but some authors maintain that it is formed from the coxosternites only. A deep median sulcus extends from the posterior region of the basisternite through the furcasternite and the entire metsternum. It marks the position of a deep infolded ridge. This infolding of the sternum is probably associated with the close approximation of the coxae, and the infolded region therefore represents the reduced mediosternite, at least in part. The external furcasternite probably represents the post-coxosternites only. The wider posterior region of the furcasternite bears two lateral process for articulation with the coxa. This second or *ventral condyle* of the leg occurs in certain insects.

The *metasternum* is greatly reduced, as the metathorax is small and the hind legs occupy most of the ventral surface. Surrounding the coxa there is a complete ring of sclerotized tissue which includes the *episternum* and *epimeron*, narrow *precoxal* and *postcoxal bridges* and a small sternal plate. The whole ring very probably represents the subcoxa. The sternal plate is even more deeply infolded than that of the mesothorax. Near the middle the plate expands to form the *ventral articulatory condyles* of the coxae. A sulcus passing through these articulatory processes divides the sternum into anterior and posterior sclerites and continues inward across the infolded portion. The *sternal apophyses* borne on the latter are apparently associated with this sulcus. The sulcus is therefore the *sternacostal sulcus* the external sclerites are the *precoxosternites* and *postcoxosternites* while the inflected plate is probably the *mediosternite*.

## THORACIC APPENDAGES

### Legs

In the vast majority of insects each thoracic segment bears a pair of legs. In the cases where legs are absent, for example, in all dipteran, and many coleopteran and hymenopteran larvae, the condition is secondary. Typically the legs are concerned with walking and running, but they may be specialized for a variety of functions some of which are described below :

### The Typical Walking Leg

The leg consists of five parts, the coxa, trochanter, femuritibia, and tarsus. Between adjacent parts is a narrow, annulated membrane, the *Corium,* and usually a mono- or dicondylic articulation. The coxa is short, thick segment strengthened at its proximal end by an internal ridge, the *Basicosta*. The coxa usually has a dicondylic articulation with the pleuron. In some orders the *Basicostal Sulcus* is U- or V-shaped over the posterior half of the coxa. The sclerite thus

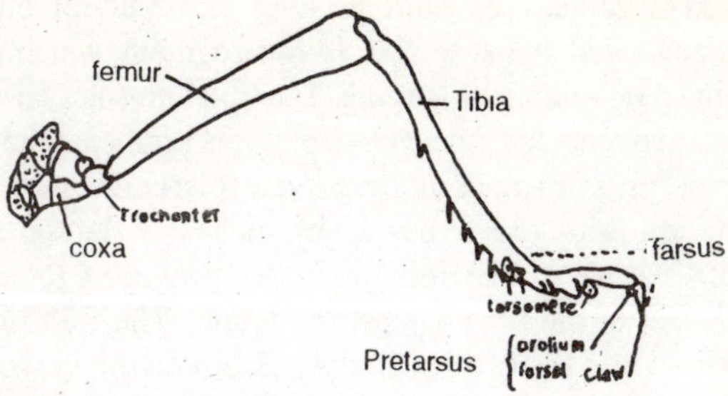


**Fig. 3.7.** A typical insect leg.

demarcated becomes thickened and is known as the *Meron*. The trochanter is a small segment. It always has a dicondylic articulation with the coxa but is usually firmly fixed to the femur, which is generally the largest leg segment. Following the slender tibia is the tarsus, a segment that is usually subdivided into between two and five *Tarsomeres* and a *Pretarsus*. The pretarsus, in most insects, takes the form of a pair of *Tarsal Claws* and a median lobe, the *Arolium*. Right angles to the body and determine the anterior, posterior, dorsal and ventral surfaces.

The pleuron and sternum surround a *coxal cavity* into which the base of the leg is set. An *articular membrane* connects the base of the leg with the rim of the coxal cavity and provides for free movement of the limb. The first or proximal segment of the leg is known as the *coxa*. Note its size and form. Find the *basicostal* sulcus which runs parallel to the proximal edge of the coxa and cuts off a narrow marginal sclerite known as the *basicoxite*. Internally the sulcus is represented by a ridge, the *basicosta*. The coxa articulates by a single hinge joint with the pleural coxal process. Note that the basicosta and basicoxite are thickened and modified to form the articular surface at this joint. A *coxal sulcus* extends from the joint to the distal edge of the coxa in *Romalea*, but is not generally present in other insects. The second segment is a short cylindrical one known as the *trochanter*. It has two lateral articulations with the coxa forming a hinge joint which permits movement in the vertical direction. The *femur* or third segment is the largest in the leg. The trochanter and femur are quite closely united in this insect and are not freely movable on each other.

The next segment, the *tibia*, is long and slender but heavily sclerotized and rigid because this is the segment which gives the propulsive push in walking or leaping. The tibial muscles lie within the femur, which accounts for the large size of this segment. Note the two rows of strong spurs or spines along the ventrolateral edge of the tibia. The articulation with the femur is by means of lateral condyles. Examine the nature of the joint and the provision to insure the complete flexion of the tibia against the femur. The fifth segment or *tarsus* is divided into three subsegments. Examine the ventral surface and note that it bears a number of swellings known as *tarsal pads*. These are paired on the first and second subsegment, unpaired on the third. The fact that the first subsegment bears three pairs of pads probably indicates that it is formed by the fusion of three subsegments, since the typical number appears to be five. The last subsegment bears a small dorsal sclerite with a distal median process, known as the *unguifer*, with which the claws articulate.

Beyond the tarsus there are several structures, collectively known as the *pretarsus*, which represent a sixth segment of the leg. Viewed dorsally the pretarsus is seen to consist of two strong curved *claws* or *ungues* and a thick median lobe, the *arolium*. On the ventral side, at the base of the arolium and between the ungues, find a quadrangular sclerite, the *planta*. Proximal to the planta, and disappearing into the cavity at the distal end of the tarsus is a second sclerite, the *unguitractor*. Cut carefully through each side of the last tarsal segment and remove the tarsal pad so as to expose the unguitractor. Observe that the end of the uguitractor is produced into a slender filiform apodeme which extends through the tarsus into the tibia. This is the *unguitractor apodeme* and is really the "tendon" of the muscle which retracts the pretarsus, that is to say, the muscle which contracts when the insect clings to an object with its claws.

**Blaberus Cranilfer or Other Cockroach**

Examine the anterior face of the coxa of the middle or hind leg, i.e., the face which is directed ventrally when the coxae are flexed against the body. Find the *basicostal sulcus* and the *basicoxite*. Note that while the basicostal sulcus runs parallel with the base of the coxa is front of the pleurocoxal articulation, it bends distally at the articulation and runs obliquely to the hind (dorsal ) margin of the coxa. The portion of the basicoxite behind the articulation is thus converted into a triangular sclerite which is now known as the *meron*. Note that the meron never takes part in the coxotrochanteral articulation.

The pleuron which consists of a fairly broad episternum and narrower epimeron separated by the pleural sulcus. The episternum is partially divided into *anepisternum* and *katepisternum*. Partly fused with the latter you will find a large triangular sclerite, traversed by a deep longitudinal groove, lying parallel to the anterior edge of the base of the coxa. This is the *trochantin*, which is larger and better developed here than in most other insects. Its ventral end makes an articulation with the coxa so that the later has two basal articulations, the *pleurocoxal* and the *trochantinocoxal*. The trochantin, like the pleuron, is derived from the supracoxal arc of the subcoxa, and in the cockroach the two remain partly united.

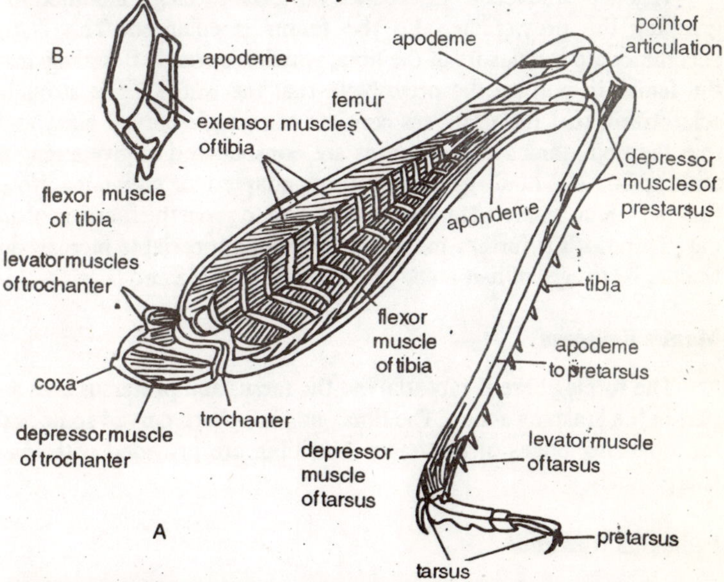


**Fig. 3.8.** Hind leg of a grasshopper showing (A) the musculature and (B) a transverse section of the femur.

## FUNCTIONAL MODIFICATIONS OF LEGS

Many insects have their legs modified to subserve functions other than walking. The following are examples of some of the more common types of adaptation. In addition to the special adaptations described, examine the various segments and note the variety of forms they may assume.

**Romalea or Melanoplus Sp**

The hindlegs are slatatorial, i.e., modified for leaping. Note the enlargement of the femur as compared with that of the other legs. This provides room for the large extensors of the tibia, the principal muscles used in leaping. The tibia is strongly sclerotized since the force of leaping is exerted against it. Note the strong spurs at the end of the tibia for gaining purchase against the substratum.

**Benacus Sp**

This is a predaceous insect with raptorial forelegs, modified for grasping the prey. Here also the femur is enlarged. This is to accommodate the flexors of the tibia, which is strongly flexed against the femur in holding the prey. Note that the whole leg is strongly sclerotized, that the pretarsus consists of a single curved claw, and that the tibia, tarsus and pretarsus are consolidated to give strength and rigidity. The hindlegs are natatorial, adapted for swimming. Note that the femur, tibia and tarsus are flattened to serve the function of an oar. The width is further increased, without appreciable increase in weight, by fringes of hair along the edges of the tibia and tarsus.

**Mantis Religiosa**

The foreleg here is raptorial but the tarsus and pretarsus take no part in the grasping action. The tibia ends in a sharp curved spine and the opposing edges of the femur and tibia are provided with long sharp spines.

**Pediculus Humanus**

The legs here are modified for clinging to the fibers of the clothing of its host. The tibia is stout and at one side bears a thumb-like process with a spine at its distal end. There is a single *tarsal segment*, to which the "thumb" is opposed, and a curved *pretarsal claw*. Tarsus and pretarsus work against the "thumb" is much the same way as the human forefinger works against the thumb in grasping an object.

**Gyrinus Sp**

The hindlegs here are even more highly specialized for swimming. The femur, tibia and the first four subsegments of the tarsus are all broad and flattened and their edges are best with flattened setae which are folded back against the leg when not in use. Note that the tarsus is articulated in such a way that when not in use it can be drawn back against the side of the tibia so that only the distal half project beyond the tibia.

**Passalus Cornutus**

The insect lives in decaying wood and uses its forelegs in helping to excavate its galleries. The foreleg is therefore adapted for digging (fossorial). Note that the segments are strongly sclerotized and rigid and that the tibia is flattened, with several tine-like processes. The tarsus is slender and normal in structure. Many soil-inhabiting insects have similar forelegs. In some of these the tarsus is permanently lost, in others it is worn off by the digging operation.

**Cicada Sp., Nymph**

The immature stages of the cicadas have a subterranean habitat and their forelegs are markedly modified for digging. The femur is enlarged and very stout and is provided with strongly sclerotized tines. The tibia too is heavily sclerotized with a strong terminal tine. Compare this leg with the corresponding one in the adult.

**Gryllotalpa Sp**

This also is a subterranean insect and the forelegs are perhaps more highly specialized than the legs of any other insect. The trochanter in some species is produced distally into a flattened spade-like structure. Apart from being very stout the femur is not greatly modified, but the tibia is short and stout, bearing distally two or three strong flattened and pointed tines. The first two segments of the tarsus are also produced into strong tines, and the tine of the first segment can work against one of the tibial tines to function as shears in cutting through fine rootlets.

**Dytiscus Sp. Male**

The forelegs are an example of marked sexual variation. The first three tarsal segments are greatly expanded to form a flat disc provided with a number of suckers on its under surface. It enables the male to cling to the female. Compare the corresponding legs in the female.

**Apis Mellifera (Honeybee)**

Examine the legs of the worker and find the specialized structures described below :

(a) The *pollen comb* for removing the pollen from the body hairs is made up of several rows of stiff hairs on the inner surface of the first segment of the hind tarsus. (b) The *corbiculae* or pollen baskets form the concave outer surfaces of the hind tibiae. (c) The tibiotarsal articulation of the hind leg is so arranged that the approximated ends of the two segments form a pair of nippers known as the *wax pincers*. (d) The apex of the middle tibia bears a strong *spur* used in scraping pollen from the baskets and also in cleaning the wings. (e) On the front leg there is the *antennal comb*, a notch in the proximal end of the flexor surface of the basal segment of the tarsus. The notch may be closed by the *velum*, a process from the adjoining end of the tibia. (f) A *brush* of stiff hairs, used in cleaning the antennal comb is to be found on the distal end of the extensor surface of the front tibia. (g) The extensor surface of the basal segment of the front tarsus bears a row of spines used as an *eye brush*.

**Some Muscles of the Leg**

**Acheta Assimilis**

The extrinsic leg muscles originate in the notum, the pleuron, the sternum, the sternal apophyses and the spina. They are inserted chiefly in the rim of the coxa but some are inserted in the trochanter. They act as promotors, remotors, rotators and adductors of the leg. The intrinsic muscles occur as pairs or paired groups within the coxa, femur and tibia and are inserted respectively on the proximal edge of the trochanter, tibia and tarsus. The extent to which muscles arise within the trochanter depends on the degree to which the latter is developed as a distinct segment. These intrinsic muscles act as flexors and extensors of the segment in which they are inserted. When the trochanter is reduced and more or less fused with the femur, the coxotrochanteral muscles are in part responsible for the movements of the femur.

The tergosternal muscle which lies internal to all of the others and which originates on the antecosta of the postscutellum and is inserted on the sternal apophysis. Its removal more fully exposes three prominent dorsoventral muscles which arise in the notum and follow a somewhat oblique course to the coxal cavity. Their origin is partly hidden by the longitudinal dorsals which should be removed. Their insertions are partly hidden by the sternal apophysis and the muscles which arise from it; these may be dissected away carefully but in doing so note that some of the muscles which arise from the apophysis are themselves inserted in the base of the leg. They are therefore part of the extrinsic leg musculature and the student may, if he desires, study their arrangement, and deduce their action before cutting them away. The anterior of the three muscles mentioned above is a notocoxal muscle which takes its origin in the scutum and is inserted in the anterior edge of the coxa. It functions as a *notocoxal promotor* of the leg, i.e., its contraction pulls the leg forward. Detach the head of the muscles from the scutum and raise the muscle. External to it a long narrow muscle will be found whose origin lies close to that of the first and which is also inserted in the anterior edge of the coxa. The two sets of muscle fibers are inserted in a common apodeme and may be regarded as one double-headed muscle. Identify the apodeme on the skeleton. It is characteristic of most of the muscles of the appendages, which have a very specific action, that they are inserted by means of tendon-like apodemes or ingrowths of the cuticle.

The posterior of the three muscles is also a notocoxal muscle which originates in the scutum immediately posterior to the anterior muscle or promotor. It is inserted by an apodeme into the posterior edge of the coxa and is the *notocoxal remotor* of the leg. It is also double-headed. Remove the promotor and remotor so as to expose the middle muscle more fully. It will be seen to originate from the scutum external to the origin of the two other muscles and to pass through the coxa to be inserted on the ventral edge of the trochanter by a flat apodeme. It also is double but the two heads are of equal size. Since the trochanter is greatly reduced and is no longer an independent segment, this muscle functions as an *extrinsic extensor of the femur*. The removal of the three extrinsic leg muscles will expose other dorsoventral muscles external to them. These are concerned in part with wing movements, but leg muscles are also included among them.

Two sets of muscles originate within the coxa, a ventral group inserted in the ventral edge of the trochanter and a dorsal group inserted on the dorsal edge. These are *intrinsic extensors* and *flexors of the femur* respectively. Two muscles originate in the femur. The large dorsal muscle has groups of relatively short fibers converging on a large flat central apodeme which arises from the dorsal edge of the tibia. This is the *extensor of the tibia* and the force of its contraction is largely responsible for the leaping of the insect. The enlargement of the hind femur in leaping insects is to accommodate this very strong muscle. The ventral femorotibial muscle is inserted on the ventral edge of the tibia. It functions as the *flexor of the tibia* and is not as strongly developed as the extensor. Within the tibia there is a dorsal *levator of the tarsus* and a ventral *depressor*. It difficulty is experienced in dissecting the tibia the dissection may be limited to the distal end where the insertion of these muscles may be seen. There is in addition a small muscle originating near the proximal end of the tibia and inserted in a long ligamentous apodeme which originates in the unguitractor and traverses the tarsus and tibia. This muscle is the *flexor of the claws*.

# 4

# Locomotion

Locomotion is the displacement i.e., the movement from one place to other. Mobility at some stage of the life history is a characteristic of all animals. They must move in order to find a mate, for dispersal and, in many cases, in order to find food. The success of insects as terrestrial animals is in part due to their high degree of mobility arising from the power of flight, but more local movements by walking or swimming are also important. Most insects move over the surface of the ground by running or hopping, the power for these movements coming from the legs. The legs move in sequences which are varied at different speeds in such a way that stability is always maintained. Co-ordination of these movements involves central mechanisms, but segmental reflexes are also important. The legs can only function in this way when the skeleton is rigid. In soft-bodied larval forms the muscles work against a hydrostatic skeleton maintained by the turgor pressure of the haemolymph. These forms crawl by extending the body anteriorly, obtaining a fresh purchase on the sustratum and then drawing up the rest of the body. Some forms have abdominal appendages which assist in crawling.

In aquatic insects legs also play an important role in movement, the *swimming legs* being modified to expose a maximum area on the power stroke and a minimum on the return stroke. With other adaptations this ensures a maximal thrust on the backstroke so that the insect swims forwards. Larvae of aquatic Diptera which are legless move by lateral flexure of the whole body and larval Anispotera use a method of jet propulsion.

## WALKING

### Movements of the Leg

In describing the movements of the legs the following terms are used (Hughes, 1952) :

*Protraction* – The complete movement forwards of the whole limb relative to its articulation with the body.

*Promotion* – The movement of the coxa resulting in protraction.

*Retraction* – The backward movement of the leg between the time it is placed on the ground and the time it is raised.

*Remotion* – The corresponding movement of the coxa.

*Adduction* – The movement of the coxa towards the body.

*Abduction* – The movement of the coxa away from the body.

Levation – The raising of the leg or a part of the leg, part of protraction.

*Depression* – Lowering the leg, or a part of the leg. The terms levation and depression are to some extent interchangeable with.

*Extension* – Extention and increase in the angle between two segments of the leg.

*Flexion* – A decrease in the angle between the two segments of the leg.

The muscles responsible for these movements fall into two categories, *extrinsic*, arising outside the leg, and *intrinsic*, wholly within the leg and running from one segment to the next. The coxa is moved by extrinsic muscles arising in the thorax and a fairly typical arrangement is shown in Fig 4.1 with promotor and remotor muscles arising on the tergum, abductor and adductor muscles from the pleuron and sternum and rotator muscles also from the sternum. The functions of the muscles may very, depending on the activities of other muscles and also on the type of articulation. In *Apis* (Hymenoptera), which has rigid pleural and sternal articulations, promotor and remotor muscles from the tergum are absent.

In the pterothoracic segments, muscles run from the coxae to the basalar and subalar sclerites. They are concerned with wing

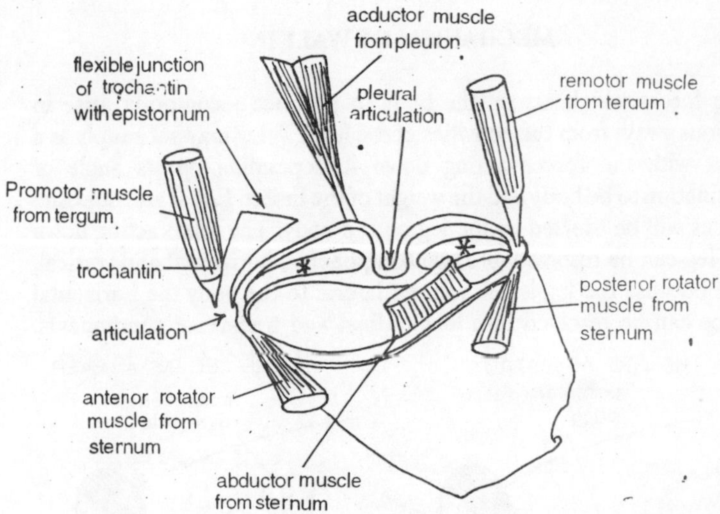


**Fig. 4.1** Inner view of coxa showing the extrinsic leg muscles which move it. Muscles arising from the points marked with asterisks and inserted at the wing base are omitted.

movements. The intrinsic musculature of the leg is much simpler than the coxal musculature, typically consisting only of pairs of antagonistic muscles in each segment. In *Periplaneta* (Dictyoptera) there are three levator muscles of the trochanter arising in the coxa and three depressor muscles, two again with origins in the coxa and a third arising on the pleural ridge and the tergum. The femur is usually immovably attached to the trochanter, but the tibia is moved by extensor and flexor muscles arising in the femur and inserted into tendons from the membrane at the base of the tibia. Levator and depressor muscles of the tarsus arise in the tibia and are inserted into the top of the metatarsus, but there are no muscles within the tarsus moving the tarsomeres.

It is characteristic of the insects that the pretarsus has a depressor muscle, but no levator muscle. The fibers of the depressor occur in small groups in the femur and tibia, being inserted into a long tendon which arises on the unguitractor. Levation of the pretarsus results from the elasticity of its basal parts.

## MECHANISM OF WALKING

The forces which act on the body to produce locomotion arise in various sways from the activities of the legs : A leg may act simply as a strut with the forces acting down it depending on its angle of inclination to be body and the weight of the insect. Equal and opposite forces will be exerted by the leg on the body. The force acting down the leg can be resolved into two components, horizontal and vertical, and because the leg is splayed out lateral to the body the horizontal force can be resolved into longitudinal and transverse components.

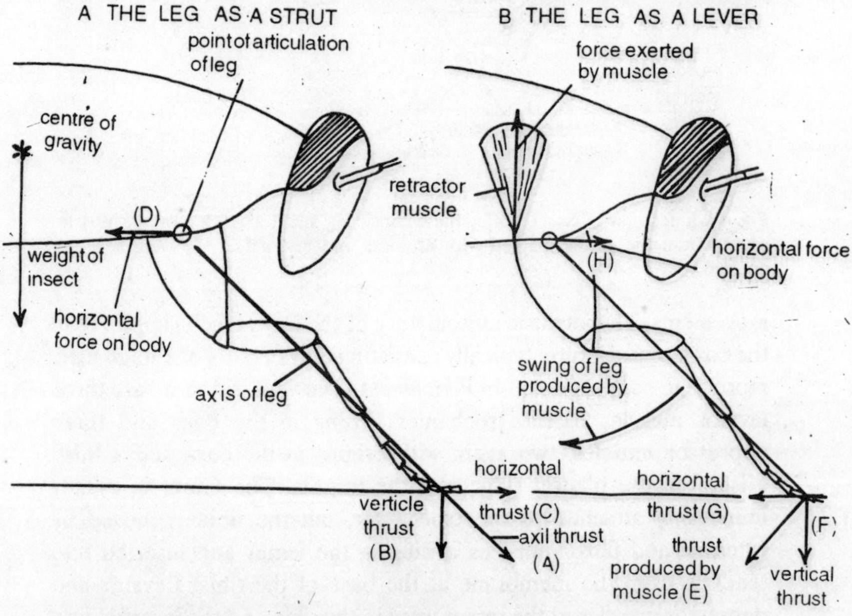


**Fig. 4.2.** A. Diagrammatic representation of a leg acting as a strut. The axial thrust (A) is exerted down the length of the leg by virtue of the weight of the insect. The size of the axial thrust depends, among other things, on how much of the weight is borne by the other legs. It can be resolved into vertical and horizontal components (B and C), but because the foot is held by friction wit the substratum it does not move. Instead, an equal and opposite horizontal force (D) acts on the body and; in this case, tends to push it back unless balanced by other forces. B Diagrammatic representation of a leg acting as a lever. Contraction of the retractor muscle tends to swing the leg back so that the foot exerts a thrust (E) on the ground. This can be resolved into vertical and horizontal components (F and G), but since the foot is held still by friction and equal and opposite horizontal force (H) acts on the body pushing it forwards. For a fuller consideration of the mechanics involved.

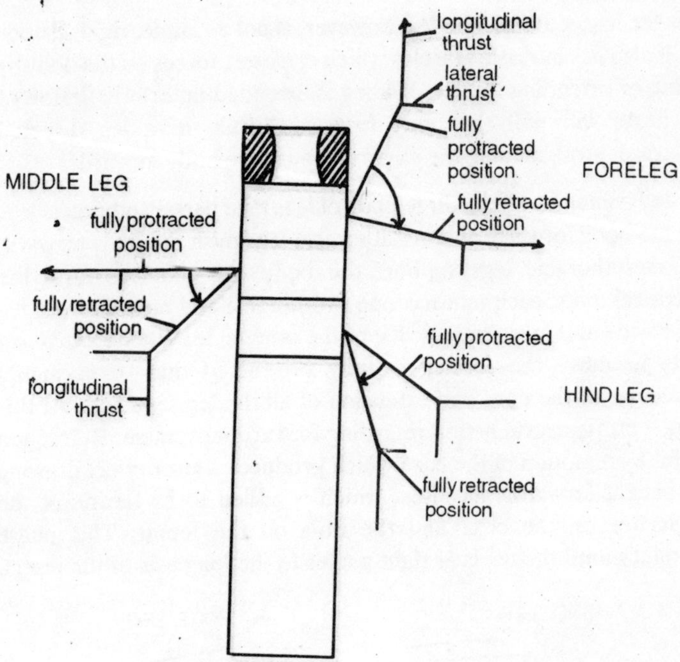


**Fig. 4.3.** Diagram to show the positions of the legs forming a typical triangle of support when fully protracted and fully retracted, together with the longitudinal and lateral components of the horizontal strut effect which the legs exert on the ground at these times. The forces acting on the body will be in the opposite directions.

The relative sizes of the longitudinal and transverse components will vary according to the position of the leg. In this diagram it is assumed that only three legs are on the ground and it is clear that for most of its movement the strut effect of the foreleg tends to retard forward movement, while that of the middle and hind legs promotes forward movement. So long as all the longitudinal and lateral forces balance each other there will be no movement, but if the forces are not balanced the body will be displaced due to a fall in the centre of gravity.

A leg can also act as a lever, that is a bar on which external work is done so that it rotates about a fulcrum. This effect is produced by the extrinsic muscles which move the leg relative to the body and so lever the insect along. The leg, however, is not a simple, rigid strut or bat. It also has intrinsic muscles which can exert forces on the body by flexing or extending the leg. If a leg is extended anteriorly flexion of the joints will pull the body forwards, while in a leg directed backwards straightening the joints will push the body forwards.

In *Periplaneta* all the three pairs of legs take part. Prothoracic legs pull the body forward, and metathoracic legs push the body forward, and mesothoracic legs support the body. All the legs form the functional units, each unit has one prothoracic and metathoracic leg of one side and mesothoracic leg of the other side. When *Periplaneta* starts to move the foreleg is fully protracted due to maximum promotion of the coxa and extension of all the leg segments. At this stage it exerts a strut action retarding forward movement. Retraction begins by remotion of the coxa which produces a lever effect drawing the animal forwards, an effect which is added to by flexion of the trochanter on the coxa and the tibia on the femur. This phase continues until the leg is at right angles to the long axis of the insect.

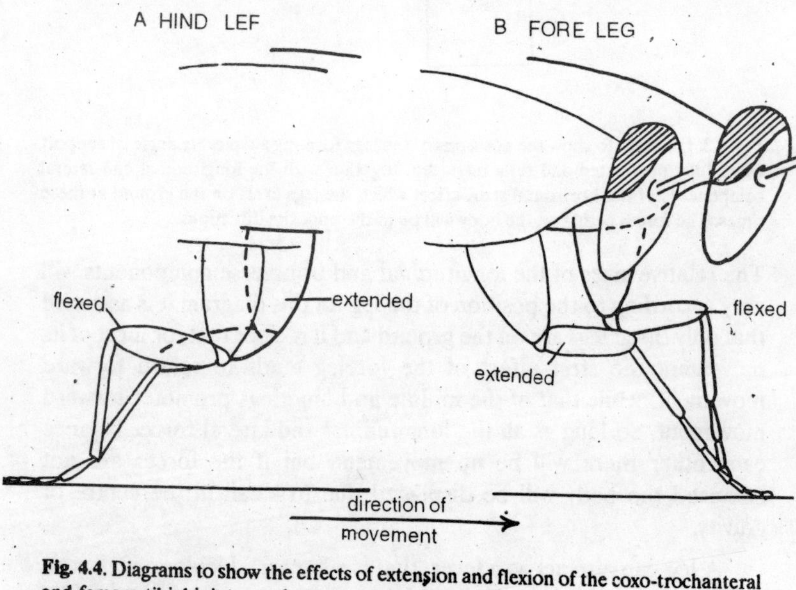


**Fig. 4.4.** Diagrams to show the effects of extension and flexion of the coxo-trochanteral and femoro-tibial joints on the movement of the body, while the feet remain still. A Extension of the hind leg pushes the body forard, B. Flexion of the foreleg pulls the body forwards.

When it has passed this position it exerts a strut effect which, aided by extension of the leg, tends to push the insect forwards.

During protraction the leg is lifted and flexed so that it exerts no forces on the body. The promotor muscle of the coxa probably starts to contract before retraction is complete so that the change over from retraction to protraction is smooth. As the leg swings forwards it extends again so that in each cycle of movement the intrinsic muscles undergo two phases of contraction and relaxation, while the extrinsic muscles only contract and relax once. The tarsi of the middle and hind legs are always placed on the ground behind their coxae so that their longitudinal strut effect always assists forward movement. The main propulsive forces of both pairs of legs are derived from extension of the trochanter on the coxa and of the tibia on the femur pushing the insect forwards. The longitudinal forces produced by these movements are mostly such that the insect moves forwards. At the same time lateral forces are produced and when, for instance, the right foreleg is on the ground it tends to push the head to the left. This is partly balanced by the other legs, but there is some tendency for the head to swing from side to side during movement.

Only *Periplaneta* has been fully studied in this way, but it is probable that the leg movements and forces involved in walking are broadly similar in other insects.

## PATTERNS OF LEG MOVEMENT

When an insect is walking its legs move in a definite sequence following two general principles. First, no leg is raised until the leg behind it is in a supporting position and, second, the movements of the two legs of a segment alternate. Usually the prothoracic legs pull the body and metathoracic legs push the body, and mesothoracic legs support the body. The pattern of leg movements and the number of legs on the ground at any one time depend on the relative times of protraction, when they are in the air, and retraction, when they are on the ground. Changes in the pattern occur automatically following changes in the relative periods of protraction and retraction [but see Wilson (1966a) for a slightly different interpretation of the manner in which these changes may be brought about]. At low speeds the

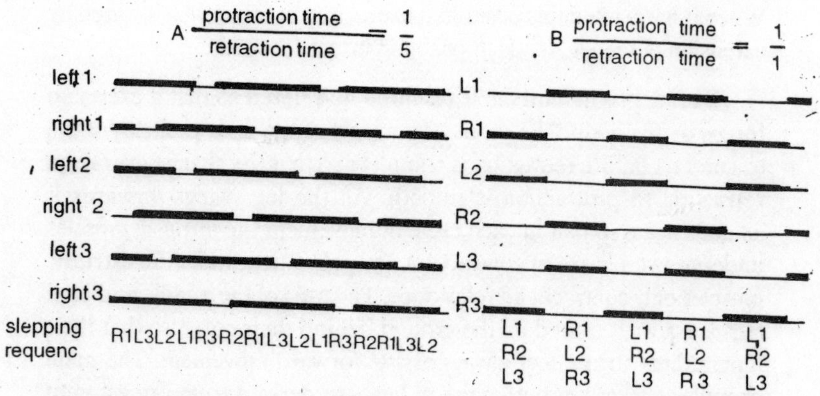


**Fig. 4.5.** Diagram showing the disposition of the legs with different protraction time: retraction time ratios. Thick lines indicate retraction with the foot on the ground, thin lines protraction with the foot in the air.

retraction time is long relative to protraction so that $\frac{\text{protraction}}{\text{retraction}}$ is low. Thus $\frac{\text{protraction}}{\text{retraction}} = 0.31$ for *Blatta* (Dictypotera) moving at 3.2 cm./sec. and at lower speeds the value is still lower. Under these circumstances most of the legs are on the ground for most of the time and the legs are protracted singly in the sequence $R_3\ R_2\ R_1\ L_3\ L_2\ L_1\ R_3$ etc. (Where R and L indicate right and left, and 1, 2 and 3 the fore, middie and hind legs). As the speed increases retraction time becomes relatively shorter and $\frac{\text{protraction}}{\text{retraction}}$ approaches unity. With $\frac{\text{protraction}}{\text{retraction}} = 1.0$ only these legs remain on the ground at any one time. The three are the fore and hind legs of one side and the middle leg of the opposite side. Between them they form a triangle of support. As they are protracted the other three are retracted and *vice versa* in the sequence $\begin{matrix} R_1 & L_1 & R_1 \\ L_2 & R_2 & L_2 \\ R_3 & L_3 & R_3 \end{matrix}$ etc., the insect thus being supported on laternate triangles of legs. This occurs in *Carausius*, but in *Periplaneta* the legs of one triangle are not all protracted simultaneously, but follow each in rapid succession. This pattern of movement based on alternating triangles of support is commonly observed. The insect

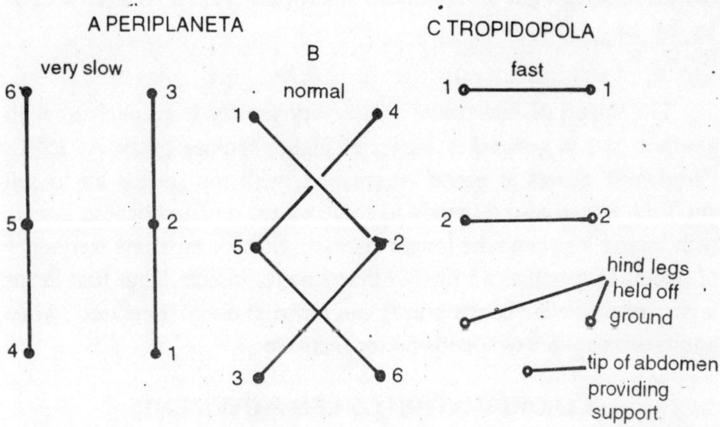


**Fig. 4.6.** Diagrams showing the order in which the feet are lifted (A) in *Periplaneta* at very low speeds, (B) in *Periplaneta* and most other insects at normal speeds, and (C) in *Tropidopola* at high speeds. Numbers indicate the sequence of stepping; points jointed by straight lines indicate legs stepping together or in rapid succession.

never has less than three legs on the ground and can stop at any point without losing its stability since the three legs enclose the vertical axis through the centre of gravity. Stability is enhanced by the fact that the body is slung between the legs so that the centre of gravity is low.

In some insect other patterns of movement occur. Thus *Petrobius* (Thysanura) moves the two legs of a segment together and this is also true of the climbing grasshopper *Tropidopola*. This insect is effectively quadrupedal, using only the anterior two pairs of legs, while the tip of the abdomen provides an additional point of support. At fast speeds the legs are moved in the sequences $\frac{L_1}{R_1}\ \frac{L_2}{R_2}\ \frac{L_1}{R_1}$ etc. At slow speeds mantids are also functionally quadrupedal using only the posterior

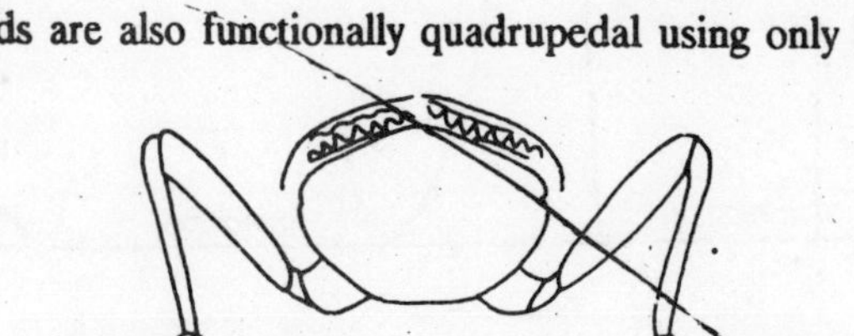

**Fig. 4.7.** Transverse section through the mesothorax of *Forficula* (Dermaptera) to show the body suspended between the legs.

two pairs of legs, but the sequence of stepping is $L_3\ L_2\ R_3\ R_2\ L_3$ etc., or
$\begin{matrix} L_3 & L_2 & L_3 \\ R_2 & R_3 & R_2 \end{matrix}$ etc.

The speed of movement varies very greatly from one insect to another, but in general is higher at higher temperatures. At 25° C. *Periplaneta* moves at about 70 cm./sec., with top speeds up to 130 cm./sec/. Speed also depends to some extent on size because insects with longer legs can take longer paces so that for the same frequency of pacing they will move further than smaller insects. Thus first instar larvae of *Blattella* (Dictyoptera) can move at about 3 cm./sec., while adults are capable of speeds up to 20 cm./sec.

## CO-ORDINATION OF LEG MOVEMENTS

Leg muscles are under the control of central nervous system. So the movements of the legs are controlled by the central nervous system and by feedback from the proprioceptors in the legs. The ganglia of

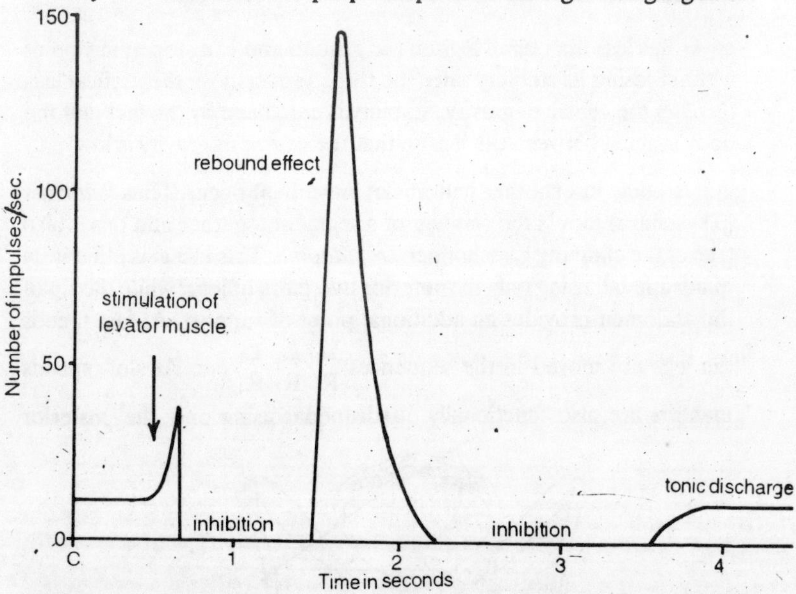


**Fig. 4.8.** The changes in the discharge to a depressor muscle following stimulation of a levator muscle. After an initial rise the discharge is briefly inhibited and then rebounds at a high frequency which might cause contraction of the muscle.

the head, influenced by the peripheral sense organs, exert on overall control, both inhibitory and stimulatory , but movements of an individual leg may depend largely on a system of reflexes. Stimulation of the campaniform sensilla of the legs causes a reflex contraction of the depressor muscles; touch on the upper side of the leg produces reflex levation. The excitation of one set of muscles inhibits the nervous discharge to the antagonistic muscles, but, if the excitation is strong, its removal is followed up by a sharp peak, known as a rebound effect, in the discharge to the antagonistic muscles which may lead to their contraction. Thus, the stimulation of one muscle could lead to the contraction of its antagonistic muscles and if the input is sufficiently strong the alternation could be maintained (Pringle, 1940). It has been shown that the rate of reaction of the sensilla and the speed of conduction through the reflex is adequate to account for the fastest movements of the legs seen in running. Thus, the reflexes almost certainly do have a role in the control of leg movements, although they may be overridden by stronger nervous outputs from other sources. Some reflex inhibition in the muscles in the contralateral leg also occurs so that, for instance, stimulation of the depressor muscles of one side inhibits the discharge to the depressor muscles on the other side of the segment. As a result of this the two legs of a segment are out of phase.

It has been clearly seen that there must also be co-ordination between the segments to maintain stepping and control timing. This might involve inter- segmental reflexes, but there is little evidence for their existence. Some measure of intersegmental co-ordination could arise from the perception by the sensilla of each leg of changing drag and gravi- tational effects resulting from the activities of the other legs, but there are good arguments suggesting that such explanations are not adequate and it, seems certain that autonomous central nervous processes are involved.

## TYPES OF LOCOMOTION

## TERRESTRIAL LOCOMOTION

### Walking and Running

Walking and running are accomplished by the six thoracic legs. Actually these limbs, unless modified for some function other than walking, serves to basic purposes: they suspend and support the insect body off the ground, and they exert the necessary forces to propel the insect. The body literally "hangs" close to the ground, resulting in a low center of gravity and thus a high degree of stability. Also, the tip of the abdomen may be in contact with the ground, providing additional stability. The use of legs as a mode of terrestrial locomotion is found in the adults of nearly all flying and non-flying insects and many nymphal and larval forms as well. In many insects walking legs are the only source of locomotion or the only source used to any great extent. For example, apterygote insects such as silverfish and proturans are solely walkers and runners; cockroaches, although they possess functional wings, seldom fly. In several beetles the elytra (forewings) are fused and hence these insects are grounded for life. Larval and nymphal forms are entirely dependent upon legs for locomotion.

Although several insects are rather sluggish in their walking and, in fact, may use their legs more as organs for clinging to a surface than for actual locomotion, many are rapid, agile runners. For example, cockroaches are well known for their running ability. The fastest speed measured for the cockroach *Periplaneta americana* is 2.9 miles per hour or 130 cm/sec. In the absolute sense this is a much slower speed than many vertebrates are able to attain. However, in relation to

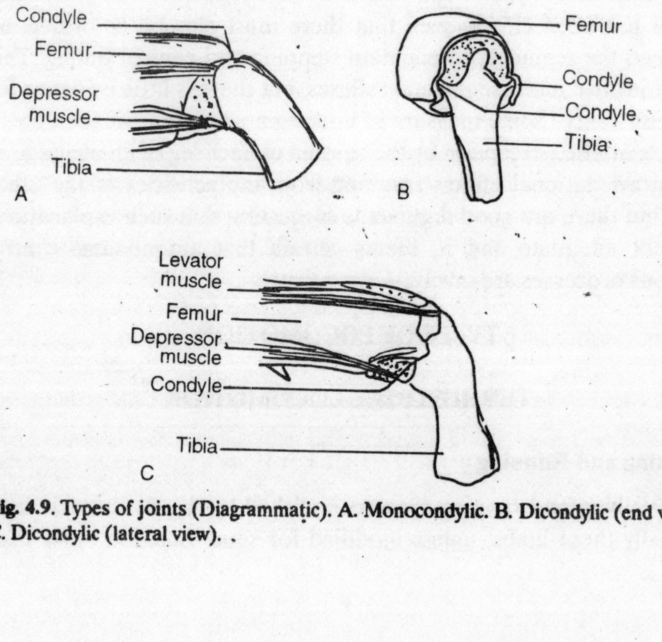


**Fig. 4.9.** Types of joints (Diagrammatic). A. Monocondylic. B. Dicondylic (end view). C. Dicondylic (lateral view).

body size, it is remarkably fast. Insects also have the ability to acclerate rapidly over short distances and to change direction suddenly. The speed of walking and running is proportional, within certain limits, to temperature.

## Coordination of Legs During Walking and Running

Insect walking and running may be resolved into two basic components, the movements of each leg and the co-ordination of movements of all the legs. It is pretty well-established that each thoracic segment contains afferent and efferent pathways necessary for the individual movement of the two legs associated with it. However, the co-ordination of all the thoracic legs in intersegmental, as evidenced by the fact that section of the longitudinal connectives disrupts walking. This co-ordination lies within the thoracic ganglia since decapitation (removal or brain and subesophageal ganglion) does not disrupt the ability of an insect to walk. Currently, it is though that locomotor rhythms are centrally generated and that these central patterns are modified, according to circumstance, via sensory feedback. Thus, changes in walking pattern with experimental amputation of different legs can be explained by the difference in the patterns of nervous signals from the remaining legs. Although, as previously mentioned, decapitation does not affect the co-ordination of walking, it may result in increased locomotor activity. As a result, the brain is thought to exert a degree of inhibitory influence on locomotor activity.

## Aids to Walking and Running

For the propulsive force applied by the legs to move an insect there must be a certain amount of friction between the leg and the substrate. Although tarsal claws generally suffic on rough or dirty surfaces, which afford ample points of grasping, there are situations in which tarsal claws fail. For example, the very smooth surface of an inclined piece of glass or a window pane. Yet many insects, such as the house fly, are able to walk with ease on such surfaces. These adhesive structures are usually covered with dense mats of tiny hairs with expanded tips. The expanded tips are covered with a secretion from glands located at the bases of the hairs and are the parts of the adhesive pads that come into direct contact with the substrate. Apparently, molecular forces among the expanded tips, the glandular fluid, and the substrate account for the adhesion of the pad. Since it is the tiny hairs that are ultimately responsible for the "clinging" ability, they are commonly reffered to as *tenent hairs*.Gillet and Wigglesworth (1932) advance the following explanation of the functioning of tenent hairs in *Rhodnius prolixus* (Hemiptera- Heteroptera; Reduviidae).

The tenent hairs that comprise the climbing organ on the distal end of the tibia are wedge-shaped at their tips and only the hindmost part comes into contact with the substrate. When the insect moves in an anterior direction (toward X), or an external force (e.g., gravity) pulls the insect in an anterior direction, the secretion between the hairs and the substrate acts as a lubricant, reducing the friction. However, if the insect is moving or being forced to move in the opposite direction (posterior or toward Y), the surfaces of that hair tips come into very close contact with the substrate and "seizure" occurs, the surfaces being held together by the adhesive forces of the molecules of the glandular secretion. Thus *R. prolixus* can walk up a pane of carefully cleaned glass at an incline of 80° but slips at an angle of 22° when walking down the same pane. The holding power of adhesive pads is well-illustrated by the fact that one bug in the same family as *R, prolixus* can support a tension of greater than 50 grams.

**Jumping**

Under this heading we include all the means, other than wings, by which insects are able to propel themselves through the air. Many insects are capable of jumping, but only in certain groups has it become a pronounced specialization. The suggestion has been made that perhaps wings first appeared on a jumping insect that used the jumping ability for a propulsive force and early wings merely as gliding surfaces. Most jumping insects use modified hindlegs in jumping although other specialized mechanisms occur in some species. Those that use their legs in jumping typically have enlarged, muscular femora on the hindlegs. Examples are the members of several orthopteran families (e.g., grasshoppers and katydids), flea beetles (Coleoptera, Chrysomelidae), and fleas (Siphonaptera). Relative to their size these insects are capable of rather astounding leaps. For example, the trajectory of a jumping flea may reach a height of 8 inches and cover a distance of 13 inches; if proportionally carried out by a human being, this would result in a jump of 800 feet high. So a flea the size of a man would have relatively much more mass per unite cross-sectional area of muscle than a normal sized flea and would thus probably still be

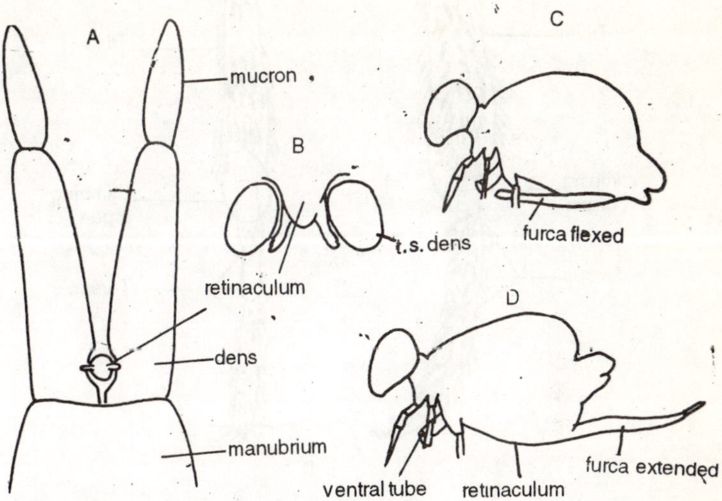


**Fig. 4.10.** Jumping in Collembola. A.Furca and retinaculum seen from below. B. Diagram showing the retinaculum holding thee dentes. C and D. Diagrams of a collembolan with the furca in the flexed and extended positions. Jumping is produced by the swing from the flexed to the extended position.

able to jump only 8 inches high. Grasshoppers have been reported to leap up to 30 cm high over a distance of 70cm. Springtails (Collembola), click beetles (Elateridae) and cheese maggots (Diptera, Piophilidae) have unique methods of jumping.

Springtils are minute wingless insects that possess a fork-shaped structure, the *furcula*, on the posterior part of the abdomen and a structure called the *tenaculum* of the anteroventral portion of the abdomen. The furcula is bent anteriorly and its tip engaged by the tenaculum. Prior to a jump, tension is built-up in the muscles of the furcula. When released by the tenaculum, the furcula "snaps" posteriorly against the substrate, propelling the insect into the air. Click beetles utilize a similar principle, an elongated spine on the posterior part of the posternum being inserted into a receptacle on the mesoternum. The jump produced by the sudden release of this prosternal spine is elicited only when the beetle is on its back and hence serves as a righting mechanism. Cheese maggots use their entire body in jumping. The last abdominal segment is held by the mouth hooks, muscular tension is built up within the body, and the last

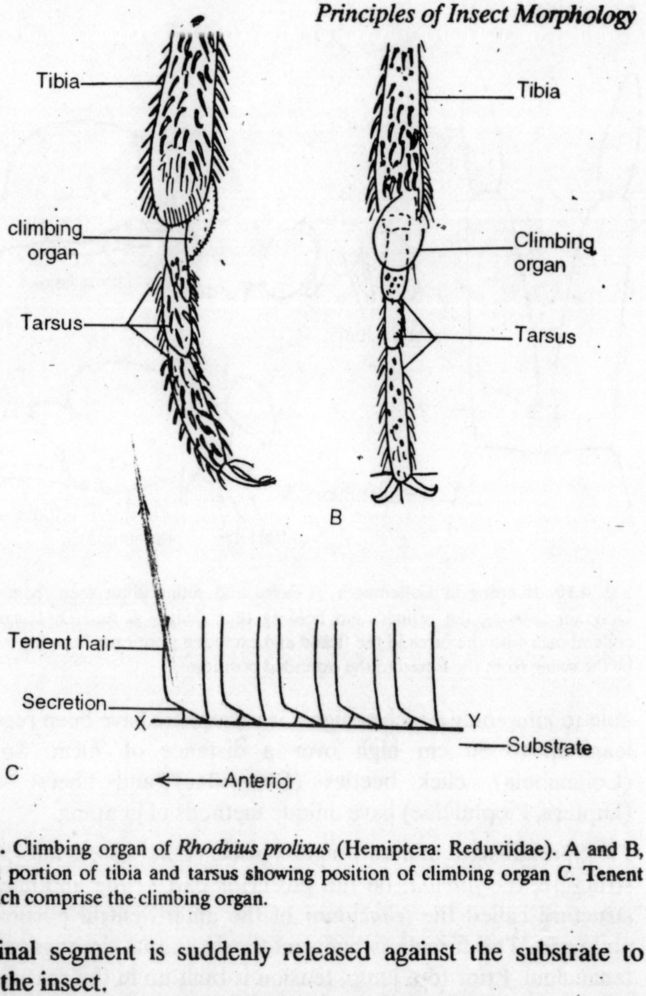


**Fig. 4.11.** Climbing organ of *Rhodnius prolixus* (Hemiptera: Reduviidae). A and B, Proximal portion of tibia and tarsus showing position of climbing organ C. Tenent hairs which comprise the climbing organ.

abdominal segment is suddenly released against the substrate to propel the insect.

It has been assumed for a long time that direct contraction of leg muscles accounts for the force released when an insect jumps. However, recently, the role of cuticular elasticity has come to be appreciated. Cuticular elasticity is associated with regions of the cuticle that contain high concentrations of rubbery protein, *resilin*. The Oriental rat flea, *Xenopsyllae cheopis*, has been studied extensively in this regard. In this flea, resilin is concentrated in the pleural arch of the metathorax. It has been known for a long time that the large metathoracic legs are the most important in jumping. The pleural arch is considered to be homologous with the wing hinge ligament of winged insects. The metathoracic leg is so constructed that it can be "cocked." When this leg is in the cocked position, the femur is raised up and engaged by a catch and the trochanter is in contact with the substrate. A catch mechanism also engages plates of the mesothorax to the metathorax. Also when the metathoracic leg is in the cocked position, the resilin in the pleural arch is compressed (i.e., elastic energy stored in it). When the catches are released, the elastic energy in the resilin is released and translated to the trochanter via a tendon. The trochanter thrusts against the substrate, accelerating the flea from the substrate. Subsequently, the thrust of the tibia against the substrate provides additional acceleration. Jumping in locusts (Orthoptera, Acrididae) and click beetles is also thought to involve energy stored in resilin.

**Crawling**

Crawling (or creeping) is generally associated with the locomotion of larva insects, particularly those that propel themselves by means other than the six thoracic legs alone. However, many larval and most nymphal insects walk like adults. Some of these hexapodous immature forms, for example, several beetle larvae, are aided in their locomotion by eversible *Pygopodia*. Which arise from the terminal abdominal segment. A pygopodium is everted by means of hemolymph pressure and aids the thoracic legs in the progression of the insect.

The immature forms of butterflies and moths (caterpillars) and certain wasps (Hymenoptera, Symphyta) bear, in addition to the three pairs of thoracic legs, accessory legs, or *prolegs*, on certain of the abdominal segments. The movement of the prolegs depends on the integrated activity of the body musculature of the abdomen. Since these insects are typically soft-bodied, hemolymph pressure is important in maintaining a "hydrostatic skeleton." The pressure of the hemolymph is maintained by the turgor muscles of the insect body. Puncturing one of these insects results in an immediate shriveling; owing to the contraction of the turgor muscles. This indicates that these muscles respond to a reduction in hemolymph pressure by contraction and gives some insight as to how they function in an intact larva. In addition to the turgor muscles, there are locomotor muscles that fail to contract when a larva is punctured and are involved with

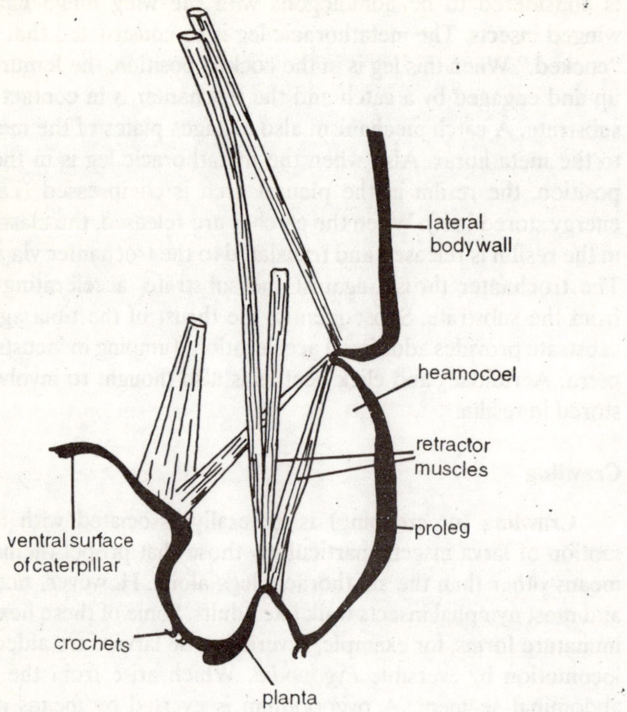


**Fig. 4.12.** Transverse section through part of an abdominal segment of a caterpillar showing the proleg.

the actual movement of the prolegs. These muscles are arranged in longitudinal, dorsoventral, and transverse patterns. Walking patterns in larvae with prolegs have been best described in caterpillars. As in walking solely with thoracic legs, the rhythms involving prolegs are probably generated in the central nervous system. Usually single peristaltic waves of contraction pass anteriorly from the posterior end of the insect. As each wave passes, the two legs of each segment always move simultaneously. Each wave involves three main phases. Hughes and Mill (1974) describe these three phases as follows :

In the first, the dorsal longitudinal and dorsal intrasegmental muscles contract, together with the large transverse muscles. This results in the segment becoming shortened dorsally and consequently

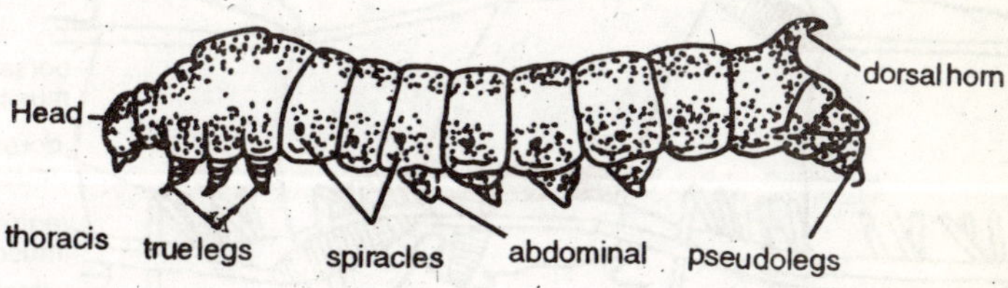


**Fig. 4.13. Caterpillar walking.**

its posterior end becomes inclined forward so that the segment behind is lifted from the ground. In the next phase the segment contracts dorsoventrally and its feet are released simultaneously from the substratum. After the legs have been moved forward, the ventral longitudinal muscles contract so as to bring the segment down toward the ground and the feet become fixed. The wave of peristalsis passing over the body is therefore not limited to a single segment but involves a simultaneous contraction of muscles in at least three segments. Contraction starts at the last segment with the release of the terminal appendages. They are lifted and placed on the ground at varying distances forward.

This description is simplified, since other sets of muscles (e.g., the transverse muscles) are no doubt involved. In addition, contractions of the turgor muscles must be in co-ordination with the contractions of the locomotor muscles. An interesting variation of this type of crawling is shown by the larvae of geometrid moths. These larvae are often referred to as *inch worms* or loopers. In these larvae much of the trunk is out of contact with the substrate as they move along. Terrestrial larvae having no legs whatsoever, for example, fly maggots, depend entirely upon peristaltic waves of contraction for their progression. Many possess spines or other structures that increase the friction between their bodies and the substrate. In most apodous

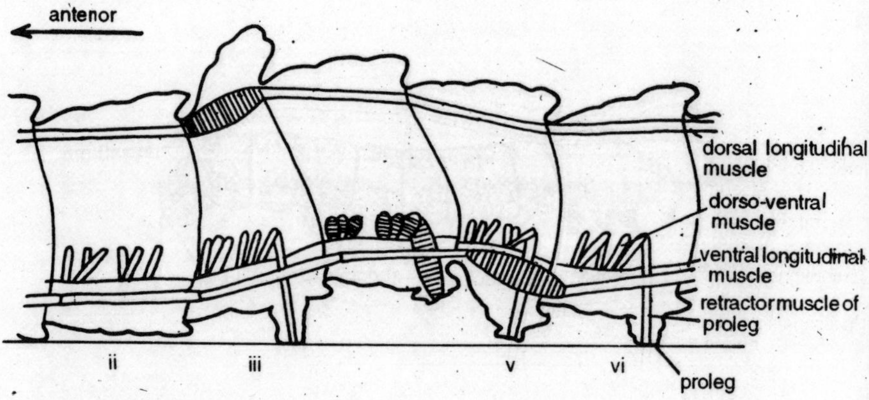


**Fig. 4.14. Diagrammatic section through a caterpillar showing a wave of contraction which passes along the body from behind forwards and produces forward movement Contracted muscles are shown hatched.**

larvae, waves of peristaltic contraction, like those in larvae with prolegs, pass from posterior to anterior, that is, in the direction of progression. However, in certain ones adapted to burrowing through the soil, for example, burrowing crane fly larvae and bibionid larvae (Diptera), the waves of contraction proceed from anterior to posterior, or in the direction opposite that of progression. In this method, the insect's posterior serves as a stable point while the anterior portion of its body is pushed through the soil.

## AQUATIC LOCOMOTION

We shall examine the locomotion of aquatic insects from the standpoint of adaptations, first, those that enable insects to propel themselves on the surface of the water and, second, those by which insects "swim" beneath the water's surface. Nachtigall (1974a) reviews the literature on aquatic locomotion.

### Surface Locomotion

To appreciate fully the various adaptations associated with insect locomotion on the water's surface, we must first consider some of the characteristics of that surface and its interaction with other surfaces. At temperatures and pressures typical of the environments in which

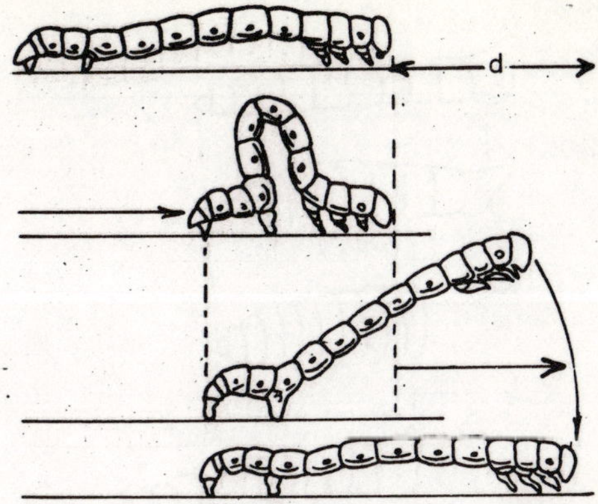


**Fig. 4.15.** Locomotion of "inchworm." geometrid moth caterpillar. *d* = distance advanced in one cycle.

insects are found, the molecules of water are strongly attracted to one another, a phenomenon known as *cohesion*. In the absence of outside forces upon the molecules, a droplet of water tends to contract so that it has the smallest possible surface area. Theoretically this would result in a sphere. The molecules of the gases that compose air have little attraction for water molecules, and little attraction for one another. Hence any surface of water in contact with air tends to contract. This force of contraction is called *surface tension*, and its presence makes the water's surface behave somewhat like a thin, elastic membrane. Some surfaces or surface coatings attract water molecules, and when this attractive force is equal to or greater than that between water molecules themselves, the water spreads out on the surface, which is said to be wettable or *hydrophilic* (water loving). Other surfaces have the opposite effect, exerting little or no attractive force on water molecules. Such surfaces are water repellent or *hydrophobic* (water fearing).

The magnitude of the forces associated with surface tension seems rather small when we think of large animals, which have a comparatively small ratio of surface to volume and hence comparatively little surface area per unit body mass. However, surface-tension forces are quite significant in small organisms, such as insects, that have relatively large ratios of surface to volume and thus a comparatively large amount of surface area peer unit body mass.

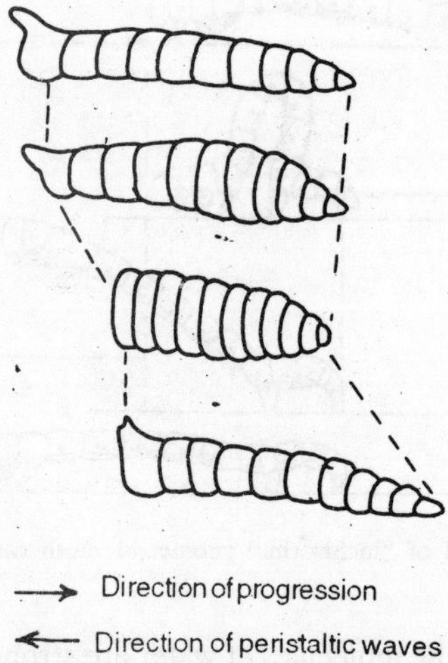


**Fig. 4.16. Locomotion in a crane fly larva.**

Since, in general, the insect cuticle is hydrophobic, many small insects are able to be supported against the force of gravity by the forces of surface tension. However, if the cuticle is made hydrophilic by the application of a "wetting agent," such as a detergent, the insect will immediately sink or become trapped by the water. Many surface-dwelling forms secrete a waxy material that makes their tarsi hydrophobic and allows these insects to walk on the water's surface. The surface-dwelling bugs (several families of Hemiptera) are able either to walk on the water as other insects would walk on the ground (e.g., the water measurer, *Hydrometra*, Hydrometridae) or to use the middle legs as synchronous oar, rowing from place to place (e.g., water striders, Gerridae and others). The *collophores* and tarsal claws of certain springtails (e.g., *Podura aquatica*) are hydrophilic while the rest of the integument is hydrophobic. This enables these insects to be anchored to the water and yet easily move from place to place by means of their furculae. Beetles in the genus *Stenus* (Staphylinidae) secrete a substance from anal glands that lowers the surface tension of the water behind the insect. As a result they are drawn forward by the effects of the contractive forces of the water in contact with their hydrophobic bodies. These beetles are able to attain speeds of 45 – 70 cm/sec. Water striders in the genus *Velia* (Veliidae) are similarly

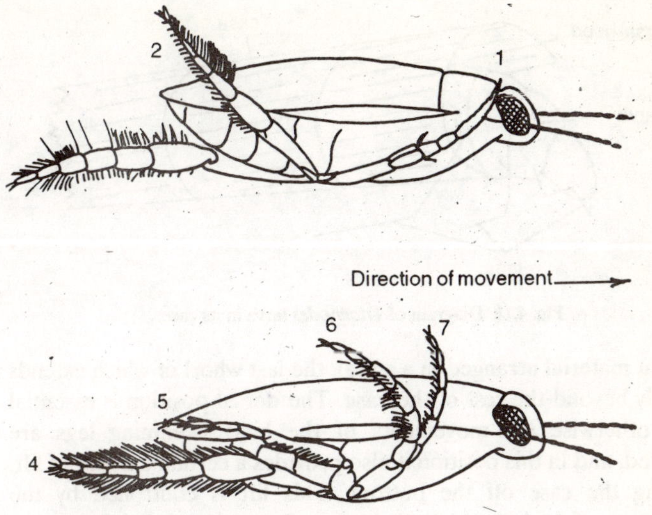


**Fig. 4.17.** Swimming movements of a predaceous diving beetle *Acilius sulcatus*. Numbers indicate the temporal sequence of leg positions during the swimming stroke. Top, thrust stroke; bottom, recovery stroke. Note the positions of the hairs at various points in the stroke cycle.

propelled by a surface-tension-reducing substance discharged through their posteriorly directed mouthparts. This occurs as an escape response (Blum, 1978).

**Bottom Dwellers**

Bottom-dwelling aquatic insects, such as *Aphelocheirus* (Heteroptera) and larval Odonata and Trichoptera, can walk over the substratum in the same way as terrestrial insects. The larva of *Limnephilus* (Trichoptera) basically uses an alternation of triangles of support, but because of the irregularity of the surface the stepping pattern tends to become irregular. The forelegs many step together instead of alternating and the hind legs may follow the same pattern. Normally the power for walking comes primarily from traction by the fore and middle legs and pushing by the hind legs, but under difficult conditions the hind legs may be extended far forwards outside the middle legs so that they help the other legs to pull the larva along. In the bottom-dwelling *Triaenodes* (Trichoptera) the larval case is built

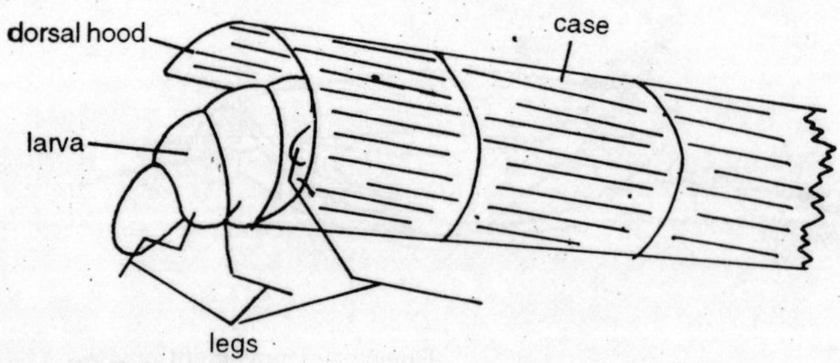


**Fig. 4.18.** Diagram of *Triaenodes* larva in its case.

of plant material arranged in a spiral, the last whorl of which extends dorsally beyond the rest of the case. The dorsal position is essential since otherwise the movements of the hind swimming legs are hindered, and in this position it also provides a certain amount of lift, carrying the case off the bottom. This lift is controlled by the movements of the legs which tend to produce a downward thrust.

Larval Anispotera can walk across the substratum using their legs, but they are also able to make sudden escape movements by forcing water rapidly out of the branchial basket so that the body is driven forwards. The branchial basket so that the body is driven forwards. The branchial is compressed by longitudinal and dorso-ventral contractions of the abdomen, the contractions being strongest in segments six to eight in which the brachial basket lies. Before this contraction the anal valves close and then open slightly leaving an aperture about 0.01 sq. mm. in area. The contractile movement lasts about 0.1 sec., and water is forced through the anus at a velocity of about 250 cm./sec., propelling the larva forwards at 30 – 50 cm. sec. As the abdomen contracts the legs are retracted so as to lie along the sides of the body, offering a minimum of resistance to the forward movements (Hughes, 1958). Co-ordination involves giant fibres running in the ventral nerve cord.

**Free-swimming Insects**

Larval and pupal Diptera, larval and adult Heteroptera and adult Coleoptera form the bulk of free-swimming insects and, apart from the Diptera, most of these use the hind legs, sometimes together with the middle legs, in swimming. The hind tibiae and tarsi, and sometimes also those of the middle legs, are flattened antero- posteriorly to form a paddle which is often increased in area by inflexible hairs or, as in *Gyrinus* (Coleoptera), by cuticular blades 1 $\mu$ thick and 30–40 $\mu$ wide. In *Acilius* (Coleoptera) the hairs constitute 69% of the total area of the hind tibiae and 83 % of the tarsi. The hind legs of these insects are relatively shorter than the hind legs of related terrestrial insects, but the tarsi are relatively longer. The point of attachment of the hind legs is displaced posteriorly compared with terrestrial insects and in dytiscids and gyrinids the coxae are immovably fused to the thorax. This limits the amount of movement at the base of the leg and the basal muscles are concentrated into two functional groups, a powerful retrac or group and a weaker protractor group. Intrinsic muscles of the legs tend to be reduced and the movements of the distal parts of the legs during swimming are largely passive. The two legs of a segment move together, contrasting with the alternating movement of the legs in terrestrial insects, but *Hydrophilus* (Coleoptera) is an exception. This beetle uses the middle and hind legs in swimming, the middle leg of one side being retracted simultaneously with the hind leg of the opposite side, but out of phase with the contralateral middle leg.

*Buoyancy*

Many free-swimming insects are buoyant and when they stop swimming come to rest at the surface of the water in a characteristic position which results from the distribution of air stores on and in the body. Most forms float head down and *Notonecta* (Heteroptera), for instance, rests at an angle of 30° to the surface. As it kicks with its swimming legs this angle is increased to 55° so that the insect is driven down, but as it loses momentum during the recovery stroke of the legs it will tend to rise again. If the driving movements of the legs are repeated rapidly, before the insect rises very much, the path may be straightened out and by controlling the rate of leg movement the insect can dive, move at a constant level or rise to the surface. The beat tends to be faster at higher temperatures so that movement becomes more uniform as the temperature rises. In *Dytiscus* the buoyancy effect is offset at faster speeds by using the middle and hind legs alternately, the while *Hydrophilus* achieves the same effect by using the legs of the two sides out of phase. Hence, these insects produce a continuous driving force which offsets their buoyancy.

*Thrust*

The thrust which pushes the insect down through the water is developed during retraction of the legs, but, because the insect is surrounded by the medium, protraction of the legs also produces forces. These tend to drive the insect backwards and if it is to move

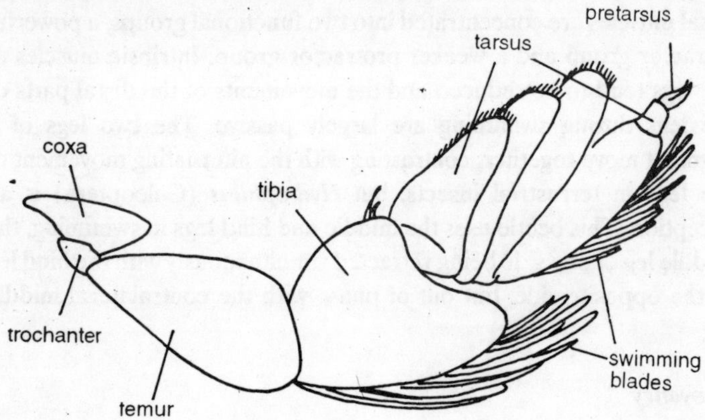


**Fig 4.19.** Hind leg of *Gyrinus*.

forwards the forward thrust produced on the backstroke must exceed the backward thrust produced on the forward, recovery stroke of the legs. The thrust which a leg exerts in water is proportional to its area and the square of the velocity with which it moves. Hence to produce the most efficient forward movement a leg should present a large surface area and move rapidly on the backstroke, while presenting only a small surface and moving relatively slowly on the recovery stroke. To achieve a large surface area during the backstroke the swimming legs of *Dytiscus* are straight with the fringing hairs, which are articulated at the base, spread to expose a maximum area. On the forward stroke, however, the femorotibial joint flexes so that the tibia and tarsus trail out behind. At the same time the tibia rotates through

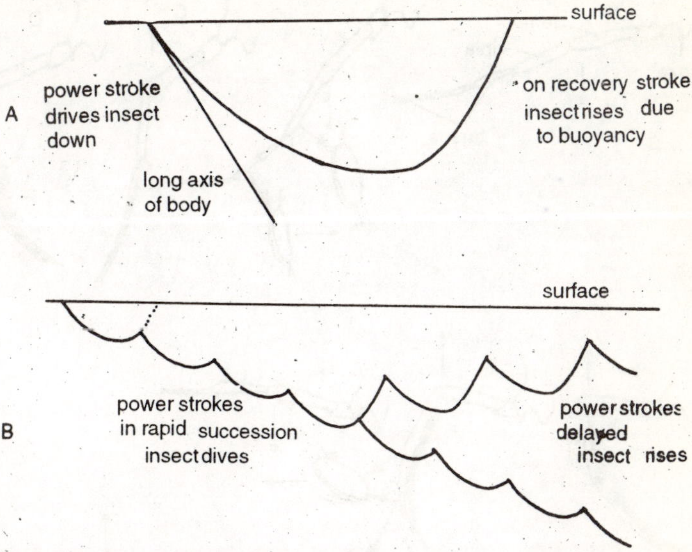


**Fig. 4.20.** Diagram of the path through water of an insect such as *Notonecta*. A. The path due to a single swimming stroke by the legs. B. Different paths produced by differences in the timing of successive strokes.

45° so that the previously dorsal surface becomes anterior and the fringing hairs fold back. The tarsus, which articulates with the tibia by a ball and socket joint, rotates through 100° in the opposite direction. These movements are passive resulting from the form of the legs and the forces exerted by the water, and they ensure that the tibia and tarsus are presented edge on to the movement, producing a minimum of thrust. Subsequently, at the beginning of the backstroke the leg and hairs extend passively to expose a maximum surface area again. There is no extensor tarsi muscle and the extensor tibiae is weak. The power for the stroke comes from the muscles moving the trochanter on the fixed coxa.

Similar device are employed by other insects for exposing a maximum leg area during the power stroke and a minimum during the recovery stroke. The swimming blades fringing the leg of *Gyrinus* are placed asymmetrically so that they open like a venetian blind, turning to overlap and produce a solid surface during the power stroke. In the

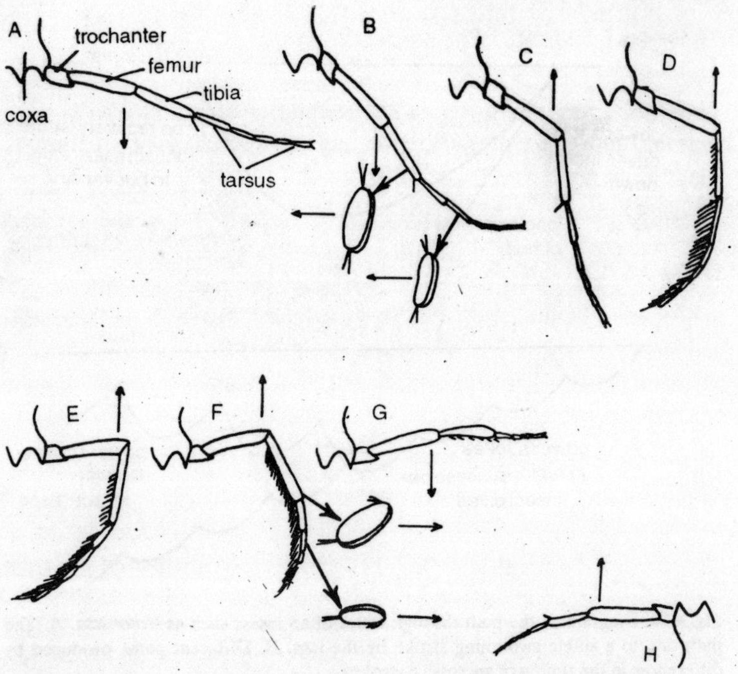


**Fig. 4.21. Diagram showing successive positions of the right hind leg of *Dytiscus* during swimming. G, H, A and B show stages of retraction, C – F stages of protraction. Insects are cross- sections of the tibia and tarsus showing their orientation during retraction (B) and protraction (F). Small arrows indicate the direction of movement of the femur relative to the body of the insect.**

recovery stroke the tarsomeres collapse like a fan and are concealed in a hollow of the tibia which in turn is partly concealed in a hollow of the femur. These changes decrease the area of the middle leg by 35 % and of the hind leg by 28%.

The relative power developed on the forward and backward strokes also depends on the relative speeds of the strokes. In *Acilius* (Coleoptera) the backstroke is faster than the forward stroke so that for a given leg area the forward thrust on the body exceeds the backward thrust. In *Gyrinus*, on the other hand, the backstroke is slower than the forward stroke so that for a given area the backward thrust is greater than the forward thrust. Hence, if the area of the legs of *Gyrinus* remained constant the insect would tend to move backwards and it is only because the reduction in area of the leg on the forward stroke reduces the backward thrust on the body that the net effect is to push the insect forwards. The legs move in an arc so that lateral thrust is produced in addition to the longitudinal thrust. In most insect, where the legs of the two sides move in phase, the lateral forces developed on the two sides balance each other out, but in *Hydrophilus*, where the legs are used alternately, there is some deviation to either side, although the lateral thrust of the hind leg on one side is largely balanced by the opposite lateral thrust of the contralateral middle leg.

Forward thrust is minimal at the beginning and end of each stroke, but when the legs are at right angles to the body the whole of the thrust developed is longitudinal. It is advantageous if the velocity of the leg is greatest at this point and is low at the beginning and end of the stroke so that the lateral forces, which are produced mainly

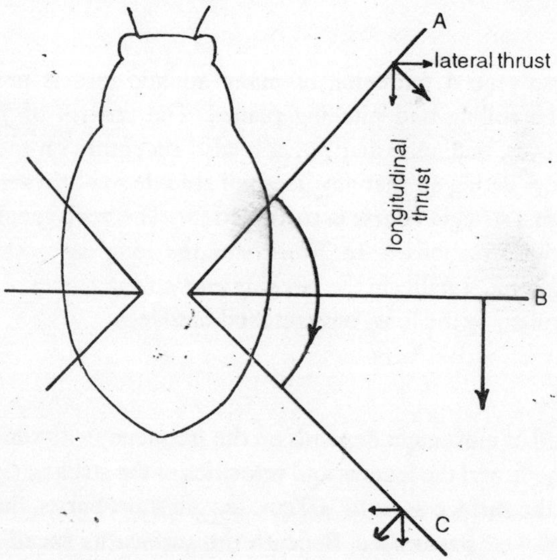


**Fig. 4.22.** Diagram of a water beetle showing the thrust exerted by the hind leg at different points of the power stroke. It is assumed that the velocity of the leg at A and C is only half its velocity at B, where, since the leg is at right angles to the body, only longitudinal thrust is produced. Equal and opposite forces act on the body.

during these phases, are kept to a minimum. This is the case, at least in *Acilius* and *Gyrinus*. In the latter the leg is moving most rapidly when it is at an angle of 90 – 135° to the body. After this the velocity rapidly falls to zero.

*Streamlining*

Most aquatic insects are streamlined and dorso-ventrally flattened so that they offer a minimum of resistance as they pass forwards through the water, *Acilius*, for instance, only creates about three times the resistance of an ideal streamlined body and deviations of up to 10° on either side do not markedly increase the resistance. There is, however, a marked increase in resistance if the insect turns broadside or ventral side to the direction of movement and this facilitates turning and braking. Turns are made by producing strokes of unequal amplitude on the two sides or, in making a sharp turn, the leg on the inside may be extended and kept still while the contralateral leg paddles.

*Stability*

The dorso-ventral flattening of many aquatic insects provides stability in the rolling and pitching planes. The control of yawing involves the eyes, antennae and possibly also receptors on the legs, these receptors acting so that any unequal stimulation as a result of deviation from a straight course is corrected for. The head ganglia are involved in these responses. In *Triaenodes* the long case acts as a rudder giving some stability in the pitching and yawing planes. Rolling may be controlled by the long, outstretched hind legs.

*Speed*

The speed of movement depends on the frequency with which the strokes are made and the lengths and velocities of the strokes. *Gyrinus* can swim on the surface at up to 100 cm./sec., in short bursts, the hind leg making 50 – 60 strokes/sec. Beneath the surface its speed rarely exceeds 10 cm./sec. *Acilius,* making 3 – 10 strokes/sec., can reach 35 cm./sec., while *Triaenodes*, with hind legs making 13 strokes sec., only moves at about 1.7 cm./sec. because of the high drag effect of the larval case.

## AERIAL LOCOMOTION

As mentioned earlier, insects, alone among the invertebrates, posses the ability to fly. This ability is perhaps one of the most important reasons for their tremendous success relative to the rest of the animal kingdom. Flight has enabled insects to take advantage of environmental situations that are virtually untouchable by their non-flying rivals. For this they develop wings and powerful flight muscles having greatest number of mitochondria.

# 5

# The Wings

Almost all insects are provided with wings with the exception of *Lepisma* – the silver fish. Some of them are poor flier as we see in the case of cockroaches while most of there have perfectly developed wings adapted for aerial life. Insects differ from the other flying animals in that their wings are structures superadded to the primitive motor equipment of their ancestors. The birds and the bats, in acquiring the power of flight, have lost the use of a pair of limbs for other modes of locomotion, since their wings are the forelegs made over for purposes of flying. Insects thus seem related to the winged creatures of fiction, though the latter, it must be observed, are given wings quite irrespective of their anatomical possibilities of using them. And yet the wings of insects, as we shall presently see, when first acquired were probably outgrowths of the back incapable of movement. Certainly their evolution into organs of flight has involved much reconstruction in the thoracic segments to contrive a motor mechanism for them.

## ORIGIN AND EVOLUTION OF THE WINGS

The oldest insects known from the fossil records lived in Carboniferous times, their remains being found in the lower beds of the Upper Carboniferous or Pennsylvanian period. These ancient insects had two pairs of fully developed wings, which differed but little in structure from the wings of modern insects. Many of the Carboniferous insects, however, had in addition to the wings a pair of small, fiat lobe projecting laterally from the tergum of the prothorax, and these lobes suggest that at an earlier period the wings themselves

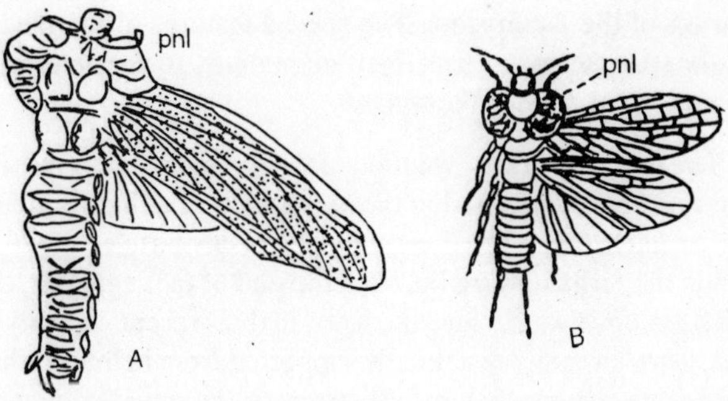


**Fig. 5.1.** Examples of fossil insects with paranotal lobes on the prothorax. A. *Stenodictya lobata.*, B, *Lemmatophora typica.*

were developed from similar tergal lobes of the meso- thorax and the metathoax. We may visualize the immediate ancestors of the winged insects, therefore, as creatures having the body already differentiated into head, thorax, and abdomen and characterized by the possession of a series of three partly overlapping, fanlike extensions, or ***paranotal lobes***, projecting on each side of the body from the thoracic terga. If the paranotal lobes, then, were the precursors of the wings, it seems most probable that they serves as gliding organs, allowing their possessors to launch themselves into the air from an elevation and to sail off to sail off to some more distant objective.

The structure and development of the wings of modern insects attest the origin of the wings from lateral folds of the tergal margins, for each wing is essentially a hollow extension of the body wall, the dorsal lamina of which is directly continuous with the tergal plate supporting it while the ventral lamina is reflected into the lateral wall of the segment. The pleural plates of the thoracic segments must have been evolved for the purpose of supporting the bases of the paranotal lobes from below. We can thereby understand why the pleura of adult pterygote insects have the same essential structure in all the thoracic

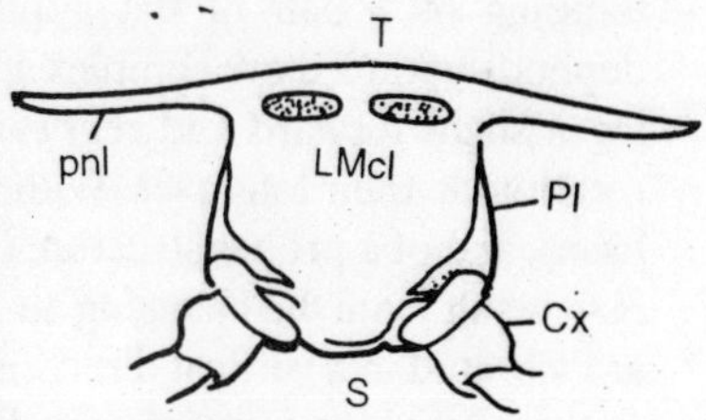


**Fig. 5.2.** Diagrammatic cross section of a thoracic segment with paranotal extensions of the tergum.

segments, and why they differ so characteristically from the primitive pleurites of the Apterygota. The special features of the pleura in the wing-bearing segments are final adaptations to the later-developed mechanism for moving the wings.

The first step in the evolution of the paranotal lobes into organs of flight must have consisted in the acquisition of a line flexibility in the base of each lobe. The longitudinal dorsal muscles could then by arching the terga upward between the end of the segments, throw the wing flaps downward, since the later, in their recent capacity of glider lobes, were already substantially supported from below on the pleura. In order to give mechanical efficiency to the muscles in arching the terga, however, movement between the tergal plates had first to be eliminated. Sclerotic continuity in the dorsum of the wing bearing region has been acquired either by a union of the successive terga or by a forward extension of the acrotergal lips of the metatergum and first abdominal tergum into the regions of the intersegmental membranes. With the downstroke of the wings produced by an upward, bend of the tergum, the upstroke must depend on some antagonistic force that will flatten the tergum. It is possible that, in the first place, the elasticity of the tergal plates sufficed to restore the normal contour of the latter; but most modern insects have special tergal depressor muscles in the two wing-bearing segments. These muscles lie laterad of the longitudinal muscles; they are attached dorsally on the anterior lateral parts of the tergum and ventrally on the sternum before the coxae. They are perhaps derived from the primitive lateral body muscles.

The wings of insects are thus movable up and down by a relatively simple mechanism. Flight, however, is not to be achieved by the mere flapping of a pair of flat appndages. Forward motion in the air depends upon a more complex movement in the motor organs, involving a slight forward and rearward action of the wings and a partial rotation on their long axes. With the wings sufficiently flexible at their bases, and of a proper structure in their distal parts, these movements may result from the changing air pressure on their surfaces when they are vibrated in a vertical direction. The stiffening of the anterior parts of the wings by a forward crowding of the veins, and the flexibility of the more weakly supported posterior areas, automatically gives a torsion to the wing planes in motion. But if the wings had to depend upon air pressure alone for the slant of their planes that gives the forward impulse to the insect in the air, it is evident that the wings on opposite sides of the body would have always approximately the same degree of movement. Controlled or differential action in the two wings of a pair would then be impossible, and the insect would have no power of directing its flight or of changing its course. Most modern insects, however, do control efficiently their motions on the wing, and many of them, besides being well able to direct their forward flight, can also fly sidewise and backward without changing the position of the body, or they can hover at on point in the air. It is possible that some insects may shift their course by altering the posture while flying, but it has been shown that steering, for the most part, is a function of the wings.

The rotary movement of the wings is produced chiefly by powerful muscles lying against the laurel walls of the wing-bearing segments. These muscles are inserted usually on small sclerites situated immediately beneath the wings, respectively before and behind the wing fulcrum or in some cases on lobes of the dorsal margin of the pleuron. The principal pair of these so-called "*direct wing muscles*" in each side of the segment take their origins ventrally on the coxa), showing that they are primarily leg muscles that have been given over to the service of the wings. When insects in this way acquired the direct action of muscles on the wing bases, they possessed a mechanism capable of controlling the movements of each wing separately and thus became endowed with the power of directive flight.

## DEVELOPMENT OF THE WINGS

Since the insect wing in a flattened, double-layered expansion of the body wall, its own walls consist of the same elements as the body wall, namely, cuticula, epidermis, and basement membrane, and its lumen contains nerves, tracheae, and the body fluid, or blood. The wings of insects with incomplete metamorphosis grow externally in the same manner as do the legs, the mouth parts, and other appendicular organs. The wing buds appear first in the second or third instar of the nymph as hollow, flattened outgrowths of the body wall along the lateral margins of the dorsum in the mesothorax and metathorax. They increase in size at each moult, without much change in structure until they assume the adult form at the transformation to the imago. In insects with complete metamorphosis the wings develop during the larval stage beneath the outer cuticula, usually within pouches of the epidermis. The time of their first formation varies in different insects from a late embryonic period to the last larval instar. The internal wing buds are normally everted from their pouches during the prepupal period of the last larval stage, but they are then still covered by the loosened cuticula of the larva. When this last larval skin is shed, and the insect enters the pupal stage, the wings are first exposed as external organs. During the pupal stage they develop rapidly and then quickly take their final form when the insect issues as an imago from the pupal skin.

In the very young wing bud the epidermis consists of upper and lower layers corresponding to the dorsal and ventral surface of the wing fold; but very soon the epidermal cells become elongate, and the inner ends of those in the opposing layers meet and unite. The fused basement membranes then become the so-called *middle membrane* of the wing. Along certain lines, however, the basement membranes do not come together; the channels thus left open, which are remnants of the primitive wing cavity, determine the courses of the future veins of the wing. The channels contain the wing nerves, tracheae and blood. In later stages of development, the epidermal cells condense along the vein channels and form here the thick cuticular layers that are to constitute the walls of the wing veins. When the wing development is completed, the epidermis has largely disappeared, and the mature wing is almost entirely a cuticular structure. Nevertheless, an active circulation of blood persists in the adult wing, observed in insects of most of the principal orders, and sense organs are of frequent occurrence on the wing surfaces. The histological changes in the growth of the wings are somewhat more complicated in the Holometabola, but the developmental pro- cesses are essentially the same as in insects with a simpler meta- morphosis.

The origin and growth of the tracheae of the wings are of much importance in the study of the wing venation, because, in many cases, the tracheation of the young wing serves as a key to the homology of the veins of the adult wing. The wing tracheae arise from a basal trachea, or two united tracheae, at the base of the wing bud, and in general are given off into the latter in two groups, one anterior, the other posterior. In insects with incomplete metamorphosis the tracheae take their places in the wings before the veins are formed and thus appear to determine the courses of the veins. In the Holometabola, however, the vein channels may be formed in advance of the tracheae, and, though in the Coleoptera, Neuroptera, and Lepidoptera each trachea is said to penetrate the vein corresponding to the one its homologue occupies in insects with incomplete metamorphosis, in the Trichoptera, Hymenoptera, and Diptera the relations between the tracheae and the veins are not so clearly preserved.

The internal wing buds of holometabolous larvae are aerated first by a few simple tracheae, but in the later larval stages they are supplied with numerous bundles of tracheoles that grow directly from the walls of the primary wing trachea. The definitive wing tracheae are finally formed during the last larval stage and become functional in the pupal stage, when the earlier tracheae and tracheoles degenerate. From the walls of the definitive tracheae, finally a second set of tracheoles is developed, and these tracheoles become functional at the change to the imago. The development of the wing tracheoles and tracheae was first studied by Gonin (1894); the details of the origin of the wing trachea from the basal trachea or tracheae of the wing in the principal groups of insects have been described by Chapman (1910), and the development of the wing veins of a cockroach by Beck (1920).

## THE STRUCTURE OF THE WINGS

In studying the wings of insects we must give special attention to three features of their structure, namely, the *articulation* to the body, the *veins*, and the differentiation of the alar surface into *wing regions*. The veins serve to strengthen the wing and to adapt it to the movements demanded of an organ of flight. The articular parts furnish the basal structure in the wing necessary for the movements of flight in the distal area and constitute also the flexor apparatus in the wing-flexing insects. The wing regions are local differentiations of the wing area partily subserving the function of flight, but largely accessory to the act of flexion.

The principal veins of the wings spring from the wing base, and most of them, except those of the posterior area, branch in varying degrees in the distal part of the wing. It is probable that all the diverse patterns of wing venation found in living and extinct insects have been derived from a single type of primitive venation; but the true primitive venation is not actually known, because the oldest fossil insects yet discovered have a highly complex system of wing veins. The venation

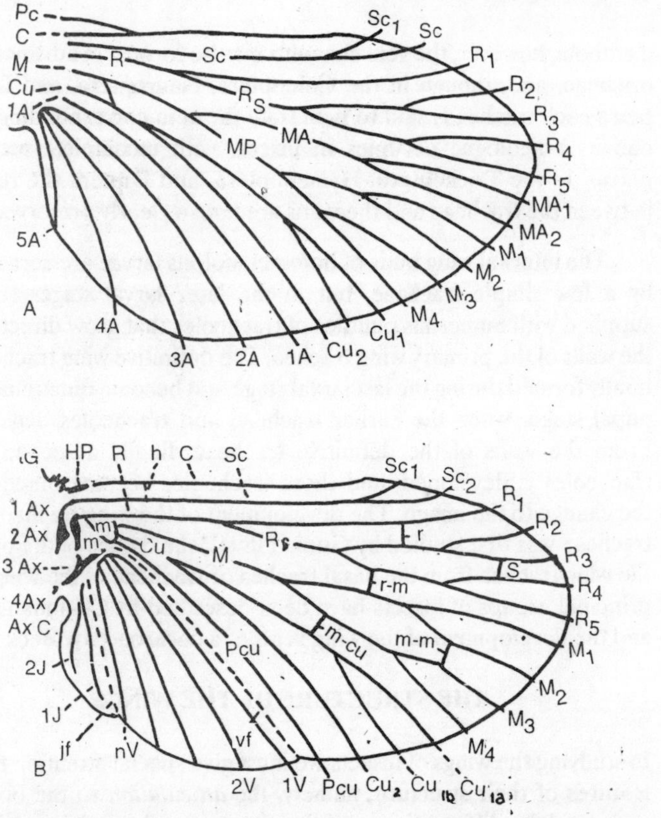


**Fig. 5.3.** Diagrams of wing venation. A, the archetype venation with veins named according to the Comstock-Needham system. B, the usual wing veins and axillaries as designated in the accompanying text. *A.* anal veins; *Ax*, axillary sclerites (first, second, third and fourth); *AxC*, axillary cord; *C*, costa *Cu*, cubitus; *h* humeral cross-vein; *HP* humeral plate; *J*, jugal veins; *if*, jugal fold; *M*, media; *m, m*, median plates; *MA*, media anterior;; *m-m* median cross-veinl; *m-cu* mediocubital cross-vein; *MP*, media posterior; *Pc*, precosta; *Pcu*, postcubitus (first anal); *Rs*, radius; *r*, radial cross-vein; *R* radial sector; *s*, sectorial cross-vein; *tg*, rudiment of tegula; *V*, vannal veins (anal veins except the first); *vf*, vannal fold.

pattern given in Fig. 5.3 A represents the plan of venation which students of the wing veins of insects regard as an ancient type from which the venation of modern insects has been derived, and it is therefore termed the *archetype* venation. The theoretically complete archetype venation includes the following veins, named according to the Comstock-Needham system of vein nomenclature; first, a small *precosta* at the base of the wings; second, a *costa*, which is usually marginal in modern insects; third, a two-branched *subcosta*; fourth, a five-branched *radius*; fifth, a six-branched *media*, sixth, a three-branched *cubitus*; and, finally, a varying number of *anal veins*.

The vein nomenclature given above is adopted in the present text with the exception that the anal veins are not recognized as a homogeneous group. The first anal vein in more generalized insects is always associated at its base with the cubitus, and in the wings of many nymphal insects it is represented by a distinct trachea. The independence of the first anal from the other anals becomes an important feature in a study of the mechanism of the wings. For this reason it is here designated the *postcubitus*. The rest of the anals, which constitute a definite functional group of veins in generalized insects, associated with the flexor scleritesof the wing base, are distinguished as *vannal veins* because the wing region containing them often forms a large fanlike expansion (vannus) of the posterior part of the wing. At the base of the wing, proximal to the vannal region there is usually a small but variously developed lobe, the jugum, which may contain one or two *jugal veins*. The second principal branch of cubitus, the postcubitus, and the vena dividens sometimes present in the fold between the postcubital and vannal veins are the first, second, and third plical veins of Forbes.

In the wings of modern insects the precosta does not appear, and the anterior fork of media is usually absent. The remaining veins in the wing are subject to many modifications in different groups of insects by the union of adjacent veins or by a partial or even complete suppression of certain veins, and the venational pattern may be further complicated by the addition of secondary veins. It often becomes, therefore, a difficult matter to identify with certainty the veins that are present, and the problem of determining the wing- vein homologies in the various orders of insects has been a major subject in entomology. A help in the study of the venation of adult insects may be derived from an examination of the basal connections or association of the veins with small sclerites in the articular region of the wing. These sclerites, the pteralia, are present in the wings of all insects, but they are particularly developed in insects that flex the wings over the back when at rest.

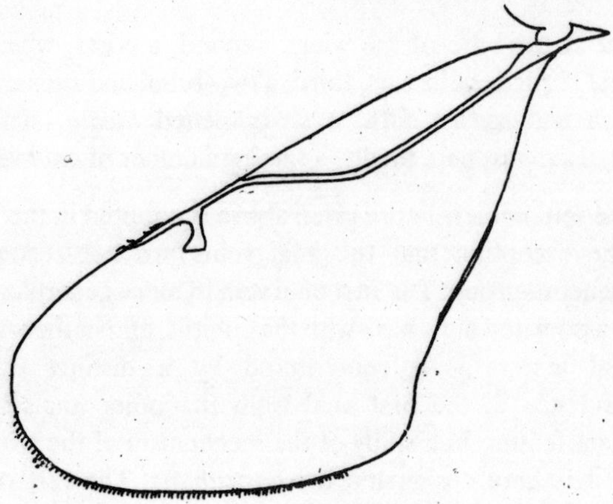

**Fig. 5.4.** Forewing of *Perilampus* (Hymenoptera) showing extreme reduction of venation.

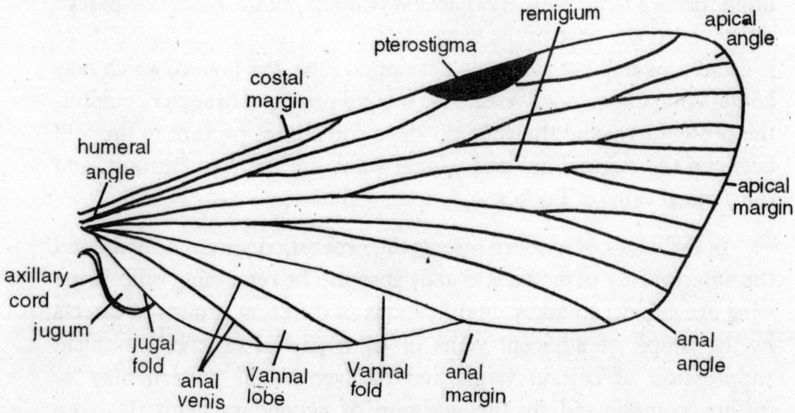


**Fig. 5.5.** Diagram illustrating some of the features of the wing.

## Areas of the Wing

In order to give Maximum efficiency and support to the wing during flight the longitudinal veins tend to be concentrated towards the anterior margin of the wing. The region containing the bulk of the veins is called the remigium and behind it the area supported only by the anal veins is called the vannal or anal region. The remigium and the vannal region are separated by the vannal fold which may lie

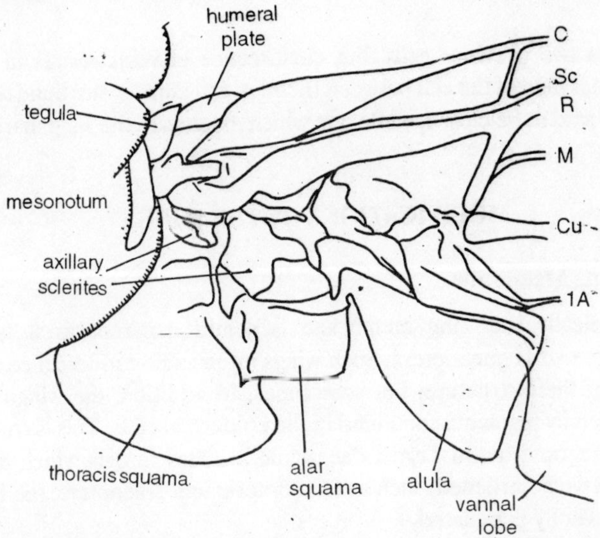


**Fig. 5.6. Base of the right wing of a tabanid showing the arrangement of the various structures.**

before or behind the vein IA. Proximal to the vannus in some insects is another membranous lobe called the jugum. In some Diptera there are three separate lobes in this region of the wing base known from proximally outwards as the thoracic squama, alar squama and alula. There is some confusion in the terminology and homologies of these lobes, but it appears that the thoracic squama is derived from the posterior margin of the scutellum, the alar squama represents the jugum and the alula is a part of the vannal region which has become separated off from the rest. Some Coleoptera have a lobe called an alula folded beneath the elytron. It appears to be equivalent to the jugum. The wing margins and angles are also named. The leading edge of the wing is called the costal margin, the trailing edge is the anal margin and the outer edge is the apical amrgir. The angle between the costal and apical margins is the apical angle, that between the outer and anal margins is the anal angle, while the angle at the base of the wing is called the humeral angle.

## Cells of the Wing

The veins devide the area of the wing into a series of cells which are most satisfactorily named after the vein forming the anterior boundary of the cell. If a vein atrophies the cell which remains includes two previous cells, but coalescence of veins results in the disappearance of the cell between them. A cell entirely surrounded by veins is said to be closed, while one which extends to the wing margins is open.

## MODIFICATIONS OF THE WINGS

### The Wing Membrane

Typically the wing membrane is semitransparent as it is in Odonata and Hymenoptera. Such wings often exhibit iridescence as a result of their structure, but sometimes, in addition, the wings are patterned by pigments contained in the epidermal cells. This is true in some Mecoptera and Trypetidae, while in many insects which have the fore wing hardened, such as Orthoptera and Coleoptera, the fore wing is wholly pigmented.

### Hairs and Scales on the Membrane

The surface of the wing membrane is often set with small non-innervated spines called microtrichia. Typically trichoid sensilla are confined to the veins, but in Trichoptera comparable hairs, known as macrotrichia clothe the whole of the wing membrane. In Lepidoptera the wings are clothed in scales. These vary in form from typical hairlike structures to flat plates and they usually cover the body as well as the wings. A flattened scale consists of two lamellae with an airspace between, the inferior lamella, that is the lamella facing the wing membrane, being smooth, the superior lamella usually with longitudinal and transverse ridges. The two lamellae are supported by internal struts called trabeculae. The scales are set in sockets of the wing membrane so that they are inclined to the surface and overlap each other to form a complete covering. In primitive Lepidoptera their arrangement on the wings are random, but in Papilionoidea, for instance, they are arranged in rows.

Pigments in the scales are responsible for the colours of many Lepidoptera, the pigment being in the wall or the cavity of the scale. In other instances physical colours result from the structure of the scale. Some specialised scales are associated with glands, while the scales may also be important in smoothing the airflow over the wings and body. On the body they are important as an insulating layer helping to

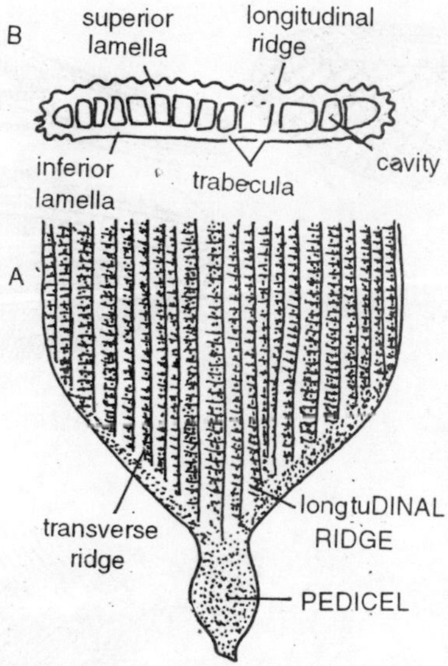


**Fig. 5.7.** A. Basal half of a typical lepidopteran scale. B. Transverse sect on of a scale.

maintain the high temperature of the thorax. Scales also occur on the wing veins and body of Culicidae and on the wings of some Psocoptera and a few Trichoptera and Coleoptera.

## Wing Form

In Odonata, Isoptera, Mecoptera and male Embioptera the two pairs of wings are similar in form, roughly shaped like elongate triangles, but in most other groups of insects one or other of the wings becomes modified from this basic form. Thus, the hind wings of Plecoptera, Dictyoptera and Orthoptera have large vannal lobes so that they are generally much more extensive that the fore wings. Sometimes the hind wings have a projection from the hind margin as in swallow-tailed butterflies and some Lycaenidae, while in the Nemopteridae the hind wings are slender ribbons, trailing out behind the insect. The hind wings are similarly modified in some Zygaenidae. Sometimes the hind wings are very small, as in Ephemeroptera. Hymenoptera and male coccids, while in some Ephemeroptera, such as *Cloëon*, and some male coccids they are absent altogether. In

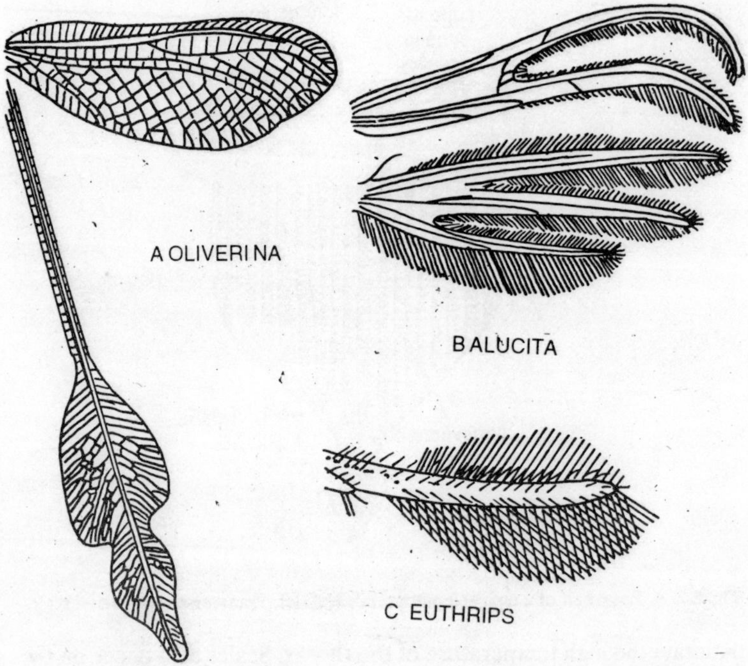


**Fig. 5.8.** A.Wings of *Oliverina* (Neuroptera). B. Wings of *Alucita* (Lepidptera). C. Forewing of *Euthrips* (Thysanoptera).

Diptera the hind wings are modified to form the halters, while in male Stepsiptea the fore wings form similar dumb-bell-shaped structures.

Sometimes the outline of the wings are irregular, as in *Polygonia c-album* (Lepidoptera) where it serves to break up the outline of the resting insect. In the plume moths, Pterophoridae and Orneodidae, the wings are very deeply cleft and divided into a number of lobes fringed with scales. Wing fringes are common in Lepidoptera and Culicidae and in some Tinaeoidae they are so extensive as to greatly increase the effective area of the wing. The wings of very small insects are often reduced to straps with one or two supporting veins and long fringes of hairs. This occurs in Thysanoptera, in Trichogrammatidae and Mymaridae amongst the Hymenoptera and in some of the small Stphylinoidae amongst the Coleoptera.

Some insects have both pairs of wings reduced and they are said to be brachypterous or micropterous. This occurs, for instance, in some Orthoptera and Heteroptera. The completely wingless, or apterous condition is also widespread. Winglessness occurs as a primitive condition in the Apterygota, while the ectoparasitic orders Mallophaga, Anoplura and Siphonaptera are secondarily wingless. Wingless species are also widespread in most other orders, but apparently do not occur in Odonata or Ephemeroptera. Sometimes both sexes are wingless, but frequently the male is winged and only the female is apterous. This is the case in coccids, Embioptera, Strepsiptera, Mutilidae an some Chalcididae. In the ants and termites only the reproductive caste is winged and here the wings are shed after the nuptial flight, breaking off by a basal suture so that only a wing scale remains. The break is achieved in different ways, but termites frequently rest the wing on the ground and then break it off by twisting the wing base. After loss of the wings the flight muscles degenerate. Quite commonly the development of the wings varies within a species either geographically or seasonally. Such wing polymorphism occurs in various groups, but is particularly well-knows in Heteroptera. For instance, *Gerris lacustris* is bivoltine in Britain and the overwintering generation is largely macropterous, that is fully winged, while the summer generation contains a relatively high proportion of micropterous individuals. In this case the wing length is determined

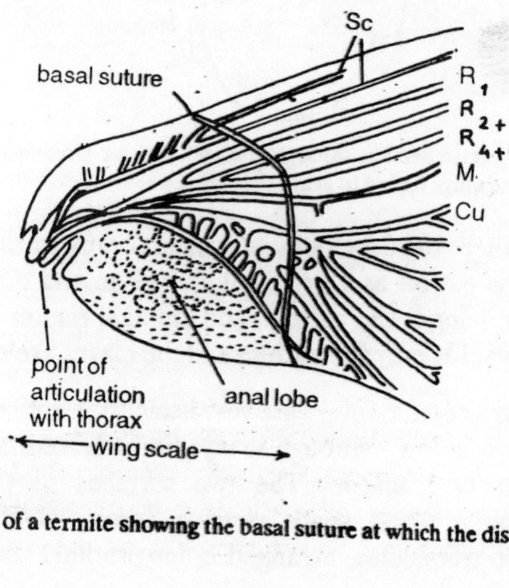


Fig. 5.9. Wing base of a termite showing the basal suture at which the distal part of the wing breaks off.

largely by the environment and to a lesser extent by genetic segregation.

**The Protective Function of the Fore Wings**

The fore wings of many insects become more fully sclerotised than the hind wings and serve to protect the latter when they are folded up at rest. Fore wings modified in this way are known as elytra or tegmina. Leathery elytra occur in Orthoptera, Dictyoptera and Dermaptera, while in Heteroptera only the basal part of the wing is hardened such wings being known as hemelytra. The basal part of the hemelytron may be subdivided into regions by well-marked veins and

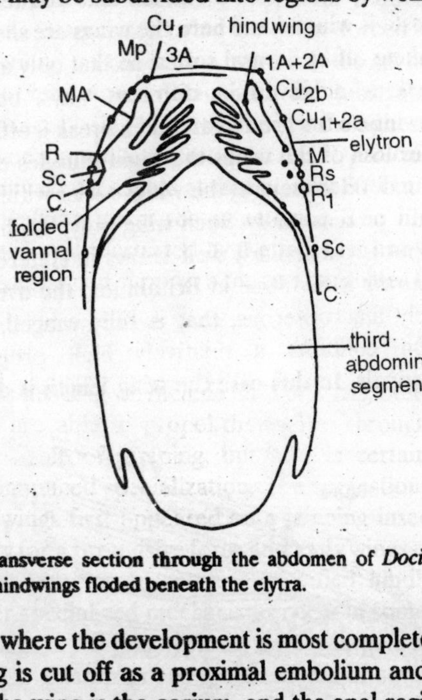


**Fig. 5.10. Transverse section through the abdomen of *Dociostaurius* (Orthoptera) showing the hindwings floded beneath the elytra.**

in capsids, where the development is most complete, the costal margin of the wing is cut off as a proximal embolium and distal cuneus, the centre of the wing is the corium, and the anal region is cut off as the clavus. In lygaeids only the corium and the clavus are differentiated.

The elytra of Coleoptera are very heavily sclerotised and the basic wing venation is lost, although it may be indicated internally by the arrangement of tracheae. The two surfaces of the elytron are separated by a blood space running across which are cuticular columns, the trabeculae, arranged in longitudinal rows and marked

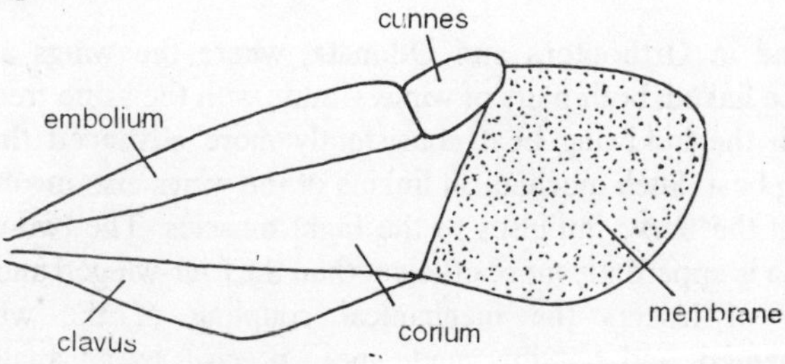


**Fig. 5.11.** Forewing of an anthocorid (Heteroptera).

externally by rows of striations. Primitively there are eight such striae, although the number may be increased in some Adephaga. The elytra of beetles do not overlap in the midline, but meet and are held together by a tongaed and grooved joint while in some Carabidae, Curculionidae and Ptinidae they are fused together so that they cannot open and in these species the hind wings are also atrophied. At the sides the elytra are often reflexed downwards, the vertical part being called the epipleuron and the horizontal part the disc.

### Sound Production

In various groups of insects the wings are modified for sound production and they may be retained for this function when they are no longer used in flight.

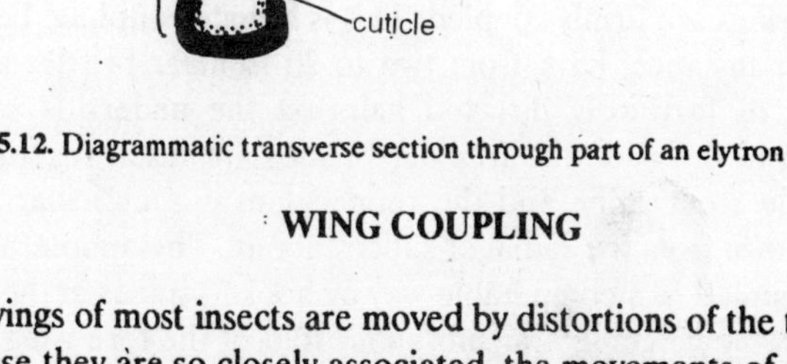


**Fig. 5.12.** Diagrammatic transverse section through part of an elytron of a beetle.

## WING COUPLING

The wings of most insects are moved by distortions of the thorax and, because they are so closely associated, the movements of each of the thoracic segments must influence the other. Hence, it is impossible for the fore and hind wings to beat completely independently of each other and in Orthoptera and Odonata, where the wings are not otherwise linked, both pairs of wings vibrate with the same frequency and with the hindwing beat consistently more advanced than the forewing beat. Such mechanical linking of the wings also involves the timing of the nerve impulses to the flight muscles. The two-winged condition is apparently more efficient than the four-winged and in the majority of insects the mechanical coupling of the wings is supplemented, and possibly made more precise, by an anatomical coupling of the fore and hind wings so that they move together as a single unit.

This wing coupling may take various forms, but in many species involves lobes or spines at the wing base. A primitive arrangement is found in some Mecoptera of the family Choristidae in which there is a jugal lobe at the base of the fore wing and a humeral lobe at the base of the costa margin of the hind wing. Both lobes are set with setae, those on the humeral lobe begins termed frenular bristles, and, although they do not firmly link the wings, they overlap suffi- ciently to prevent the wings moving out of phase. From this the types of coupling occurring in other Mecoptera, Neuroptera, Trichoptera and Lepidoptera can be derived. In some of the older forms of Trichoptera only the jugum is present on the fore wing. It lies on top of the hind wing so that it is not a very efficient coupling mechanism, but the Hepialidae have a strong jugal lobe which lies beneath the costal margin of the hind wing so that this is held between the jugum and the rest of the fore wing. This is called jugate wing coupling. In Micropterygidae the jugum is folded under the fore wings and holds the frenular bristles. This is jugo-frenate coupling.

Many other Lepidoptera have the frenulum well-developed and engaging with a catch or retinaculum on the underside of the fore wing so that the wings are firmly coupled. This is frenate coupling. Female noctuids, for instance, have from two to 20 frenular bristles and a retinaculum of forwardly directed hairs on the underside of the cubital vein; in the male the frenular bristles are fused together to form a single stout spine and the retinaculum is a cuticular clasp projecting down from the radial or subcostal vein. Thysanoptera have the wings coupled in a comparable way by hooked spines at the base of the hind wing catching a membranous fold of the fore wing. The wings of the Paplionoidea and some Bombycoidea are coupled by virtue of an extensive area of overlap between the two. This is known as amplexiform wing coupling. A similar arrangement occurs in some Trichoptera, often together with some other method of coupling.

Other insects have the wings coupled by more distal modifications which hold the costal margin of the hind wing to the anal margin of the fore wing. Thus Hymenoptera have a row of hooks, the hamuli, along the costal margin of the hind wing which catch into a fold of the fore wing; Psocoptera have a hook at the end of $Cu_2$ of the fore wing which hooks on to the hind and Heteroptera have a short gutter edged with a brush of hairs on the underside of the clavus which

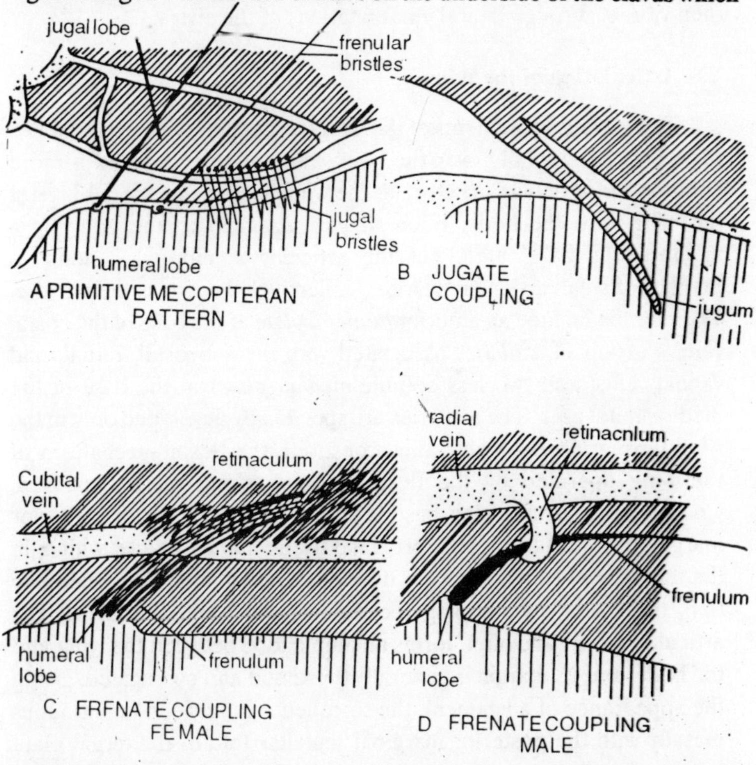


**Fig. 5.13.** Wing coupling mechanisms involving the jugal and humeral regions of the wings. A. Primitive mecopteran pattern in *Taeniochorista* (Mecoptera). B. Jugate coupling in *Charagia* (Lepidoptera). C. Frenate coupling in female *Hippotion* (Lepidoptera). D. Frenate coupling in male *Hippotion* (Lepidoptera). All diagrams represent the mechanisms as seen from below with the attachment to the thorax immediately to the left. Membrane of the forewing shown diagonally hatched, that of the hindwing with vertical hatching.

holds the costal margin of the hind wing. Homoptera exhibit a variety of modifications linking the anal margin of the fore wing to the costal margin of the hind wing. Other insects have become functionally two-winged by the reduction or complete loss of one pair of wings. In Diptera and some Ephemeoptera the fore wings alone function as propulsive organs, while in the Coleoptera the hind wings provide the power for flight. The elytra are not vibrated, but are extended at 30 – 45° above the horizontal by tonic contractions of the fight muscles aided by self-locking apodemes at the base. In Cetoniinae the elytra are raised only sufficiently to allow the wings to unfold and the wings then vibrate through lateral emarginations of the elytra.

**The Articulation of the Wings**

The various movements of the wings, especially in insects that flex the wings horizontally over the back when at rest, demand a more complicated articular structure at the wing base than a mere hinge of the wing with the body. Each wing is attached to the body by a membranous basal area, but the articular membrane contains a number of small articular sclerites, collectively known as the *pteralia*. The pteralia include an anterior *humeral plate* at the base of the costal vein, a group of *axillaries* associated with the subcostal, radial, and vannal veins, and two less definite *median plates* at the base of the mediocubital area. The axillaries are specifically developed only in the wing-flexing insects, where they constitute the flexor mechanism of the wing operated by the flexor muscle arising on the pleuron. Characteristic of the wing base is also a small lobe on the anterior margin of the articular area proximal to the *humeral plate,* which, in the forewing of some insects, is developed into a large, flat, scale-like flap, the *tegula*, overlapping the base of the wing. Posteriorly the articular membrane often forms an ample lobe between the wing and the body, and its margin is generally thickened and corrugated, giving the appearance of a ligament, the so-called *auxiliary cord*, continuous mesally with the posterior marginal scutellar fold of the tergal plate bearing the wing.

The articular sclerites, or pteralia, of the wing base of the wing-flexing insects and their relations to the body and the wing veins, shown diagrammatically in Fig. 5.13, are as follows:

### *The Humeral Plate*

Usually a small sclerite on the anterior margin of the wing base, movably articulated with the base of the costal vein; greatly enlarged in Odonata.

### *The First Axillary*

This sclerite is the anterior hinge plate of the wing base. Its anterior part is supported on the anterior notal wing process of the tergum; its posterior part articulates with the tergal margin. The anterior end of the sclerite is generally produced as a slender arm, the apex of which is always associated with the base of the subcostal vein, though it is not united with the latter. The body of the sclerite articulates laterally with the second axillary.

### *The Second Axillary*

This sclerite is more variable in form than the first axillary, but its mechanical relations are no less definite. It is obliquely hinged to the outer margin of the body of the first axillary, and the radial vein is always flexibly attached to its anterior end. The second axillary presents both a dorsal and a ventral sclerotization in the wing base; its ventral surface rests upon the fulcral wing process of the pleuron. The second axillary, therefore, is the pivotal sclerite of the wing base, and it specifically manipulates the radial vein.

### *The Third Axillary*

The third axillary sclerite lies in the posterior part of the articular region of the wing. Its form is highly variable and often irregular, but the third axillary is the sclerite on which is inserted the flexor muscle of the wing. Mesally it articulates anteriorly with the posterior end of the second axillary, and posteriorly with the posterior wing process of the tergum or with a small fourth axillary when the latter is present. Distally the third axillary is prolonged in a process which is always associated with the bases of the group of veins in the anal region of the wing here termed the vannal veins. The third axillary, therefore, is usually the posterior hinge plate of the wing base and is the active sclerite of the flexor mechanism, which directly manipulates the vannal veins. The contraction of the flexor muscle revolves the third axillary on its mesal articulations and thereby lifts its distal arm; this movement produces the flexion of the wing.

### *The Fourth Axillary*

This sclerite is not a constant element of the wing base. When present it is usually a small plate intervening between the third axillary and the posterior notal wing process and is probably a detached piece of the latter.

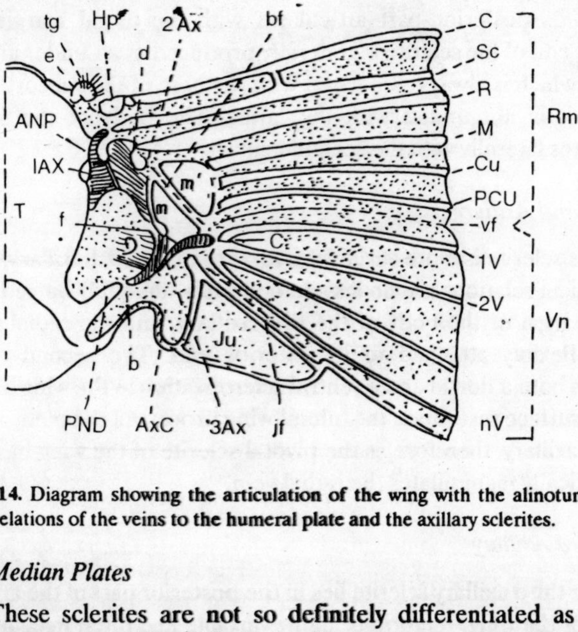


**Fig. 5.14.** Diagram showing the articulation of the wing with the alinotum, and the basal relations of the veins to the humeral plate and the axillary sclerites.

### *The Median Plates*

These sclerites are not so definitely differentiated as specific plates as are the three principal axillaries, but nevertheless, they are important elements of the flexor apparatus. They lie in the median area of the wing base distal to the second and third axillries and are separated from each other by an oblique line which forms a prominent convex fold during flexion of the wing. The proximal plate is usually attached to the distal arm of the third axillary and perhaps should be regarded as a part of the latter. The distal plated is less constantly present as a distinct sclerite and may be represented by a general sclerotization of the base of the mediocubital field of the wing. When the veins of this region are distinct at their bases, they are associated with the outer median plate.

## The Wing Base of Ephemerida

The mayflies, when at rest, bring the wings together vertically over the back, but they do not flex the wings in the sense of folding them horizontally. A flexor mechanism, therefore, is not developed in the bases of the wings; and yet the structure of the articular areas of the wings is not radically different from that of the wing-folding insects. At the base of each wing of the mayfly there is a small humeral plate intermediating between the head of the costal vein and a small tergal lobe of the body segment supporting the wing. In the axillary region there is a group of weakly defined sclerites which in their arrangement and relations to the vein bases give a suggestion of the axillaries of the wing-folding insects. The posterior part of the axillary membrane in the mayfly has the usual form of a fold bordered by a corrugated by a corrugated thickening, or axillary cord, continuous with the posterior margin of the tergum.

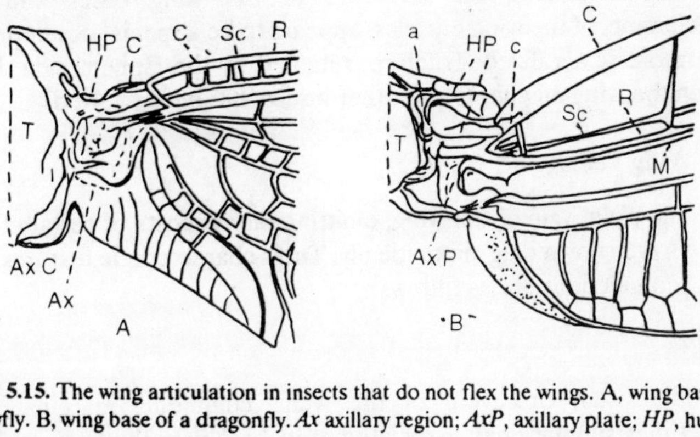


Fig. 5.15. The wing articulation in insects that do not flex the wings. A, wing base of a mayfly. B, wing base of a dragonfly. *Ax* axillary region; *AxP*, axillary plate; *HP*, humeral plate.

## The Wing Base of Odonata

The articular region of the wing of a dragonfly contains two large, strongly sclerotized plates. The anterior plate supports the costal vein by a small intermediate sclerite at the base of the latter and thus corresponds to the humeral plate of Ephemerida and the wing-flexing insects. The great enlargement of the humeral plate in Odonata is evidently a specialized feature of the flight mechanism in this group. The posterior plate of the dragonfly wing carries the four basal shafts of the postcostal veins and hence may be termed the *axillary plate*, since it corresponds in position to the group of sclerites in the ephemerid wing that appear to represent the axillaries of the wing-flexing insects. The humeral plate of the odonate wing is hinged to the anterior half of the lateral edge of the tergum of the segment supporting the wing, or in some species to a distinct sclerite of the tergum. The axillary plate is articulated to the posterior half of the lateral tergal margin opposite a deep membranous area of the latter. The pleural wing process supporting the wing has two arms, one applied to the humeral plate, the other to the axillary plate. The basal plates of the dragonfly's wing turn up and down on the fulcral arms when the wings are lifted or depressed. The two plates, however, are slightly movable on each other, and, since the costal veinis doubly hinged to the humeral plate by a small intermediary piece at its base, the costal area of the wing can be quite freely deflected independent of the rest of the wing area, which is solidly supported on the axillary plate by the veins attached to the latter. The flight mechanism of the Odonata, including the structure of the wing bases and the attachments of the wing muscles, appears to be a special development of a more generalized structure, retained by the Ephemerida, from which the wing mechanism of other insects has been evolved.

## The Wing Veins

The usual vein of the wing, omitting the precosta of certain fossil insects are shown diagrammatically. Their characteristic features and basal connections are as follows:

### *Costa*

The usual first vein of the wing, commonly marginal, but sometimes submarginal; associated at its base with the humeral late. The trachea of the costal vein is perhaps a branch of the subcostal trachea.

### *Subcosta*

The second vein of the wing, typically forked distally into two short breaches; associated at its base with the distal end of the neck of the first axillary.

### *Radius*

The third an generally the strongest vein of the wing. Toward the middle of the wing it forks into a first undivided branch and a second branch, called the *radial sector,* which subdivides dichotomously into four distal branches. Basally the radius inflexibly united with the anterior end of the second axillary.

### *Media*

The fourth vein of the wing. In the archetype pattern the media forks into two main branches, a *media anterior* which divides into two distal branches, and a *median sector,* or *media posterior,* which has four terminal branches. In most modern insects the media anterior has been lost, and the usual "media" is the four-branched media posterior with the common basal stem. In the Ephemerida, according to present interpretations of the wing venation, both branches of the media are retained, while in Odonata the persisting media is the primitive anterior branch. The stem of the media is often united with the radius, but when it occurs as a distinct vein its base is associated with the distal median plate or is continuously sclerotized with the latter.

### *Cubitus*

The fifth vein of the wing, primarily two branched. The primary forking of the cubitus takes place near the base of the wing, forming the two principal branches. The anterior branch may break up into a number of secondary branches but commonly it forks into two distal branches. The second branch of the cubitus in Hymenoptera, Trichoptera, and Lepidoptera was mistaken by Comstock and Needham for the first anal, as has been shown by Tillyard, Lameere, Tanaka, Imms and others. Proximally the main stem of the cubitus is associated with the distal median plate of the wing base.

### *Postcubitus*

This vein is the *first anal* of Comstock and Needham, except where these writers wrongly identified the second branch of cubitus as the *"first anal."* The postcubitus, however, has the status of an independent wing vein and should be recognized as such. In nymphal wings, as amply shown by Comstock (1918), its trachea arises between

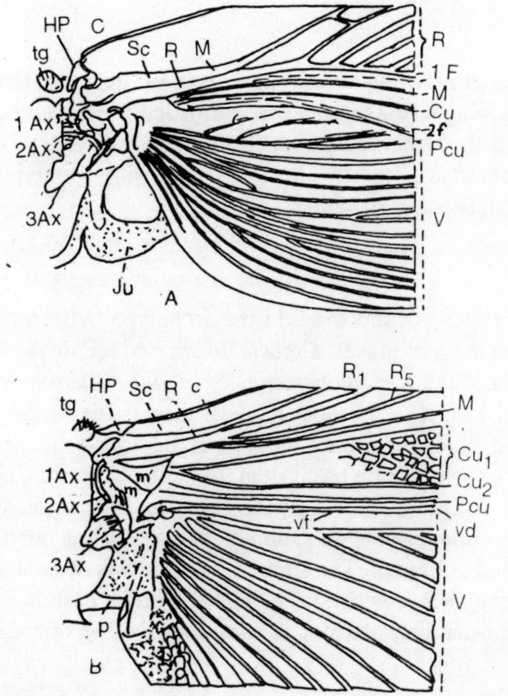


**Fig. 5.16.** Bases of the fore and hind wings of *Periplai.acta americana*.

the cubital trachea and the group of vannal tracheae. In the mature wings of more generalized insects the postcubitus is always associated proximally with the cubitus and is never intimately connected with the flexor sclerite of the wing base. In Neuroptea, Mecoptera and Trichoptera the postcubitus may be more closely associated with the vannal veins, but its base is always free from the latter. The postcubitus is usually unbranched; according to Lameere (1922) it is primitively two branched. In a dormer paper the writer (1930) called this vein *"second cubitus"* and mistakenly regarded it as representing the second branch of cubitus in Orthoptera.

## *Vena Dividens*

This is apparently a secondary vein present in the hind wing of some Orthoptera, developed in the fold that sets off the vannal region from the wing region before it.

### *Vannal Veins*

The vannal veins are the anal veins that ar immediately associated with the third axillary, and which are directly affected by the movement of this sclerite that brings about the flexion of the wings. In number the vannal veins vary from 1 to 12, according to the expansion of the vannal area of the wing. The vannal tracheae usually arise from a common tracheal stem in nymphal insects, and the veins are regarded by Lameere (1922) and Tanaka (1926) as branches of a single anal vein. Distally the vannal veins are either simple or branched.

### *Jugal Veins*

The jugal lobe of the wing is often occupied by a network of irregular veins, or it may be entirely membranous ; but some times it contains one or two distinct small veins, the *first jugal vein*, or *vena arcuata*, and the *second jugal vein*, or *vena cardinalis.*

### *Cross-veins*

All the veins of the wing are subject to secondary forking and to union by cross-veins. In some orders of insects the cross-veins are so numerous that the whole venational pattern becomes a close network of branching veins and cross-veins. Ordinarily, however, there is a definite number of cross-veins having specific locations as indicated at B of Fig. 5.13. The more constant cross-veins are the *humeral crossvein (h)* between costa and subcosta, the *radial cross-vein* between and the first fork of *Rs*, the *sectorial cross-vein* between the two forks of *Rs* the *median cross-veins* between $M_2$ and $M_3$, and the *mediocubital cross-vein* between media and cubitus. The veins of the wings appear to fall into an undulating series of *convex veins* and *concave veins*, according to whether they have a tendency to fold up or down when the wing is relaxed. The basal shafts of the veins are convex, but according to Lameere (1922) each vein forks distally into an anterior convex branch and a posterior concave branch. Thus, the costa and subcosta are regarded as convex and concave branches of a primary first vein, *R,* is the concave branch of the radius, *MP* the concave branch of the media, $Cu_1$ and $Cu_2$ are respectively convex and concave, while the primitive postcubitus and the first vannal have each

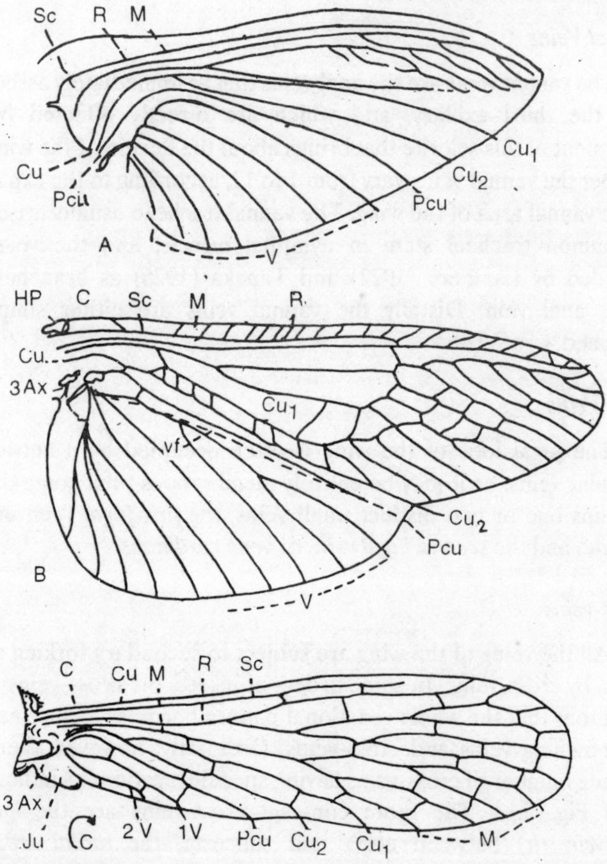


**Fig. 5.17.** Examples of wing venation. A, hind wing of nymph of *Scudderia. B, hind wing of a plecopteron*, *Isogenus*, C, forewing of *Panorpa*.

an anterior convex branch and a posterior concave branch. The convex or concave nature of the veins has been used as evidence in determining the identities of the persisting distal branches of the veins of modern insects, but it has not been demonstrated to be consistent for all wings.

## The Wing Regions

In the wings of all insects we must distinguish a basal *articula area* from true *ala*, or distal expanse of the wing containing the veins. The wing base of Odonata, as we have observed, contains two large plates, an anterior humeral plate supporting the costal vein, and a posterior axillary plate supporting the other veins. The structure here is probably a specialized development of a more generalized structure of the wing base in primitive insects. In the Ephemerida the humeral plate is small, as it is in insects generally, and the axillary region contains a group of indistinctly differentiated sclerites. In the wing-flexing insects the axillary sclerites are well defined and individualized; the area containing them is a definite feature of the wing base and may be termed the *axillary region* of the wing.

The true alar area of the wing is always more or less asymmetrical in form. The contour of the front margin is different from that of the hind margin, and the pattern of the anterior venation never matches with that of the venation of the posterior part of the wing. There is a tendency for the anterior veins to become thickened and crowded toward the forward margin in such a manner as to give greater rigidity to the front half of the wing, while the weaker posterior veins are more widely spaced and give flexibility to the rear half. The alar area thus becomes differentiated into an anterior region, which is actively effective in flight, and a posterior more passive region. The anterior rigid part of the wing may be termed the *remigium*. In more generalized slow-flying insects, the posterior flexible part of the wing is often enlarged to form a fanlike expansion of the wing and hence may be termed the *vannus*. In the more specialized swift-flying insects the vannus is reduced; but since it contains the veins connected with the flexor sclerite of the wing base and is therefore an essential part of the flexor apparatus, the vannus is seldom entirely obliterated. Finally, there is often developed at the base of the wing proximal to the vannus a membranous lobe of the wing, the *neala* of Martynov (1925), commonly called the *jugum* because that of the forewing in some insects serves to yoke the two wings on each side with each other. At the posterior angle of the wing base there sometimes occurs a membranous lobe, or pair of lobes, known as the *alula*, or *calypter*.

The three regions of the alar surface are commonly separated by lines of folding in the wing membrane. This is true particularly when the wings are wide and cannot be placed flat over the body in the flexed position. There occurs then between the remigium and the vannus, or approximately separating these regions, a *plica vannalis*, or

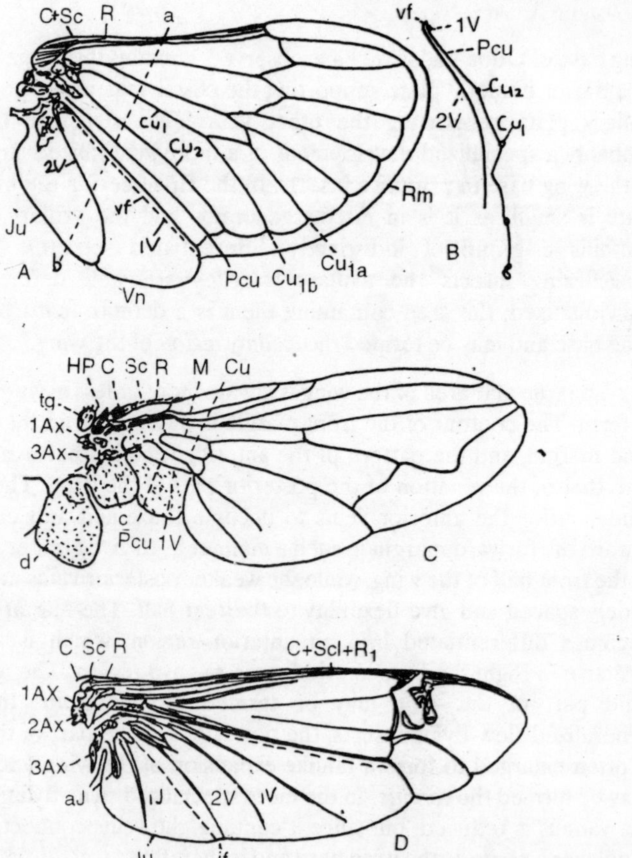


**Fig. 5.18.** Examples of wing venation. *A, Magicicada septendecim*, hind wing, extended. B, section of same along line *a-b* when folded. *C, Musca domestica*, wing and calypteres. *D, Epicauta peennslvanica.*

*vannal fold*. This fold either allows the vannus to take a horizontal. The outer squama arises from the wing base behind the third axillary sclerite and evidently represents the jugal lobe of other insects; the larger inner squama arises from the posterior scutellar margin of the tergum of the wing-bearing segment and forms a protective, hoodlike canopy over the halter. In the flexed wing the outer squama of the alula is turned upside down above the inner squama, the latter not being affected by the movement of the wing. In many Diptera a deep incision of the anal area of the wing membrane behind the single vannal vein sets off a proximal alar lobe distal to the outer squama of the alula.

**The Wing Muscles**

The movements of the wings in the majority of insects are accomplished by five pairs or paired sets of muscles in each alate segment. These muscles are the *dorsal muscles*, the *tergosternal muscles* the *axillary muscles*, the *basalar muscles*, and the *subalar muscles*. The dorsal and tergosternal muscles are often called the *"indirect wing muscles,"* the axillary and epipleural muscles the *"direct wing muscles,"* but, strictly speaking, only the axillary muscles in most insects are attached directly on the wing bases. In addition to the muscles listed above as specific wing muscles, it is probable that most of the segmental and intersegmental muscles of the pterothorax that are not leg muscles have some action in relation to the wing movements. Particularly the tergopleural muscles, which extend from the tergum to the basalare, to the wing process, or to the epimeron, must exert some controlling influence on the movement of the tergum. Those inserted on the basalare undoubtedly have a direct action on the wings; a large tergobasalar muscle present in some Diptera sharply extends the wing in a horizontal plane. Since however, the tergopleural muscles are highly variable and are not constantly present in the wing-bearing segments, they will not be considered in the following general discussion of the wing muscles.

**The Dorsal Muscles**

These muscles are the ordinary longitudinal muscles of the back, which, in the usual secondary segmentation of the body, extend from the antecosta of one tergum to that of the next. In the wing-bearing segments of most insects the dorsal muscles are differentiated into *median longitudinal muscles* and *lateral oblique muscles*. The median dorsal muscles are usually greatly enlarged in the wingbearing segments, and their expansion is accommodated by the development of phragmatal lobes on the antecostae of the mesotergum, the metatergum, and the first abdominal tergum, but frequently also their dorsal fibers encroach upon the postcostal surface of the alinotum and on the precostal surface of the postnotum. The longitudinal dorsal muscles are the principal depressors of the wings, since, by their contraction they arch the wing-bearing terga upward between the ends of the segments and thus deflect the wings on the pleural fulcra. The action of the dorsal muscles as wing depressors, however, depends on the obliteration of the dorsal intersegmental membranes, a condition that has been brought about either by a fusion of the consecutive terga or by a forward extension of the precostal lips of the metathoracic and first abdominal terga to form the phragma-bearing postnotal plates of the mesothorax and metathorax, respectively. The dorsal muscles are most highly developed in the segment bearing the principal pair of wings; they are usually reduced or absent in a segment of which the wings are small or are used for other purposes than that of flight. With insects such as Isoptera, Blattidae, and Gryllidae having weak powers of flight, the dorsal muscles are very small in both segments of the pterothorax, but they are also reduced in the strong-flying Odonata, in which the wings are moved by the lateral thoracic muscles.

The Isreral oblique dorsal muscles arise on the posterior part of the scutum and are inserted posteriorly on the succeeding phragma laterad of the bases of the median dorsals. Usually these muscles are relatively small, and they are not always present; in function they

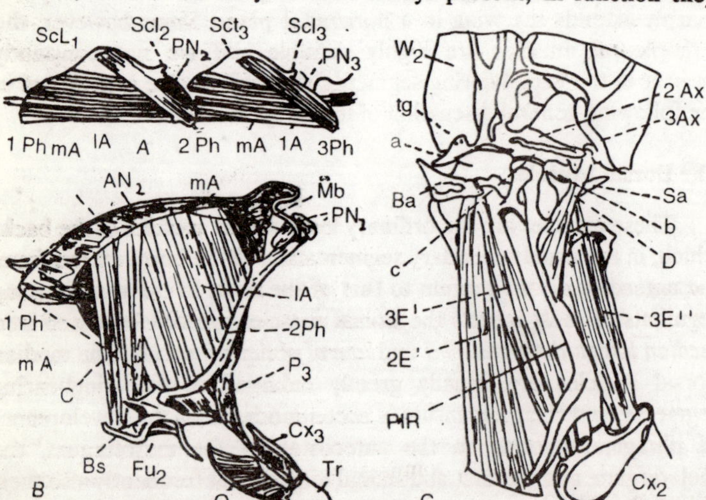


**Fig. 5.19.** The wing muscles. A, dorsal muscles of pterothorax of *Panorpa*, lateral view. B, mesothorax of *Magicicada*, median dorsals (*mA*) removed, showing almost vertical Postonotum of large lateral (*lA*). C, mesothorax of *Dissosteira* with basalar (*E'*), subalar (*E*) and wing flexor muscles (*D*) of right side, mesal view.

probably supplement the tergosternals in their downward pull on the tergum. In the mesothorax of the cicada and in the Diptera the oblique dorsal muscles are unusually large and assume a position so nearly vertical, by reason of the great size of the second phragma, that they become powerful adjuncts of the tergosternal muscles as depressors of the tergum.

### The Tergosternal Muscles

These muscles lie to the sides of the median dorsal muscles in the anterior part of the segment. They are attached dorsally on the anterior lateral areas of the tergum, and ventrally on the basisternum before the coxae. There may be one or several pairs of them in each segment. Functionally the tergosternal muscles are antagonists of the longi- tudinal dorsals, since by contraction they depress the tergum and thereby elevate the wings on the pleural fulcra. These muscles have no representatives in the prothorax, and they may be absent in the pterothorax of insects of weak flight. In the Diptera a third pair of muscles, lying between the anterior tergosternals and the posterior oblique dorsals, becomes secondarily levators of the wings. The muscles of this pair are the normal tergal remotors of the middle legs inserted on the meral lobes of the coxae. In the higher Diptera, however, the mesothoracic meron is detached from the rest of the coxa and becomes solidly incorporated into the lateral wall of the thorax. The tergal remotor of the coxa is thus anatomically transferred from the leg and given over functionally to the service of the wing, since, by the loss of movement at its lower end, it becomes a depressor of the tergum.

### The Axillary Muscles

The only muscles attached directly on the wing bases, in insects other than Odonata, are muscles arising on the pleuron and inserted on the first and third axillary sclerites. A muscles of the first axillary is known to occur only in Diptera. In a syrphid fly this muscle consists of two parts, one arising on the episternum, the other behind the pleural ridge, both inserted on the inner margin of the first axillary. A pull on the muscle turns the axillary upward on its tergal articulations, which is the usual action of the first axillary during flexion of the wings. The muscle of the third axillary is present in all the wing- flexing insects, since it is the effector of the flexion movements of the wing. The muscle arises on the pleuron, but it is variable in size and distribution. Typically it consists of a single bundle of fibers attached on the pleural ridge, but it may comprise several branches arising on the episternum, the pleural ridge, and the epimeron. Distally the flexor muscle is inserted on the base of the third axillary sclerite. Its contraction revolves the third axillary dorsally and inward on the proximal articulations of the latter and thus turns the alar area of the wing posteriorly on the axillary region by a flexure along the line of the plica basal is. A muscle corresponding to the wing flexor is well-developed in each wing-bearing segment of zygopterous Odonta. It arises on the pleural ridge and is inserted Posteriorly on the axillary plate.

**The Basalar Muscles**

The muscles of the basalar sclerites, or of the basalar lobe of the episternum usually include two muscles on each side, but sometimes three are present, or again only one. The first of the potential three muscles of t˙ ;s group arises on the episternum, the second arises on the sternum or the precoxal bridge of the pleuron or occasionally on the trochantia (*Gryllus*), the third (93*E'*) arises on the outer rim of the

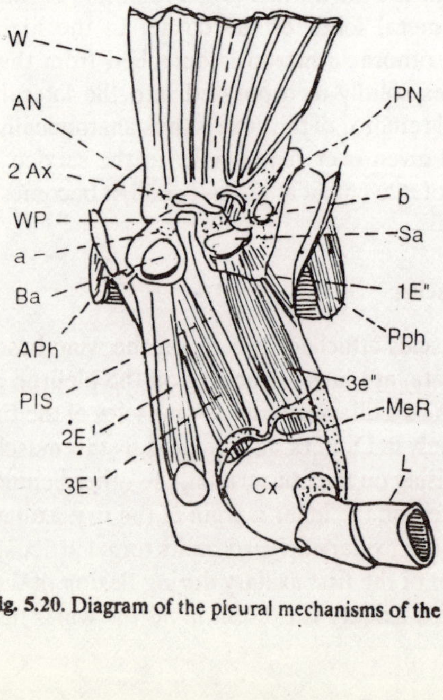


**Fig. 5.20.** Diagram of the pleural mechanisms of the wing.

coxa anterior to the pleural coxal articulation. The last muscle appears to be a pleural leg muscle that has secondarily become a wing muscle in the adult by reason of the intimate connection of the basalare with humeral angle of the wing. The basalar muscles of the adult winged insect function as depressors of the costal margin of the wing during flight and as extensors of the flexed wing, for which reasons they may be termed the *pronator-extensor* muscles of the wing. In Odonata there are two anterior wing muscles arising on the lower edge of the episternum and inserted by long tendons directly on the large humeral plate of the wing base. There are no epipleural sclerites in the dragonflies, and no pleurocoxal muscles are associated with the wing mechanism.

**The Subalar Muscles**

In most insects there is but a single subalar muscle, usually of large size, lying against the epimeral wall of the pleuron, on each side of each wing-bearing segment, which is attache ventrally on the meron of the coxa. Associated with this muscle, however, there is sometimes, as in Gryllidae, Trichoptera, and Lepidoptera, a second muscle arising on the epimeron and inserted on the posterior part of the subalare or on a distinct second subalar sclerite (*Gryllus*). This muscle is a counterpart of the first basalar muscle arising on the episternum. A subalar muscle corresponding to the sternal muscle of the basalare is known to occur only in Ephemerida; it is here a large muscle of the mesothorax arising medialy on the sternum behind the coxae and inserted dorsally on the subalar region of the pleuron. In the mesothorax of the higher Diptera the single large subalar muscle arises on the lower part of the epimeron dorsal to the meron, but this muscle is probably the usual subalar-coxal muscle transposed from the displaced meron to the pleural wall. The subalar muscles serve to extend and to depress the wing because of the close connection of the subalar sclerite with the second axillary sclerite of the wing base. They may be called therefore, the *depressor-extensor muscles* of the wing. In the Odonata two posterior pleural wing muscles take their origins on the ventral edge of the epimeron in each alate segment and are inserted directly on the axillary plate of the wing base.

## MOVEMENT AND CONTROL OF THE WINGS

The up and down movements of the wings are produced by direct and indirect wing muscles, but they also involve the elasticity of the thorax, the wing base and the muscles themselves.

### Movements Produced by the Muscles

In all insects the upward movement of the wings is produced by indirect dorso-ventral muscles inserted into the tergum of the segment bearing the wing. By contracting they pull the tergum down and hence also move down the point of articulation of the wing with the tergum. The effect of this is to move the wing membrane up with the pleural process acting as a fulcrum. The muscles producing this movement are not always homologous. In many insects they arise on the sternum or the coxae, but in Auchenorrhyncha and Psyllidae the tergosternal muscles are small and are funcitonally replaced as wing elevators by the oblique dorsal muscles. These arise on the postphragma and so are normally obliquely longitudinal but in the groups mentioned the phragma extends ventrally carrying the origins of the muscles with it so that they come to exert their pull vertically instead of horizontally (see Pringle, 1957).

The downward movement of the wings in Odonata and Blattaria is produced by direct muscles inserted into the basalar and subalar sclerites which are connected to the axillary sclerites by ligaments. Hence contraction of these muscles exerts a pull on the wings outside the fulcrum of the pleural process and so pulls the wings down. In Diptera and Hymenoptera the downward movement is produced by the dorsal longitudinal indirect muscles. Because the dorsum of the pterothorax is an uninterrupted plate, without membranous junctions contraction of the dorsal longitudinal muscles cannot produce a telescoping of the segments as in the abdomen. Instead, the center of the tergum becomes bowed upwards so that the tergal articulation of the wing is also moved up and the wing membrane flaps down. At the same time the anterior and posterior notal processes become approximated because of the hinging of the scutellum to the scutum, and this also assists in the movement of the wing.

In Coleoptera and Orthoptera the downward movement is produced by the direct and indirect muscles acting together. The

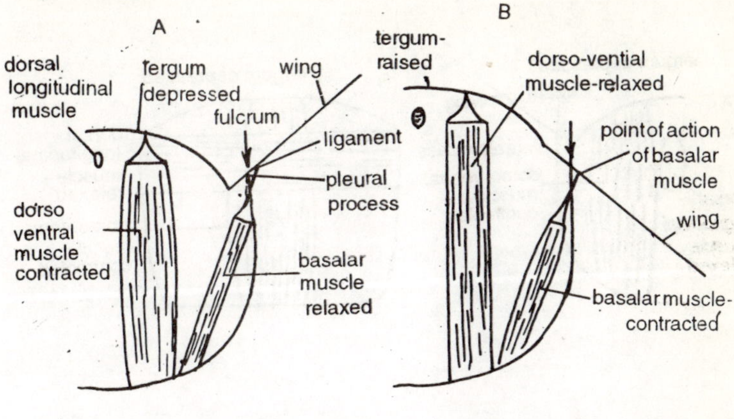


**Fig. 5.21. Diagrammatic cross-section of the thorax illustrating the wing movements in an insect, such as a dragonfly, in which the direct wing muscles cause depression of the wings.**

direct muscles are then dual purpose since they are also concerned in twisting the wing during the course of the stroke . A number of the muscles moving the wing arise in the coxa which is itself movable. Whether these muscles more the legs or the wings appears to be determined by the position of the appendages; if the wings are closed the muscles move the legs, but in flight with the legs in the flight position the wings are moved.

**Movement Due to Elasticity**

In *Schistocerca* (Orthoptera), and probably in other insects, much of the energy involved in the upstroke is stored as elastic forces for use in the downstroke. This is possible because the aerodynamic forces produced at this time act in the same direction as the wing movement so assisting its movement. Thus, the muscles have only to overcome the forces of inertia of the wing and elasticity of the wing base, and as a result some 86% of the energy they produce is stored for use in the downstroke. The elasticity of the system results partly from the pad of resilin which forms the main wing hinge, The elastic properties of this pad are almost perfect so that it absorbs less than 3 % of the energy imparted to it when it is stretched in the upward movement of the wings. The remaining energy is available for pulling the wing down. The elasticity of the flight muscles is also important. These muscles

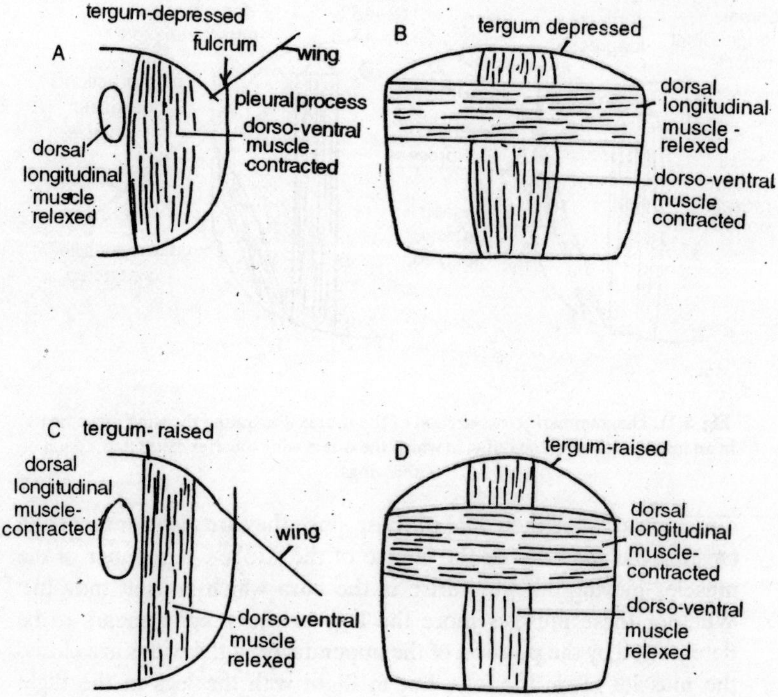


**Fig. 5.22. Diagrams illustrating the movements of the wings in an insect, such as a fly, in which both up and down movements of the wing are produced by indirect muscles. A and C, Cross-sections of the thorax. B and D. Views of the wing-bearing segment from the inside showing, in D, the shortening and bowing of the tergum produced by contraction of the dorsal longitudinal muscles.**

are characterised by a greater resistance to stretch compared with other muscles due to the elastic properties of the contractile system. The sarcolemma seems to add little to the elasticity of the muscle.

Contraction of the indirect flight muscles distorts the thorax so that the elastic properties of the thorax as a whole are also significant factors in wing movement. Figure 5.23 shows diagrammatically the manner in which the movement of the wings involves a lateral movement of the wall of the thorax. This movement is resisted by the elasticity of the thorax which is largely due to the sternopleural

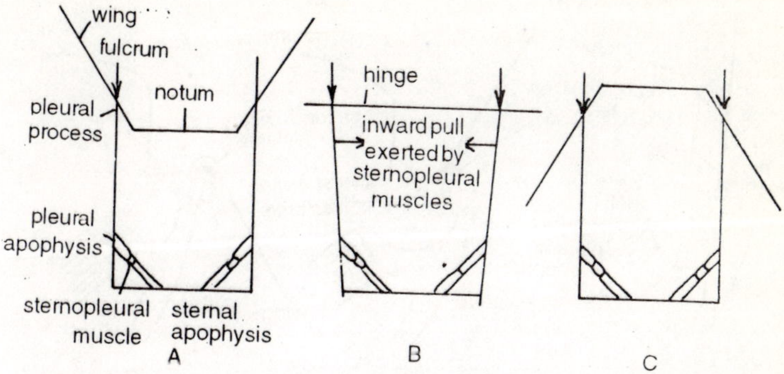


**Fig. 5.23.** Diagrammatic cross-section of the insect thorax illustrating the distortion of the thorax produced by wing movement. A Wings stable in the up position. B. Unstable position due to the inward pull of the sternopleural muscles. C. Wings stable in the down position.

articulation and, to a lesser extent, the tergopleural articulation. In the mesothorax of Coleoptera and the metathorax of Hymenoptera the pleural and sternal apophyses are fused so that lateral stiffness of the thorax is considerable and constant, but in other insects the apophyses are joined by a muscle, alterations in the tension of which can regulate the lateral stiffness.

As a result of this lateral stiffness the position of the wings is unstable for much of the stroke and they will tend to return automatically either to the fully up or fully down position, these being the only stable positions. Thus in flight the wings are moved by the muscles to the position of maximum instability and then they will swing into the up or down position as a result of the thoracic elasticity. This arrangement is called a 'click' mechanism. In an insect the wing articulation is more complex than in the diagrams, but the method of working is basically the same. Thus in *Sarcophaga* (Diptera) both dorsoventral and dorsal longitudinal indirect flight muscles produce a lateral extension of the notum, exerting forces outwards, while the pleural process is pulled inwards by the sternopleural muscle. Hence, the system XYZ in Figure 5.24 is only stable at the extreme ends of the stroke.

Contraction of the dorsal longitudinal muscles of *Sarcophaga* lowers the scutellum which is hinged to the scutum. This raises the

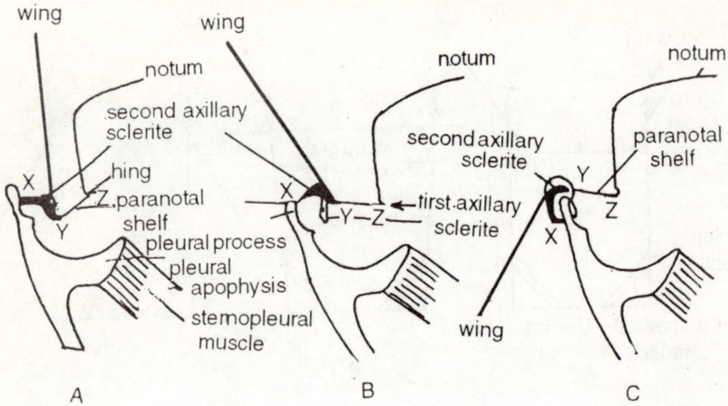


**Fig. 5.24.** Diagrammatic section through the wing base of a fly illustrating the click mechanism in the wing movement. A. Wing stable in the up position. B, Unstable position with the system XYZ in a straight line and under pressure between the arrows. C. Wing stable in the down position.

anterior end of the scutellar lever, which arises from the side of the scutellum, and this pushes the first axillary sclerite up until it reaches the unstable position when the forces exerted at X and Z cause the wing to click into the stable down position. In raising the wings the scutellar lever pulls the first axillary sclerite down to the position of maximum instability and then the wing automatically clicks up. Click mechanisms have been demonstrated in Orthoptera, Diptera, Coleoptera and possibly Odonata. At the end of each wing-stroke the movement produced by the click is limited by a stop. This gives greater efficiency in flight since it eliminates the necessity of doing work in stopping the, movement. Thus in *Sarcophaga* the upstroke is limited by the scutellar lever hitting against the pleural sclerites, but the lower limit is more variable. It is determined by the nature of the articulation between the first and second axillary sclerites which can be varied to some extent by the direct muscles altering the lateral tension. In *Schistocerca* the movement of the wings may be stopped by the contraction of antagonistic muscles before the wingstroke is complete. Thus, the dorsal longitudinal muscles may start to contract before the end of the upstroke. This requires a greater muscular output than is necessary in *Sarcophaga*, but in this way the amplitude

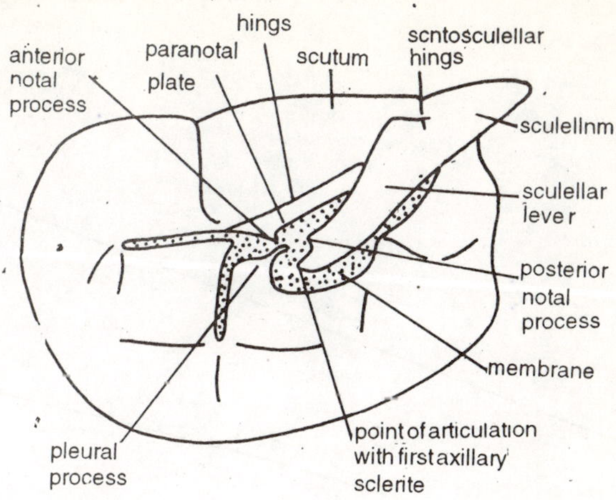


**Fig. 5.25.** Lateral view of the thorax of *Sarcophaga*.

of the wing beat is reduced, enabling the wings to be driven at a higher frequency.

### Frequency of Wingbeat

The frequency with which the wings vibrate varies considerably in different insects. Thus in butterflies the wingbeat frequency is 4 – 20/second, in *Schistocerca* it is 15 – 20/second, while in *Apis* and *Musca* the frequency is about 190/second. *Forcipomyia* (Diptera), a very small insect, has a wingbeat frequency of about 1000/second. The lower wingbeat frequencies of 30/second and less are associated with tubular or close packed flight muscles which show a normal 1 : 1 relationship with the nervous input. That is, each nerve impulse produces a single muscle contraction. At the higher wingbeat frequencies, however, this is not the case. These frequencies are associated with fibrillar flight muscles in which each nerve impulse leads to a succession frequency with which the wings vibrate in this case depends on the natural frequency of the thorax and its muscles. Even within a species the frequency of wingbeat may vary, being generally higher in smaller individuals. The frequency is greater in males than in females, although this may only reflect the generally smaller size of male insects. Frequency also varies with age, being higher in older insects, while it tends to decrease as the insect fatigues.

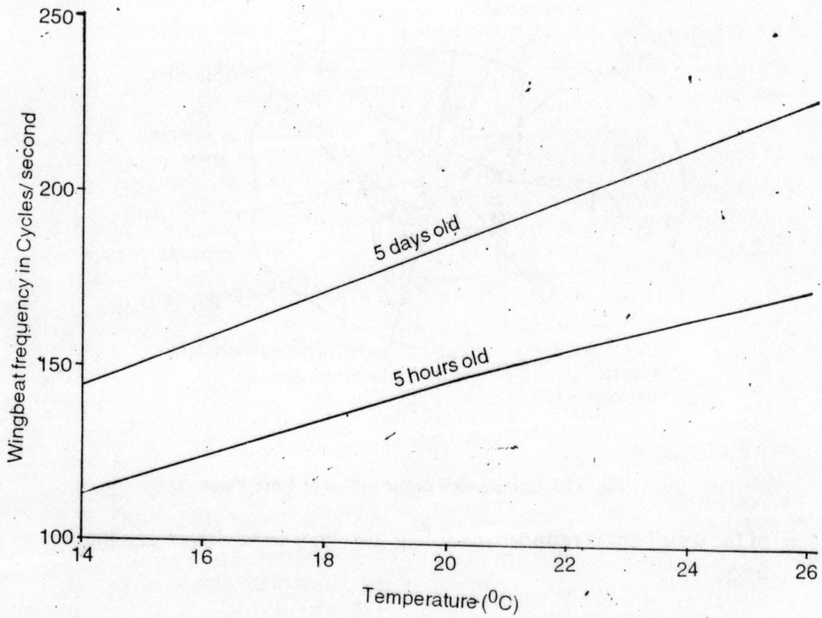


**Fig. 5.26.** Variation with age and temperature of the wingbeat frequency of *Drosophila*.

In *Drosophila* (Diptera) the wingbeat frequency increases with temperature, but in Hymenoptera temperature has no effect. The wingbeat frequency of *Schistocerca* is constant within the normal flight range of 25 – 35° C., but alters at lower and higher temperatures. Frequency can be controlled to some extent and in *Schistocerca* a higher frequency is used to increase lift. This involves an increased rate of firing of the motor nerves to the muscles moving the wings, but in Diptera frequency can be modified by alterations in the tension of certain muscles, such as the sternopleural muscles, which alter the natural frequency of the thorax.

The wings do not move at a constant speed throughout the stroke, the downward movement of the wings being rather slower than the upward movement.

**Stroke Plane**

The wings do not make simple up and down movements, but in the course of each cycle of vibration they also move backwards and forwards to some extent. As a result, the tip of the fore wing of *Schistocerca* moves in an ellipse relative to the body, moving forward and down on the downstroke, up and back on the upstroke. In some other insects, such as bees and flies, the wing tip traces a figure-of-eight relative to the body. When the insect is moving the path of the wing tip through the air follows a rather irregular path. The plane in which the wings vibrate relative to the body is called the stroke plane and in *Schistocerca* this is at a more or less constant angle of 30° to the long axis of the body. In *Apis* the plane is variable and when a bee hovers the stroke plane almost coincides with the horizontal plane through the long axis of the body. Differences in the plane of movement of the two sides produce turning movements.

### Amplitude of Wingbeat

The amplitude of the wing stroke, measured in the stroke plane, varies from one species to another. Thus in *Aeschna* (Odonata) the amplitude is about 70°, while in *Lucanus* (Coleoptera) it is 160°. In *Schistocerca* where the wings are not linked anatomically the forewing

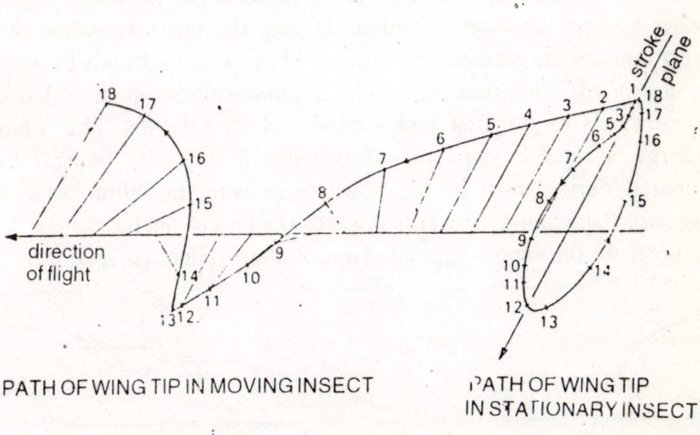


**Fig. 5.27.** Movement of the tip of the forewing of *Schistocerca*. The ellipse on the right shows the movement relative to the body of the insect, while the irregular curve shows the path of the wing-tip as the insect moves through the air. The numbers indicate the positions of the wing at regular time intervals throughout the stroke and the lines joining the wing-tip path to the flight axis show the angle which the long axis of the wing makes with the body at different stages of the stroke.

amplitude is 60 – 70, while that of the hind wing is 110. Variation of the amplitude of wingbeat on the two sides of the body may be used in steering, the insect turning away from the side of greatest amplitude. A reduction in amplitude may be associated with a higher wingbeat frequency.

**Wing Twisting**

In addition to variations in the form of the wingbeat the wing may twist in different ways in different phases of the stroke thereby altering the forces which it exerts. In many insects the twisting is produced by two direct flight muscles, the basalar muscle which pronates the wing, that is it causes the ventral surface to face downwards by pulling down the leading edge, and the subalar muscle which supinates the wing, causing the ventral surface to face obliquely forwards by pulling down the trailing edge. Since these muscles also act as direct wing depressors they are active only during the downstroke when the balance between them determines the degree of pronation of the wing. In addition there is some passive bending of the wing as a result of its flexibility and this is evident in the hind wing of the locust which always assumes a smooth camber. During the upstroke, when the direct muscles are inactive, the twisting of the wing is entirely passive. In *Sarcophaga* wing twisting results automatically from the relative move- ments of the first and second axillary sclerites. The wings undergo a regular sequence of changes in twisting, being fully pronated during most of the downstroke with, in *Schistocerca*, a posterior flap coming forwards towards the end of the movement. In the upstroke the wing is supinated and Z-shaped in cross-section.

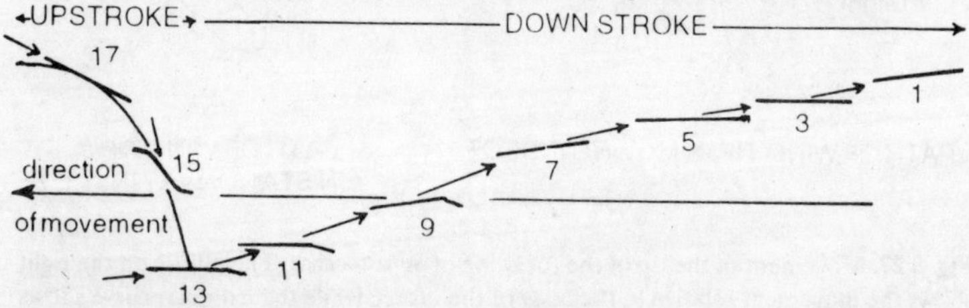


**Fig. 5.28.** Changes in the twisting of the mid-section of the forewing of *Schistocerca* in the course of a single stroke. The relative size of the wing is greatly exaggerated for clarity. Short arrows indicate the direction of the relative wind and the numbers correspond with those showing the wing-tip positions in figure 5.27.

## AERODYNAMICS

### Flapping Flight

The forces which keep an insect airborne and propel it along are produced by the movements of the wings. They vary throughout the wingbeat due to changes in the twisting of the wings and changes in the velocity of the relative wind. The relative wing is the movement of the air relative to the wing and it has two major components; one due to the airspeed of the insect and a second due to the velocity of the wing in the stroke plane. The relative wind may be regarded as the resultant of these two forces. The forces which the relative wind exerts on the wing depend on the angle at which it strikes the chord of the

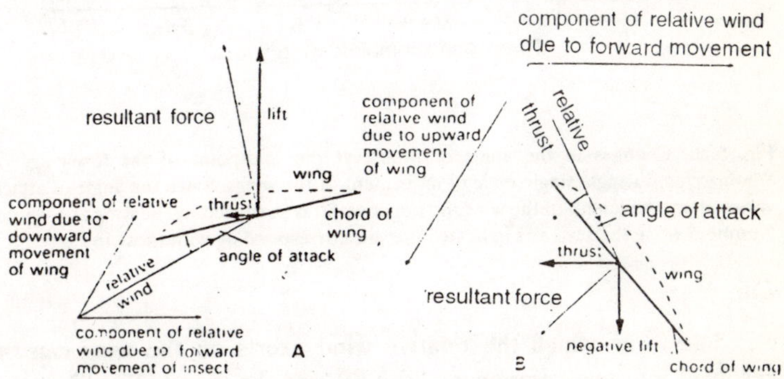


**Fig. 5.29.** Diagram illustrating the forces acting at the mid-point of the wing at different phases of the wingbeat corresponding roughly with positions and 15 in figure 5.28. A Downstroke of wing with positive angle of attack. B. Upstroke of wing with negative angle of attach. The lengths and thicknesses of the arrows have no significance in indicating the strengths of the forces.

wing. This angle is known as the angle of attack and it may be either positive or negative. Because of the twisting of the wing the angle of attack varies along its length, while the twisting also modifies the angle of attack at any one point in the course of the wingbeat. Thus in *Schistocerca* the angle of attack at the mid-point of the wing is positive and fairly constant throughout the downstroke, but becomes negative during the upstroke.

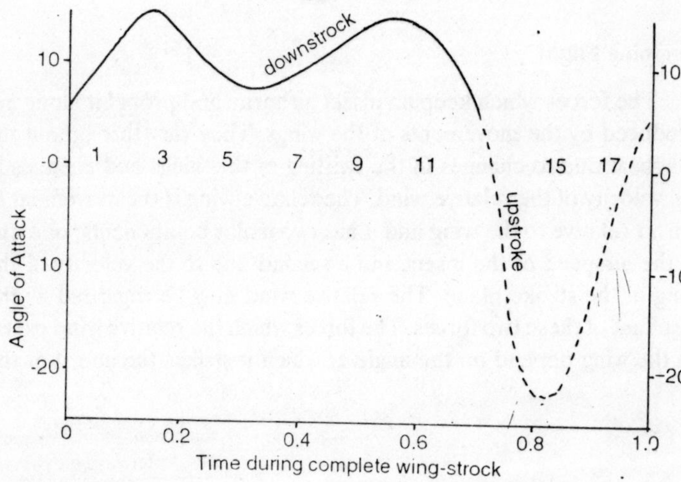


**Fig. 5.30.** Changes in the angle of attack at the mid-point of the forewings of *Schistocerca* during a single cycle of movement of the wings. Since the angle of attack varies along the length of the wing, no negative lift is produced by the wing as a whole. Numbers along the zero axis indicate intervals corresponding with those in Fig 5.28.

## *Lift*

The force which the relative wind exerts on the wing can be resolved into two components, the lift and the thrust. The lift is the vertical force produced and to keep the insect steady in the air the lift force must roughly equal the weight of the insect. Lift becomes minimal during the upstroke of the wings, but, in *Schistocerca*, because of the adjustment of the angle of attack, it never becomes negative, that is the wings as a whole never produce a force pushing the insect down. Because of their bigger area and greater amplitude of movement the hindwings produce more lift than the forewings, about 71% of the total, and because of the adjustment of the angle of attack 20% of the total lift is produced during the upstroke of the wings. The resulting variation in lift in the course of a wingbeat leads to a cycle of vertical displacements of the insect as it flies along so that instead of following a horizontal path it loops up and sown. Lift is also produced

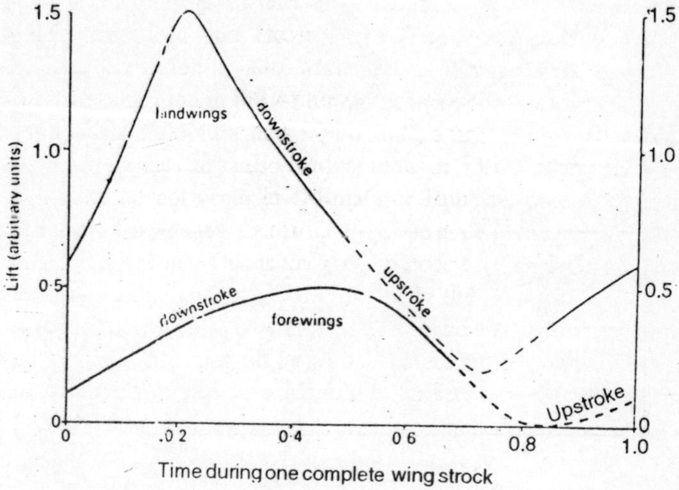


**Fig. 5.31.** Changes in the lift produced by the wings of *Schistocerca* during a wing-stroke.

by the action of the relative wind on the body as distinct from the wings, but the force is insignificant compared with that produced by the wings, amounting to less than 1/20th of the total lift in *Schistocerca*. In *Drosophila* lift is proportional to the body angle.

*Thrust*

In order to move forwards the insect must also produce a horizontal force known as the thrust. This must be sufficiently great to overcome the drag forces which resist the motion of the insect through the air. Drag results partly from the profile of the insect, that is the area which it presents to the air, but largely from an induced drag due to the development of vortices at the wing tips which dissipate much of the kinetic energy of the wings as heat. These vortices result from the maxing of air at different pressures from the two sides of the wing. The thrust, like lift, varies in the course of a wingbeat and the hind wing of *Schistocerca* produces thrust maximally in middle of the

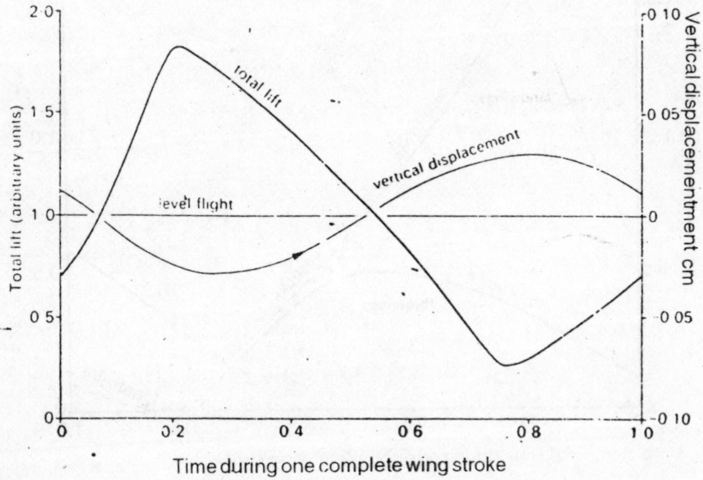


**Fig. 5.32.** Changes in the total lift and associated vertical displacements of *Schistocerca* during a wing-stroke. There is a time lag between the period of maximum lift and the corresponding rise of the insect.

downstroke and again in the upstroke. This variation results in slight changes in the forward speed of the insect in the course of a wingbeat.

Propelling the insect forwards requires relatively little energy compared with that needed to keep it in the air and in *Schistocerca* the average thrust is only 7% of the average lift.

## Gliding

Occasionally insects are seen to glide with the wings outstretched, the lift forces produced by rising air being sufficient to keep them airborne without any wing movements. Thus, during a glide an insect expends very little energy. Various Lepidoptera, locusts and Odonata have been seen to glide and it is suggested that the inability of dragonflies to fold their wings is a secondary adaptation to this, since wings locked in a stretched position are an asset in gliding. Locusts are also able to lock their fore wings in the outstretched position.

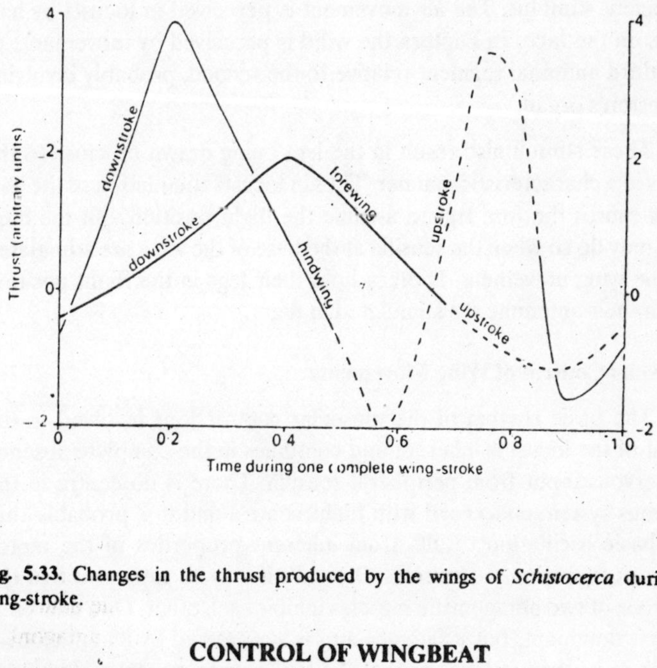


**Fig. 5.33.** Changes in the thrust produced by the wings of *Schistocerca* during a wing-stroke.

## CONTROL OF WINGBEAT

### The Intiation of Wing Movements

In most insects the wing start to beat as a result of loss of tarsal contact with the substratum and in the locust this occurs when the insect jumps into the air. When the legs are touching the ground movement of the wings is inhibited, the contact probably being perceived through the proprioceptors of the legs. The wings may also be induced to beat by various shock stimuli.

### Maintenance of Wing Movements

The loss of tarsal contact with the substratum is sufficient to maintain the movement of the wings of *Drosophila* as well as initiating it, but in most other insects flight soon stops unless the insect receives further stimulation. This is provided by the movement of wind against the head. A wind speed of only 2.m/sec., is sufficient to maintain the wing movements of *Schistocerca* and since this is less than the flight speed of the insect the relative wind produced in flight will provide sufficient stimulus. The air movement is perceived in locusts by hair beds on the face. In Diptera the wind is perceived by movements of the third antennal segment relative to the second, probably involving Johnston's organ.

These stimuli also result in the legs being drawn up close to the body in a characteristic manner. Thus in locusts stimulation of the hair beds causes the fore legs to assume the flight position, but the hind legs only do so when the sensilla at the base of the wing are stimulated by the wing movement. Diptera hold their legs in the flight position when their antennae are stimulated in flight.

### Nervous Control of Wing Movements

The basic rhythm of the muscular contractions involved in the flight of the locust is inherent and continues in the complete absence of nervous input from peripheral sensilla. There is no centre in the nervous system concerned with flight control and it is probable that the basic oscillation results from inherent properties of the motor neurons supplying the muscles. Basically it may be supposed that the neurons of two antagonistic muscles inhibit each other. One neuron is at first dominant, but it fatigues and is suppressed by its antagonist, which in its turn fatigues so that the first neuron becomes dominant again. Experimental work indicates that such simple reciprocal inhibition cannot account for the observed flight pattern in which each muscle unit receives only one or two impulses in each cycle with a relatively long time interval between successive firings. Hence, it is suggested that the oscillatory cells have follower cells which fatigue rapidly and so produce only one or two impulses. In this way, the oscillation of nerve impulses and hence of muscle contractions might be maintained.

This central oscillation is modified by the input from peripheral sensilla amongst which the stretch receptors at the base of the wing of the locust are particularly important. Each stretch receptor fires one, two or three times towards the top of the upstroke, the first impulse always being produced when the wing is in a particular position. The number of spikes produced is a measure of the velocity and amplitude

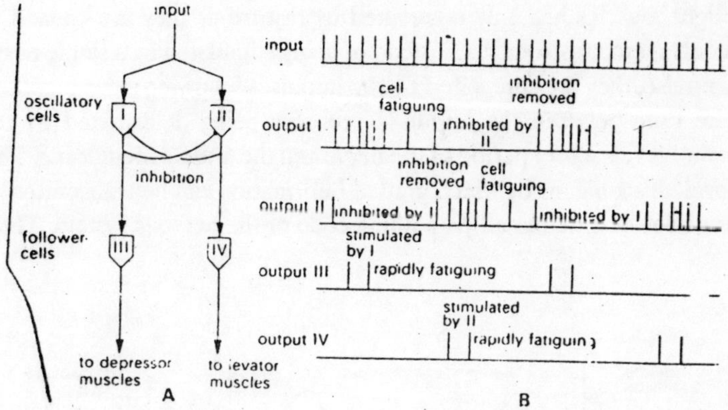


**Fig. 5.34.** A. Hypothetical arrangement of motor neurons which could account for the observed activity in locust flight motor neurons. B. The supposed activity of the cells in A. Each vertical line represents a nerve impulse. The input may be regular or irregular, but the output of cells I and II is not altered. For explanation see text.

of the wing movement. Hence, the stretch receptors transmit some precise information on the wingbeat, but it appears that this information is lost in the central nervous system and is not made use of by the insect. Instead the impulses from the receptors appear to summate over several cycles of wing movement.

A major effect of the input from the stretch receptors is to double the basic rate of oscillation of the wings, but apart from this they may tend to keep the frequency more or less constant by means of a negative feedback mechanism. As wingbeat frequency increases the amplitude of the wingbeat is reduced due to muscular braking. This possibly results in less input from the stretch receptors so that stimulation of the central oscillators is reduced and the wingbeat frequency declines. Despite the complexity of the flight movements of the locust each of the muscles involved comprises only a number of units and for instance, the first basalar muscle consists of only a single unit, while the subalar muscle consists of two muscle units. Each unit has its own motor fibre and so can act independently.

The force exerted by a muscle can be increased either by increasing the number of units which are active or by increasing the strength of the pull exerted by each unit. Although the units of the flight muscles are only innervated by fast axons they are caused to contract more strongly it, instead of being stimulated by a single nerve impulse, they are stimulated by two impulses following close together, the time between the impulses probably being determined by the relative refractory period of the nerve and the muscle membrane. This provides a means by which graded information can be transmitted to muscles despite the all-or-nothing code of the nervous system. Thus,

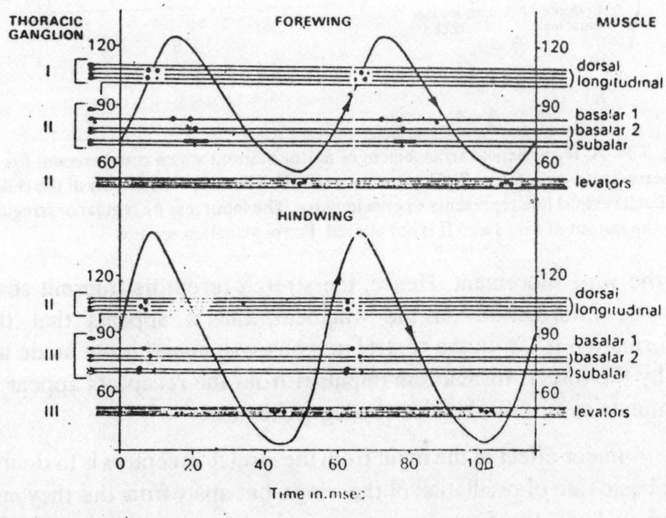


**Fig. 5.35.** Diagram illustrating the timing of firing of motor neurons to the flight muscles of fore and hind wings in relation to the wingbeat cycle. Each neuron is shown by a horizontal line with its origin in the appropriate ganglion on the left-hand side. Each dot on a line represents a nerve impulse occurring at that time; a small dot indicates that an impulse may or may not occur, a large dot that it always occurs. The heavy bar on the motor neuron to the forewing subalar muscle indicates that firing occurs within this period, but not at a precisely fixed time as with the other units. The heavy curve and the numbers along the ordinate indicate the angular displacement of the wings in degrees; 90° indicates wing horizontal, above 90° wings up, below 90° wings down.

when the insect is producing only low lift forces the second basalar muscle and some units of the dorsal longitudinal muscle of the hind wing may be inactive, whereas in producing high lift forces all the units come into action and the forces exerted by the individual units are increased by double firing of the motor neurons. .

The twisting of the wings by the controller muscles are precisely timed by the pattern of motor impulses to the muscles. Only in the case of one muscle, the mesothoracic subalar muscle, is the firing of the motor nerve very variable in its timing and this is the muscle which varies the twisting of the fore wing to control lift. The precise co-ordination of the other neurons does not arise from a fixed pattern of connections between them since they can function in other sequences than that involved in flight. The problem of control of the wingbeat is different in insects in which the wingbeat is myogenic. Here also the muscles must act in a precise sequence, but this sequence is not directly related to nervous input and the timing of firing of the motor neurons does not coincide with a particular phase of the wingbeat cycle. The nervous input to the flight muscles serves only as a general stimulator maintaining the muscle contractions. Control of the movements is exerted by the muscles which control the mechanical properties of the thorax; an increase in the lateral stiffness of the thorax produces an increased wingbeat frequency, while a decrease in stiffness leads to a reduced frequency. These changes only take effect over a number of cycles of wing movement; there is no stroke by stroke control and such precise controls is unnecessary because the insect only travels a very short distance in the course of a single wingbeat (Wilson and Wyman, 1963).

## STABILITY IN FLIGHT

Because of the variations in the forces acting on an insect during flight it has a tendency to deviate from a steady path. This instability may involve rotation about any of the three major axes passing through the centre of gravity of the body. Rotation about the long axis of the body is called rolling, rotation about the horizontal, transverse axis is pitching, and rotation about the vertical axis is yawing. Deviations from a steady path are perceived by various sensilla and the nervous input from these exerts a controlling influence on the wingbeat cycle so that the deviation is corrected and the steady path maintained. The

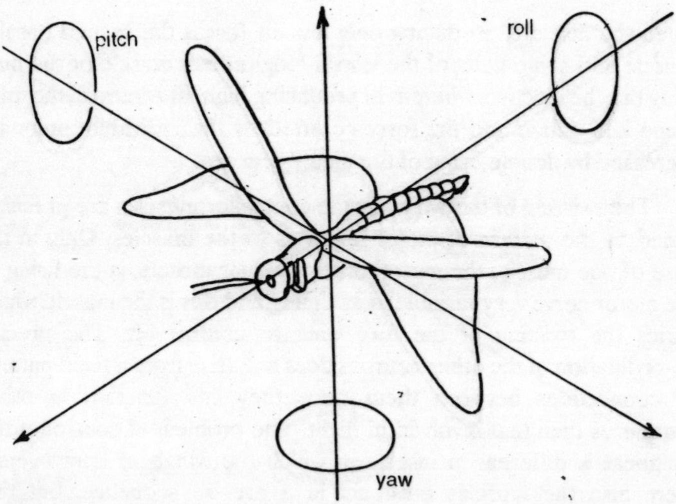


**Fig. 5.36. Diagram showing the main axes about which an insect may rotate due to its instability in flight.**

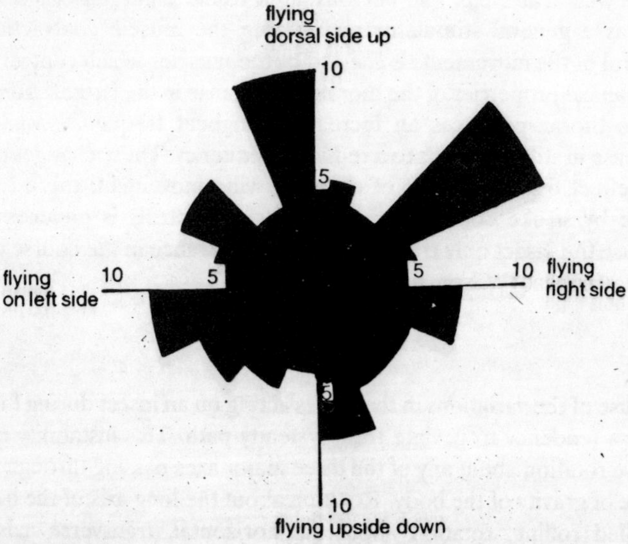


**Fig. 5.37. Diagram showing the orientation of a locust in a series of observations made when it was flying in complete darkness. The figures indicate the percentage frequency with which orientation occurred in each sector.**

sensilla at the base of the wings and in the halteres of Diptera are of particular importance in this control and they are considered separately.

*Rolling*

Vision plays an important part in the control of rolling. Odonata and Orthoptera, and probably also other insects, have a dorsal light reaction by which they align the head so that the dorsal ommatidia receive maximal illumination. To produce a dorsal light reaction a number of ommatidia must be illuminated, but the response does not depend on stimulation of a particular part of the eye since it is still apparent if the most dorsal ommatidia, which are normally concerned in the response, are covered. The response is improved by stimulation of the ocelli, although they do not play a direct rôle in orientation.

Since normally most light comes from the sun or the sky overhead the dorsal light reaction ensures that the head is usually held in a vertical position. In Odonata, where the head is loosely articulated to the thorax, it also tends to keep in a vertical plane due to its own inertia, but this is not the case in the locust where the head and thorax are broadly attached so that rolling by the thorax is immediately transmitted to the head. As a result, a locust flying in complete darkness is unable to orientate in this plane and will fly upside down or at any other angle.

# 6

# The Abdomen

As stated already that the body of an insect is divisible into three parts : head, thorax and abdomen. Embryological studies show that in all insects except the Protura and Collembola the primitive number of abdominal segments is twelve, eleven true metameres and a small terminal telson through which the anus opens. There is, however, a general tendency toward reduction of the number of segments. In no insect is a definite telson retained during postembryonic life. The metameres also show a tendency toward reduction by coalescence or atrophy at the posterior end. In generalized insects the eleventh metamere generally persists but is unsclerotized ventrally and consists of a dorsal plate, the epiproct, and two lateroventral plates, the paraprocts. The anus opens between these plates. In higher insects the eleventh metamere has disappeared as a distinct segment and the tenth also may be lot. In the generalized female the eighth segment bears first valvifer to which the first valvula is attached and the ninth segment bears the second valvifer to which the second and third valvulae are attached. The valvifers and valvulae constitute the ovipositor or external female genitalia. The ninth segment in the male bears the penis.

The pregenital or visceral segments have the simplest anatomic structure of any in body, and for this reason they are selected first for study. Each segment consists of a dorsal sclerite, the tergum, a ventral sclerite, the sternum, and between them an unsclerotized pleural membrane. The anatomical simplicity of the abdominal segments does not, however, represent a primitive condition and, from the standpoint of the morphologist, the structure of these segments is

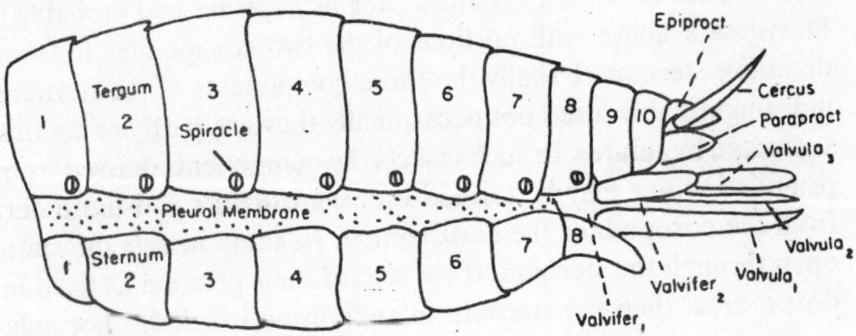


**Fig. 6.1.** Lateral view of the abdomen of a generalized insect.

relatively complex and presents problems which are not yet certainly solved.

The segmentation is *secondary*, the insertion of the longitudinal muscles being always behind the intersegmental membrane and at the anterior end of the definitive segment. Definite acrotergites and acrosternites are not usually developed and the antecostal suture is not always distinct. The antecosta may be a distinct ridge or a very slightly specialized marginal area to which the longitudinal muscles are attached.

It will be remembered that a segment consists primitively of a dorsal, a ventral and two pleural areas, the limb bases being inserted in the last. The tergum, sternum and the pleural membranes of a sclerotized abdominal segment are seldom, if ever, coincident with these primary areas. The spiracles are primitively dorsal in position since they originate lateral to the limb buds in the embryo, and their position has been used to determine the degree of correspondence between the primary dorsum and the definitive tergum. Actually, in adult insects, spiracles may be borne on the tergum, the pleural membrane or the sternum, and this is taken as indicating not only that the limits of the definitive tergum and sternum are different from those of the primitive dorsal and ventral areas, but also that these sclerites differ morphologically in different insects. The ventral area, between the leg bases, becomes sclerotized to form the primitive sternum. In some primitive insects such as *Nesomachilis* (Thysanura), in which the *coxopodites*, or basal segments of the legs, persist as independent structures, this primitive ventral plate is the only sternal element, but in other Thysanura such as *Lepisma* and probably in all Pterygota it unites with portions of the two coxopodites to form the definitive sternum. Usually the three components of the sternum are indistinguishably fused but occasionally they are partly or completely separated by sutures. In such insects the components derived from the primitive venter may be termed the *ventrosternite* and those derived from the coxopodites the *coxosternites*. In some insects the spiracles open through the sternum. If we regard their position as fixed in the dorsal area, then the sternum in such insects includes not only the components just discussed but also the lateral parts of the dorsal area.

## STRUCTURE OF THE ABDOMEN

### Calosoma Sycophanta, Larva

In this case the number of segments in the abdomen is only ten, the last being small and conical. The eleventh segment has disappeared unless it is represented by the small membranous area through which the anus opens. Note the two spine-like processes borne on the dorsal side of the ninth segment. These are the *urogomphi* and must not be confused with the cerci of other insects, which are true appendages of the eleventh segment.

The abdomen in many larval insects is wholly unsclerotized. In others, especially those which lead an active life, these are various degrees of sclerotization. In *Calosoma* there are several sclerites in each segment but the primary segmentation is not obscured. Examine a segment near the middle of the abdomen. Note the large dorsal unpaired sclerite lateral to which on each side, is a small spiracle in the membrane. Lateroventral to each spiracle there are two small sclerites of unequal size, the larger lying anterior to the smaller. Ventral to these sclerites there is a distinct groove in the membrane which runs longitudinally along the side of the abdomen and passes forward into the thorax where its course lies dorsal to the pleura. This is the *dorsopleural line*, occasionally present in insects, which is presumably the dividing line between the dorsal and pleural regions. The area dorsal to it is therefore the primitive dorsum. The sclerites found in this area collectively constitute the *tergum* and each is a *tergite*. The large median sclerite is the *mediotergite* and the small lateral ones the *laterotergites*.

Ventral to the dorsopleural line you will find two pairs of sclerites similar in arrangement to the laterotergites, and ventral to these a second longitudinal groove, the *pleuroventral line*.The two grooves presumably mark off the primitive dorsal, pleural and ventral areas. The sclerites in the pleural area may represent the remains of the coxopodites or basal leg segments. If this is so each group of two sclerites may correctly be called the *pleuron*, a term which, in the thorax, is reserved for lateral sclerites derived from the coxopodites. The individual sclerites would therefore be *pleurites*. They may, however, be sclerotizations in the pleural area having no connection with the legs and are best referred to as *pleural sclerties*. The ventral region of all except the three most posterior segments bears five sclerites – one large anterior one and four small posterior ones. These collectively constitute the *sternum*, which in this insect is a apparently derived entirely from the ventral area. The individual sclerites are known as *sternites*. The typical condition in the abdomen of adult insects is the presence of a single dorsal and ventral sclerotization; a true pleuron distinct from the other sclerites is rarely present.

**Acheta Assimilis (Field Cricket) Female**

Whenever, the speciman is viewed from the dorsal side only ten distinct terga are visible, the tenth being compound. Turning to the ventral side, however, you will find only eight distinct sterna, because the ventral parts of the segments behind the eighth are greatly modified.

**The Pregenital Segments**

These are segments 1 to 7 and are very similar in structure. Note that in each there is a *tergum* a *sternum* and a wide *pleural membrane*. The structure of the tergum and sternum is essentially the same. Each consists of a single undivided plate. The pleural membrane extends from the first to the eight segment and is not superficially segmented, except insofar as the overlap of terga and sternal extends for some distance into the pleural membrane. Find a small roundish *spiracle* in the membrane on each side of the first eight segments. This is the typical distribution of the abdominal spiracles in adult insects; that is to say, they are found in segments one to eight. Since the spiracles lie in the pleural membrane the latter is probably derived in part from the lateral portions of the primitive dorsum.

In a specimen which has not been distended by injection you will observe a distinct *lateral groove* extending through the full length of the pleural membrane. This has been interpreted as the dorsopleural line, but is in fact a temporary fold formed by the contraction of muscles inserted in minute sclerites along its length. In the fully distended abdomen the groove has disappeared and the sclerites are visible as minute black lenticular or spindleshaped bodies. There are two in each segment and they may be termed, for descriptive purposes, the "anterior and posterior ventral sclerites" of the pleural membrane to distinguish them from the "anterior and posterior dorsal sclerites." The latter lie one in front and one behind the spiracle and serve for the insertion of muscles to be examined later. They are extremely small, appearing like short black lines, but they are very variable and may not be seen in some specimens.

If a living insect is examined under the low power of the dissecting microscope and note the alternate dorsoventral contraction and expansion of the abdomen can be seen. These are respiratory movements made possible by the flexibility of the pleural membrane and caused by the contraction of the muscles to which we have referred. Observe how with each contraction the pull of the muscles causes the *"dorsopleural"* groove to become deeper and narrower, while on relaxation of the muscles it becomes broader and more shallow.

### The Genital Segments

The eighth and ninth segments give rise to the valves of the ovipositor and are, therefore, called the *genital segments*. In insects with an ovipositor these segments are always modified, especially as regards their ventral parts. In the cricket the terga are distinct and only slightly modified in form as compared with those of the pregenital segments. Examine the last ventral sclerite in front of the ovipositor. This is known as the *subgenital plate* since the female genital opening lies above it. It forms a backward- projecting flap, free at its posterior end, and is usually regarded as the eighth sternum; but there is some evidence that the latter is reduced and that the subgenital plate is a secondary sclerite developed from a fold in the membrane between the seventh and eighth sterna. The venter of the ninth segment is greatly reduced and modified and cannot be recognized by superficial examination.

### The Postgenital Segments

The tenth and eleventh segments are modified in most insects. To study these terminal structures in the cricket, cut through the abdomen in front of the ninth segment and pin the extremity in a dissecting tray with the cut surface down. The long median spear-like process is the *ovipositor* and the two lateral filiform appendages are the *cerci*. With the aid of the dissecting needle find the *anus* which opens at the posterior end dorsal to the ovipositor and between three valve-like terminal sclerites, one dorsal and two lateroventral. Examine the dorsal sclerite and note that it forms part of a plate which extends anteriorly to the ninth tergum, but is divided by a shallow groove into an anterior transverse region and a posterior triangular region. The anterior region is the *tenth tergum*. It extends laterally to the bases of the cerci, a narrow extension of it curving around the anterior and lateral sides of the bases of these appendages. The posterior valve-like plate lying dorsal to the anus is the *epiproct* and represents the eleventh tergum or a apart of it. The two lateroventral valves are the *paraprocts*. They too represent parts of the eleventh segment. Note that the *cerci* are almost surrounded by the tenth tergum, the ring being completed posteriorly by the base of the paraproct. They are, however, the segmental appendages of the eleventh segment, serially homologous with the legs and mouthparts.

There is a membranous fold between the paraprocts and the base of the ovipositor. This membrane represents, in part at least, the ventral region of the tenth segment which is unsclerotized except for a very small plate in front of the paraprocts.

### Acheta Assimilis, Male

The only important differences from the female are to be found in the genital segments. The eight is normal i.e., is resembles the anterior segments in all essentials since it is not associated with genitalia. The sternum of the ninth is well-developed and except that it is larger, resembles the subgenital plate in the female. It is sometimes referred to as the *hypandrium*. The male genitalia lie concealed in a chamber on the dorsal side of the hypandrium.

**Melanoplus, Romalea, or Other Grasshopper**

On comparing the abdomen we can note the following differences from *Achetar*. The *terga* and *sterna* are more extensive laterally, the spiracles being in the lateral margins of the former; the *pleural membrane* is reduced to a narrow inflected band between terga and sterna; the eleventh tergum or *epiproct* is distinctly separated from the tenth; the shape of the ovipositor is quite different. In the male there are apparently ten sterna; the apparent tenth, however, is really a subdivision or an outgrowth of the ninth, probably representing the fused bases of the segmental appendages of this segment.

**Acroneuria or Other Plecopteran Nymph**

In this insect the pleural membrane is absent except in the second segment an there is no demarcation between terga and sterna, the annuli being evenly sclerotized. The sternum of the first segment is fused with the metasternum the tergum being free. As external genital organs are undeveloped, the eighth and ninth segments are unmodified. The tenth segment is also completely sclerotized. Rudiments of the eleventh segment are present. The *epiproct* is probably represented by the membranous fold beneath the posterior lobe of the tenth tergum. the *paraprocts* are small but distinct and sclerotized. They are closely associated with the prominent jointed *cerci*.

**Musca Domestica, Female (Diptera)**

Whenever, a freshly killed specimen, or one that has been fully relaxed is studied. You will note that, on superficial examination, the abdomen appears to consist of only four segments. Examine these segments and note that the four *tergal plates* extend down the sides to the lateroventral line. The first tergal plate is a compound one consisting of the fused terga of the first and second segments. This is evidenced by the presence on it of two pairs of spiracles and by the two sterna associated with it. The last visible segment is therefore the fifth. There are five small *sterna*, the first crescent-shaped, the others elongated plates in the median area of the venter. Between terga and sterna are the broad *pleural membranes*. In addition to the five visible segments there are four others which are slender and cylindrical and, when at rest, are telescoped into each other and retracted within the fifth, so that only the extreme tip is visible. They form a slender tube known as the *pseudovipositor*. These ca best be examined by holding a living fly between the thumb and forefinger, or between the tips of blunt forceps, and exerting gentle pressure on the abdomen. When the pseudovipositor is extended, the insect should be pinned in the dissecting tray and the pseudovipositor should be grasped with a pair of fine forceps and pulled gently until fully extended, when it will be as long as the reminder of the abdomen. Note that the segments of the pseudovipositor are separated by extensive intersegmental membranes which make possible the telescoping of the segments. Each segments is largely membranous but is provided with one or two narrow tergal and sternal sclerites. The ninth segment bears a pair of minute lateral appendages.

**Euschistus or Anasa (Hemiptera)**

If we observe the *tergum, pleural membranes* and *sternum* from a segment near the middle of the abdomen. We find that both the tergum and pleural membranes are restricted to the more or less flattened dorsal side of the insect, while the sternum occupies not only the whole of the convex ventral side but is bent sharply at the lateral margins of the abdomen so that it extends as a pair of lateral strips on the dorsal side also. Find the *spiracles* on the ventral side of the sternum near the lateral margins of the abdomen. It is evident that the sclerotic material in this insect is distributed quite differently from that in the cricket or grasshopper. If we regard the position of the spiracles as fixed in the primitively dorsal region, then the definitive sternum includes the primitive venter, the pleural areas and part of the dorsum, the pleural membranes would also be part of the primitive dorsum.

**Apis Mellifera – Honeybee, Worker**

The clistogastrous Hymenoptera, of which the honeybee is an example, show profound modifications in the anterior segments of the abdomen. The first segment or *propodaeum* is fused with, and has become a functional part of the thorax. The second segment is greatly narrowed anteriorly (sometimes for its entire length) forming a *petiole* or narrow stalk connecting it with the propodaeum. There is thus a deep constriction between the propodaeum and the functional abdomen, or *gaster*, which allows for freer movements of the latter than is possible in other insects. It will be seen that the functional abdomen is not the same as the morphological abdomen.

When the thorax of a honeybee is observed under the microscope after removing hair. We find the deep cleft in the abdomen, at the bottom of which is the short petiole, connecting the gaster with the functional thorax. In front of this cleft, and forming the posterior portion of the dorsum of the functional thorax, find a large plate limited in front by a distinct sulcus which passes behind the attachment of the hind wings. This is the *tergum of the propodaeum*, the first segment of the morphological abdomen. The lateral edges of the plate extend from behind the bases of the hind wings to the articulation of the hind legs. Find the *spiracles* near the anterior margin. There is a very narrow transverse sclerite immediately behind the articulation of the hind legs and separated from the petiole by a distinct intersegmental membrane. This is the *sternum of the propodaeum*.

Behind the second abdominal segment (the apparent or functional first segment) there are five visible segments, the terminal one being the morphological seventh. Each is provided with a distinct tergum and sternum, and each tergum bears the usual pair of spiracles. The eighth, ninth and tenth segments are greatly modified and retracted within the seventh, where the eighth and ninth form part of the mechanism of the sting.

### Muscles of the Abdominal Body Wall

The body-wall muscles of insects are derived from those of their annelidan ancestors. In present-day annelids there are two layers of muscle fibers, an internal longitudinal and external circular, which are not organized into distinct individual muscles. These are adequate for the generalized movements of the worms. In the insects these sheets of muscle fibers are for the most part broken up into discrete muscles whose contraction is responsible for the specific movements of the appendages and other organs. The musculature is therefore most complicated in the head, the thorax, and the genital segments where the most active movements take place. The pregenital segments, especially in adult pterygote insects, have a relatively simple musculature. In the course of evolution many muscles have changed their direction and attachment, some have disappeared and new ones have arisen to meet new needs. It is not yet possible to trace the homologies of the muscles in all insects and only a few of the important muscles in different regions of the body will be studied in this course. Muscles may be named according to their function, e.g., adductor of the mandible; according to their position, e.g., longitudinal ventral; or according to their attachments, e.g. tergosternal.

**Acheta Assimilis**

The bandlike muscles in the living insect are transparent and difficult to distinguish, but are more easily seen in properly preserved specimens. In either case a brief flooding of the specimen with methylene blue or other suitable stain will increase the optical differentiation of the tissues and facilitate the removal of the fat. The *dorsal muscles* are very simple, consisting of a thin band of longitudinal fibers, divided into two groups. The *median dorsals* lie toward the mid-dorsal line. They originate in the antecosta of one segment and are inserted in the antecosta of the next posterior segment. The *lateral dorsals* lie lateral to the medians. They are greatly shortened, having their origin in the tergum some distance behind the antecosta and their insertion in the antecosta of the segment behind. In many insects there are external and internal layers of dorsal muscles. The single layer in the cricket probably corresponds to the external dorsals of other insects. The contraction of the longitudinal dorsals pulls the posterior tergum, in which they are inserted, forward. Acting in concert with the corresponding ventral muscles they telescope the segments; acting alone they bend the abdominal segments dorsally. The *ventral muscles* are somewhat more complex. We can see a prominent *transverse ventral* muscle in the anterior portion of each segment, having its attachments near the lateral edges of the sternum.

There is a relatively heavy *internal longitudinal ventral* which originates in the base of the metasternal apophysis and runs obliquely backward to the antecosta of the fifth sternum. In the second and third segments it gives off slips which are inserted on the third and fourth antecostae. External to this muscle in the fourth segment you will find a more delicate *lateral external longitudinal ventral* which originates in, or just posterior to, the antecosta of the fourth segment and is inserted in the antecosta of the fifth segment. A similar muscle is also present in the segments behind the fourth in which the internal ventral is not developed.

Near the mid-ventral line find the *median external longitudinal ventral* which originates near the middle of the sternum and is inserted in the antecosta of the succeeding segment. It will be observed that some of the longitudinal muscles, both dorsal and ventral, have changed their origin but not their insertion. This is a very common phenomenon but change in insertion is less frequent. The *dorsoventral* group of muscle includes those which lie chieefly in the pleural area and which extend in the general direction dorsum to venter. Following the general procedure in comparative anatomy they are named according to their attachments and fall into the three subdivisions described below.

An *anterior* and a *posterior tergosternal* will be found in the corresponding regions of the segment, having their attachments near the edges of the tergum and sternum. A *segmental* and *intersegmental notopleural* originate together from the notum or tergum, behind the origin of the posterior tergosternal. They diverge in their course, the segmental muscle to be inserted in the posterior ventral pleural sclerite of the same segment, and the intersegmental to be inserted in the anterior ventral pleural sclerite of the segment behind.

Five *sternopleurals* originate along the lateral edge of the sternum. The *first* is inserted in the anterior ventral pleural sclerite, the *second* in the anterior dorsal pleural sclerite, the *third* in the closing apparatus of the spiracle, the *fourth* in the posterior dorsal sclerite and the *fifth* in the posterior ventral sclerite. The third sternopleural is the *dilator of the spiracle*. In contracting, it opens the spiracular orifice and allows the exit and entry of gases to the respiratory tubes. The other dorsoventrals are *compressors of the abdomen* and are concerned chiefly with the respiratory movements. They draw the dorsum and venter toward each other, compressing the abdomen and thus increasing the blood pressure. This pressure is imparted to the tracheae and air is forced out of the spiracles. When the muscles relax, the elasticity of the body wall and the increased blood pressure force the dorsum and venter apart, the body cavity expands, pressure is decreased and air is drawn into the tracheae through the spiracles. The increased hemolymph pressure caused by the contraction of the abdominal or other muscles is used by many insects in bringing about the movement or expansion of various organs by the transmission of pressure from one region to another.

With the appearance of the sclerotized cuticle the limb took on a dorsoventral articulation with the body wall which allowed for protraction and retraction of the limb as a whole. Such limbs could be used in swimming, or could pull the body along, but could not raise it above the substratum. Later the limb may have been divided into a proximal *coxopodite* and a distal *telopodite*, having an anteroposterior articulation with each other, and could now raise the body above the substratum. Still later, further segmentation of the telopodite into *trochanter, femur, tibia tarsus* and *pretarsus* increased the efficiency of the limb in walking or climbing. At some time the coxopodite divided into a *subcoxa* and *coxa*, but the subcoxa was incorporated into the body wall leaving the coxa as the basal segment of the six-segmented leg. It is not certain however, that the subcoxa was ever a functional segment of an ambulatory leg. It may represent detached basal portions of the coxopodite incorporated in the body wall.

## ABDOMINAL APPENDAGES

### Nesomachilis (Thysanura)

Whenever the speciman is viewed from its ventral side, we find on each of the first seven segments a pair of large flattened plates, each plate bearing distally a small cylindrical tapering *stylus* and a short blunt *contractile vesicle*. The large plate are the *coxopodites* of the abdominal appendages. They are attached proximally to a narrow triangular sclerite, the *sternum*, which in this insect consists of the *ventrosternite* only. The stylus is a secondary structure articulated to the coxopodite and movable by muscles inserted in its base. The vesicle, which is retractile and therefore not usually visible, is probably a device for absorbing moisture.

### Heterojapyx

There are two styli attached to a single ventral plate indicating that the coxopodites and ventrosternite have united to form the definitive sternum. Here the sternum of the anterior abdominal segments consists of a single plate as in the Pterygota and there is no evidence of its composite origin. The sterna of the posterior segments, however, bear styli as in *Heterojapyx*.

**Corydalis Larva**

The integument of the abdominal segments consists of unsclerotized rings. The first eight bear tapering lateral processes, attached to a bulbous swelling or tubercle of the lateral body wall. This tubercle has been interpreted as the *coxopodite* but this is by no means certain. If it is, the abdominal segment illustrates what must have been the structure of a segment of the unsclerotized ancestral arthropod, namely, a dorsal region, a ventral region, and two pleural regions largely occupied by the bases of the appendages. Note that the spiracles are borne on the dorsal areas above the "leg base." The tapering lateral process of the tubercle is the *stylus*. Ventral to the stylus the tubercle bears a tuft of filamentous *gills*; remove these gills and note that they are borne on a trilobate outgrowth of the tubercle. The musculature of this outgrowth resembles that of the retractile vesicle, but the two structures are probably not homologous.

In the appendages of the tenth abdominal segment. There is a definite basal segment which bears a stylus essentially like that of the anterior appendages. On the distal extremity of the basal segment, there is a short blunt lobe bearing a pair of curved claws. The musculature of this lobe is somewhat similar to that of the gill-bearing lobe and the two may be homologous.

**Larva of Protoparce or Other Lepidopteron**

Caterpillars bear several pairs of abdominal appendages, commonly referred to as "prolegs." Examine any common species with well-developed prolegs. Note that the appendage is, in most species, attached to a lobe of the body wall. The first region of the proleg is in the form of a short cylinder which is partly sclerotized, but the extent of sclerotization differs in different species. On the assumption that the proleg is a true segmental appendage, this segment is known as the *coxa*. It is followed by a short membranous area, which in turn is followed by a distal plate known as the *planta*. This also is partly membranous, partly sclerotized. It bears a series of heavily sclerotized hook-like structures known as *crotchets*.

**Larva of Sialis**

In this specimen the first eight abdominal segments bear leg-like segmented appendages. These segments should not be regarded as equivalent to the segments of the thoracic legs. The basal segment has been interpreted as the coxopodite and the remainder as a segmented stylus but this is open to considerable doubt.

It may be stated here that other larval insects bear a variety of leg-like outgrowths which are not homologous with the legs but are adaptive larval structures, and that some structures, such as styli, crotchet-like hooks, etc., borne on abdominal appendages may not be homologous with similar structures of other insects but may be convergent adaptive structures.

**Acheta Assimilis**

There is a long tapering *cerci* at the posterior end of the abdomen and note that they are inserted in a membrane between the tenth tergum, the epiproct and the paraproct. The cerci are append- ages of the eleventh segment but owing to the partial or complete atrophy of this segment in many insects they appear to be more closely associated with the tenth. They are frequently lost. In *Acheta* a very narrow basal ring is marked off by a suture from the remainder. This ring has been interpreted as the coxopodite but this is very doubtful; the suture more probably mark is the position of a ridge for the attachment of muscles. The cerci vary in form and degree of development in different species.

# 7

# Male Genitalia

Insects are usually unisexual with a clear cut sexual dimorphism. This differentiation includes the external genitalia. The genital orifice of the male lies in the posterior region of the ninth abdominal sternum, except in Collembola and Protura where it opens in the fifth and the eleventh segment respectively. Primitively each gonoduct opened independently to the exterior, but the double opening is retained only in the Protura, Ephemerida and some Dermaptera. All other insects have a single median orifice which lies typically at the extremity of the *penis* or *phallus*, the intromittent organ used to introduce the sperm into the genital chamber of the female. In some orthopteroids it lies in the ventral body wall but is surrounded by structures which together are the homologue of the penis of other insects. Collembola and Odonata lack the penis, the so-called penis of the latter being a solid structure borne on the second abdominal segment and having no relationship to the true penis. The penis is paired in the Protura, Epemerida and Dermaptera. In other insects there is a single median organ to which Snodgrass would restrict the term "phallus."

In Ephemerida and Dermaptera there is a *double penis* originating as a pair of tubercles or penis valves on the posterior margin of the ninth sternum. The mesodermal gonoducts or *vasa deferentia* terminate blindly in a pair of ampullae lying against the base of the tubercles. The tubercles increase in size with successive moults and invaginate at the tips to form the ectodermal *ejaculatory ducts* which grow inward and make contact with the mesodermal ducts. In the Dermaptera each penis valve divides distally into a lateral and a median lobe and the ejaculatory duct invaginates in the median lobe which, with the basal portion, forms the tubular penis. The lateral lobe remains attached to, and forms a part of, the penis and is known as the *paramere*. The penes of the Ephemerida may become partially or completely united but the two orifices or *gonopores* and ejaculatory ducts persist. In *Hemimerus* and in the Forficuloidea among the Dermaptera the penes are completely fused. The two gonopores persist in the former but in the Forficuloidea the left duct atrophies and the corresponding gonopore disappears leaving a single median orifice. The *median penis or phallus* in its early development resemble the double penis, originating, like it, as a pair of tubercles or *phallomeres* on the posterior border of the ninth sternum, but the ampullae lie in close contact, one on each side of the median line, and the ejaculatory duct is formed as a single median invagination of the sternum between, and independent of, the phallomeres. This is the only important difference between the two types of penes and it is probable that the one is a modification of the other and that the penes are homologous in all insects.

The *phallomeres* in Thysanura are undivided and later unite along their dorsal and ventral edges to form the simple tubular penis of this group. In other insects they divide distally so that each consists of a basal portion on which a lateral and a median lobe are borne. In some Orthoptera and in Hymenoptea there are three lobes, lateral, dorsal and ventral. The phallomeres develop independently in the Mantidae and Blattidae. They do not unite to form a tubular penis, and the ejaculatory duct opens between their bases. In other insects the undivided bases, of the phallomeres unite dorsally and ventrally to from the *phallobase* or *caulis*, while the median lobes (or dorsal lobes if there are three pairs) unite similarly to form the *aedeagust* the two forming a continuous tube. The cavity of this tube communicates proximally with the gonopore; distally it opens to the exterior through the *phallotreme*, the orifice of the aedeagus.

The lateral lobes of the phallomeres develop into *parameres* which in the adult may remain attached to the phallobase, but in some insects migrate to the body wall where they form a pair of clasper-like organs. The ventral lobes of Hymenoptera develop into a pair of movable lobes known as the *volsella*. This uniform method of development has been observed in several groups but in others it is more complicated and what is here regarded as the primitive method is obscured. The phallus will, of course, have an outer and an inner well-known respectively as the *ectophallus* and *endophallus*. The endophallus which is usualy membranous, sometimes extends far into the body cavity and is eversible. In some insects the base of the aedeagus is retracted into a cupshaped pocket of the phallobase known as the *phallotheca*. The outer wall of the phallotheca is known as the *theca*, the inner wall as the *endotheca*. The latter also is frequently membranous and eversible.

The morphology of the penis is obscure. There can be little doubt that the true penes of all insects are homologous, but there is still some uncertainty as to whether the penis is a new structure, i.e., an organ *sui generis*, or a derivative of some antecedent structure. Various authors have tried to derive it form the appendages of the ninth or tenth segment or from the retractile vesicles of the Thysanura, but the evidence is inconclusive. A pair of claspers borne on the ninth segment of Ephemerida are universally interpreted as segmental appendages. Somewhat similar claspers are borne by many other male insects and these, for the most part, have been regarded as homologues of those in the Ephemerida. In many of these insects, however, it has been shown that the claspers originate as lateral lobes of the phallomeres and Snodgrass now believes that they are all true parameres.

The male genitalia may include other structures, in addition to the penis, known collectively as *periphallic structures*. These are usually processes of various kinds which may be borne on any of the posterior segments from the eighth to the eleventh. Only a few examples of male genitalia can be studied here but these will give some indicaion of the great variety in the structure of these organs. They are of primary importance in taxonomy and the student of any group must make himself thoroughly acquainted with the structure of the genitalia of the group.

## THE PENIS AND ASSOCIATED STRUCTURES

### Thermobia Domestica

The male can be distinguished by the absence of the ovipositor and by the fact that styli are borne only on the eight and ninth sterna. The eighth sternum otherwise resembles those anterior to it. The venter of the ninth segment is membranous and bears a pair of coxopoditesor *gonocoxites*. Cut away the eight sternum to expose the gonocoxites completely. Find the relatively stout median *penis* in the form of a simple tube growing out of the mebranous-ventral region of the ninth segment and extending posteriorly dorsal to the gonocoxites. Note the wide orifice, and the dorsally bilobate distal margin. The male can be distinguished by a pair of segmented appendages borne on the ventral side attached to the distal end of the ninth sternum.

If the insect is viewed from the dorsal side the last obvious tergum is seen which is the tenth and the lateral edge of this tergum extend to this side. Attached to each of these lateral edges there is an unsclerotized plate the two meeting in the mid-ventral line. These plates are probably the *paraprocts*, in which case they belong to the eleventh segment. Note the long, slender, multiarticulate *cerci* borne behind the tenth tergum. Dorsally, between the bases of the cerci, find a small papilla bearing a short median process. The papilla represents the reduced *epiproct* and the process is the *caudal filamental or telofilum*.

The five segmented, mesally curved *claspers* or *harpagones* borne on the posterior edge of the ninth seternum. The basal segments of the two claspers are completely united to form a continuous plate which has been often wrongly interpreted as a tenth sternum. These united segments probably represent the coxopodites of the appendages, while the distal segments represent the styli. The coxopodites in many species of mayflies are free from each other.

The bases of the two penes lie dorsal to the coxopodites. In some species of *Hexagenia* they are only slightly curved and extend posteriorly as far as the end of the second segment of the claspers. In others they are shorter and curved like a hook. Their enlarged *bases* can be seen extending laterally beyond the edges of the coxopodites. The bases of the penes are greatly enlarged and extend laterally to the posterior portion of the lateral edges of the ninth tergum, with which they are articulated. The bases are joined mesally by a narrow bridge.

**Anisolabis**

The terminal region is greatly modified. On the dorsal side the last well-developed tergum is the tenth. Behind this are the large asymmetrical forcipiform *cerci*. On the ventral side the large ninth sternum forms the *hypandrium*. The penes are concealed in the genital chamber dorsal to this plate. Cut way the hypandrium to expose more fully the terminal ventral sclerites. Between the bases of the cerci you will find two small median sclerites. Their homologies are somewhat obscure. The posterior plate is undoubtedly the *epiproct*or part of it, and the anterior plate is probably also a part of the epiproct since the cerci articulate with it. Lateral to the epiproct and mesal to the tenth tergum are two triangular sclerites which are probably the *paraprocts*. Cut away one paraproct to expose the base of the cercus and note how the latter articulates with the epiproct.

The double penis will be seen in the region from which the ninth sternum has been removed. Each consists of an elongated, partially sclerotized basal region and two distal lobes. The outer lobe which is lightly sclerotized and clasps the others is the *paramere*. The inner lobe is membranous and tubular, its cavity continuous with the slender *edaculatory duct* which extends through the basal tubular region. The inner lobes are asymmetrical, the left being longer than the right and, when at rest, bent back against the basal tube. The basal regions of the two penes may be together homologized with the phallobase of the median penis and the two inner lobes with the aedeagus. It has also been seen that proximally the two penes are united by a weakly sclerotized apodeme which projects into the body cavity.

**Blaberus Craniifer**

The tip of the abdomen has the bilobate *epiproct* and asymmetric *hypandrium*. The latter is probably formed as the result of the fusion of the ninth sternum and the gonocoxites; it bears a pair of minute *styli*. The posterior part of the abdomen has the *paraprocts* and protruding portions of the phallus. The *paraprocts* lie transversely beneath the epiproct. They are asymmetrical, the right one bearing a strongly sclerotized hook. Note the relations between the epiproct, paraprocts and cerci.

There is no tubular penis. The gonopore opens in the body wall and is surrounded by three independent lobes, a *median ventral* and a *right* and *left dorsolateral*. Qadri describes the development of the phallus in two other cockroaches, *Blatta arientalis* and *Periplaneta americana.* Development starts in the usual way, two undivided phallomeres being formed, but later each phallomere is divided into dorsal and ventral lobes. The cleft in the right phallomere descends to the base dividing in into two separate lobes. One of these takes up a position ventral to the gonopore while the other shifts dorsally. These are represented by the median ventral and the right dorsolateral lobes in *Blaberus*. The lobes of the left phallomere, the left dorsoventral in *Blaberus*, retain their common base and do not become separated from each other. The extent of pahllic structures exposed on lifting the epiproct will vary in individual specimens. The end of the median ventral lobe contains a strongly sclerotized rod which is partly surrounded by two rows of minute spines. The right dorsolateral lobe may also be partially exposed as the mouth of a membranous sac with a strongly sclerotized hook like structure protruding from it. The left dorsolateral will probably be invisible as it is concealed beneath the left paraproct.

There is a wide *ejaculatory duct* which opens through the gonopore dorsal to the median ventral lobe. The *median ventral lobe* is partly membranous, and concave dorsally, with two rows of minute teeth or spines along the lateral and posterior edges of the concavity. On the floor of the concavity there is a strongly sclerotized ridge, irregular in form, which is continued as an apodeme into the body cavity on the ventral side of the ejaculatory duct. The *right dorsolateral lobe* is in the form of a long, tapering, tubular sac terminating in a sclerotized hook. It projects into the body cavity and the distal half is invaginated into the proximal half, but the end of the hook may protrude slightly through the mouth of the sac. With the points of a pair of fine forceps grasp the hook and pull gently to evert the sac. The *left dorsolateral lobe* is thick and blunt and irregular in shape. It is largely membranous but bears several sclerites. There is a shallow cleft in it but the two lobes are attached to the common base.

### Gyrinus

In the species the posterior extremity of the male develops the genital chamber which lies dorsal to the seventh sternum. The tips of three sclerotized processes can be seen just within the orifice of the chamber or protruding slightly from it. Their outline can be seen through the semitransparent sternum. (There are only two processes in the female and they do not extend as far forward). Ventral to the digestive tract and testes there is a pair of wide tubes which converge and unite to form a more slender median tube. The latter is the *ejaculatory duct* and should be left intact.

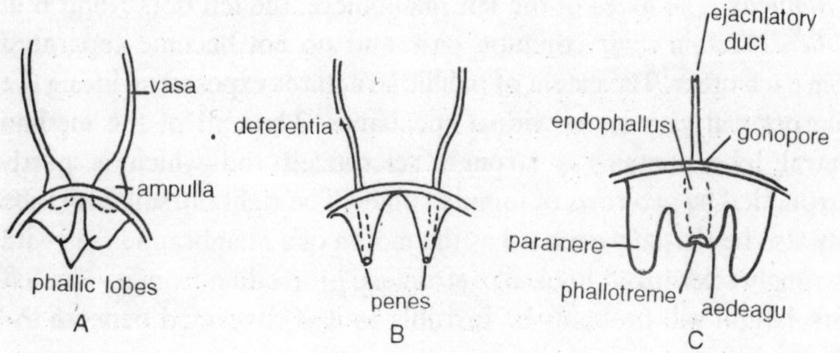


**Fig. 7.1.** Origin and development of the phallic organs. (A) Primary phallic lobes. (B) paired penes of Ephemeroptera, and (C) formation of the aedeagus.

In the abdominal segments fifth to the seventh there is a relatively large median *phallus*, into the proximal end of which the ejaculatory duct passes. There are three processes mentioned earlier; the median one is the *aedeagus*, the lateral ones the *parameres*. Anteriorly they are enclosed in a sheath, the edges of which do not quite meet on the dorsal side. This is the *phallobase*; note its oblique orifice at the anterior end. The aedeagus and parameres are retracted within the phallobase and their common base, into which the ejaculatory duct disappears, can be seen through the proximal orifice of the phallobase. The phallobase therefore forms a *phallotheca*, and its outer wall, largely sclerotized but partly membranous, is the *theca*.

Regarding the muscles of the phallus, the chief of which are as follows: (a) Muscles originating near the lateral ends of the antecostae of the seventh tergum and sternum and inserted in the common base of the aedeagus and parameres. (b) Muscles with a similar origin inserted in the anterior end of the theca. (c) Muscles originating in the posterior region of the seventh tergum and sternum and running forward to be inserted in the anterior end of the theca. (d) Intrinsic muscles originating in the posterior end of the theca and inserted in the base of the aedeagus and parameres. (e) Transverse muscles between the edges of the theca on the dorsal side. Determine the action of these muscles. On the side of aedegus there is a shallow median groove into which the *phallotreme* opens.

**Pteronidae**

Each phallomere in the Hymenoptera divides into three lobes. The lateral lobes form the parametres, the dorsal lobes unite to form the aedeagus an the ventral lobes :rom the volsella.

On the posterior end of the abdomen there is a large *hypandrium*, the ninth sternum, which forms, the floor of the genital chamber. The eighth tergum is well-developed, with a posterior median lobe which partly overlaps the reduced ninth and tenth terga. The tenth tergum is the slightly sclerotized terminal plate beneath which the anus opens. It bears a pair of small pygapods at its lateral angles. The ninth forms a narrow arch in front of the tenth, its lateral ends in contact with the ninth sternum. In the Tenthredinidae and some other Hymenoptera the phallus has rotated through 180 degrees so that the morphologically ventral side comes to lie dorsally. This condition is said to be "strophandrious." The terms "dorsal" and "ventral" as used here will refer to the definitive or secondary, and not the primary, condition.

There is a large ovoid phallus which is a compact structure and consists of a proximal *phallobase* to which is attached distally a median *aedeagus* and two large lateral *parmeres*. The *phallobase* is not continuously sclerotized but consists of five sclerites and a median dorsal membranous region. The sclerites form a basal ring, two parameral plates and two volsella plates. The *basal ring* surrounds the anteroventral orifice through which the ejaculatory duct enters the phallus. Its lateral edges can be seen on the dorsal side forming the margin of the proximal third of the phallus. Mesal to the basal ring on the dorsal side are the *parameral plates* which extend posteriorly to the outer edges of the bases of the parameres. Follow these plates around to the ventral side where they form two large sclerites behind the basal ring. Each sclerites is divided into two lobes connected by membrane. The lateral lobe and the membrane articulate with the large movable *paramere*, which forms the lateral portion of the posterior end of the phallus. The mesal or *parapenial lobe* curves dorsally and can be seen on the dorsal side as a triangular process lateral to the base of the aedeagus. On the dorsal side, mesal to the edges of the parameral plates and partially over- lapped by them, there are two long narrow sclerites, the *volsellar plates*. Each plate terminates posteriorly in two lobes. The ventral lobe or *cuspis* is a continuation of the sclerite, but the dorsal lobe or *digitus* is atriculated to the sclerite and therefore movable. These sclerites and lobes together form the *volsella*. The wall of the phallobase is completed dorsally by a membrane between the mesal edges of the volsellar plates. Ventrally it is completed by membrane between the parameral plates, but the degree of approximation between the edges of the plates varies in different species.

The *aedeagus* is a well-developed median organ borne distally by the phallobase. It is in the form of a gutter, the concavity being on the dorsal side, but he edges may approach each other fairly closely. The aedeagus is membranous but each edge is strengthened by two narrow sclerites which terminate in free spine-like processes. The outer sclerite is short but the inner is continued anteriorly as a long apodeme, the end of which is visible in the phallobase canal. These sclerites are known as *penis valves* but this term is used for the phallomeres and is unacceptable here.

**Magicicada Septendecim**

The ninth segment consists, in part, of a continuous sclerotized ring, there being no pleural membrane. The tergal region is completely sclerotized and terminates in a large hood-like spine which projects above the tenth and eleventh segments. The posterior portion of the ventral region is invaginated into the segment to form the large genital chamber within which the phallus lies. The tenth tergum is short, largely sclerotized, and its lateral edges are continued into a pair of large sclerotized hook-like processes which project ventrally and curve anteriorly. These processes or *claspers* grasp the end of the phallus, the tip of which can be seen protruding between them. The eleventh segment is small and only partially sclerotized; the anus opens between two membranous flaps at its extremity.

The genital chamber opens ventrally in the posterior region, and continues into the anterior region of the ninth segment as a *blind pouch*. The floor of this pouch is formed by two sclerites. The posterior sclerite is narrow and transverse. Attached to its anterior end is a larger sclerite which runs in an anterior direction then curves ventrally. These sclerites constitute the *basal plate* of the phallobase.

There is a tubular *intromittent organ* which is attached proximally to the anterior end of the basal plate. It runs anteriorly for a short distance, then makes a loop and runs posteriorly within the mid-dorsal groove of the genital chamber. It is held in place by the claspers on the tenth segment. Its diameter gradually increases toward the posterior end. At the distal end it bears a pair of sclerotized *serrate processes* which extend ventrally. A membranous fold formed from its inner wall projects between the serrate processes.

It is seen that the narrow posterior sclerite gives off an *apodeme* which passes into the base of the intromittent organ. The *ejaculatory duct* enters the organ ventral to the apodeme. The homologies of the components of the phallus are difficult to interpret. Snodgrass believes that the intromittent organ is the elongated *phallotheca*, its inner wall being the *endotheca*. He thinks the aedeagus is rudimentary and represented only by the *apodeme*. The commoner cicada, *Tibicen linnei*, shows some differences from the periodical cicada. The two ventral processes of the tenth tergum are united to form a single process. The "phallotheca," which is longer and more slender, and lacks the serrate processes, bends ventrally parallel to the ventral process of the tenth segment, and is held in place by a pair of small claspers borne on this segment. The apodeme and the sclerite on which it is borne are absent. Apart from these differences the general structure is similar to that of the periodical cicada and this species may be substituted if the other is unavailable.

# 8

# Female Genitalia

The external genital organs in the female consist typically of three pairs of elongated processes which, with certain basal sclerites in the body wall, constitute the ovipositor (Fig. 8.1). The ovipositor serves for the outward passage of the eggs and for their deposition in suitable situations; therefore it is frequently adapted for digging or boring in order to deposit the eggs in the soil, wood, leaf tissues or other objects. A true ovipositor is not found in all insects but the Thysanura, which lie near the ancestral stem of the Pterytota, possess an ovipositor which, though generalized, is built on the same plan as that of the Pterygota therefore it is assumed that when the ovipositor is absent in the latter it has been lost. This is borne out by the fact that some individual species of an order that normally has an ovipositor either lack the ovipositor or have it in a rudimentary condition and, conversely, in certain orders which characteristically lack an ovipositor there are a few species which possess one.

The morphology of the ovipositor in the Thysanura provides a very definite clue to the interpretation of the pterygote organ. In the Machilidae each abdominal segment from the second to the ninth bears a pair of appendages, such as were deseribed earlier, consisting of a flattened coxopodite and a terminal stylus. The appendages of the eighth and ninth segments do not differ from those of the anterior segments in general appearance, except perhaps in minor details of size and shape, and are obviously serially homologous with them. The coxopodite of each of these two appendages, however, bears from its anteromesal angle and elongated endite process or *apophysis*. The four apophyses are concave on their inner surfaces and are associated to form an elongated, backward-projecting tube which constitutes the *ovipositor*. The two pairs of apophyses because of their association with the genitalia are known as *gonapophyses*. There is a tendency for the ninth coxopodite to be elongated. The ovipositor in the Pterygota is formed from three pairs of elongated sclerotized processes known as *valvulae* (Figure 8.1). In Orthoptera the three pairs are associated to form the egg tube but in other orders only two pairs are thus associated, the third having some accessory function. The *first valvula* is borne on a small sclerite, the *first valvifer*, which originates from the eighth segment but is often displaced posteriorly. The *second valvula* and the *third valvula* are borne on a common basal sclerite known as the *second valvifer* which originates from the ninth segment.

The development of the ovipositor has been followed carefully in several species of Orthoptera. The first valvulae originate as a pair of small papillae on the posterior edge of the eighth sternum. They increase in size at each moult and differentiate into basal valvifers and distal valvulae. The second and third valvulae of each side originate as a bilobate papilla. The outer lobe, which will develop into the second valvifer and third valvula, may bear a stylus at its distal extremity and there can be little doubt that it represents the coxopodite. The stylus is lost before the insect matures. It appears, therefore, that the first valvifer of the pterygote ovipositor is the reduced coxopodite of the eighth segment and the first valvula is the gonapophysis borne on this coxopodite. The second valvifer and third valvula in the Orthoptera form a continuous structure which is obviously the greatly elongated coxopodite of the ninth segment. In other insects the third valvula is atriculated to the end of the second valvifer and it has been claimed that the valvula is the persistent stylus borne on the end of the coxopodite. In some Coleoptera, however, in which the two are distinct there is some evidence of the presence of a minute vestigial stylus at the end of the third valvula. The second valvula is obviously the gonapophysis of the ninth segment.

This is the most generally accepted interpretation of the ovipositor, but many early writers maintained that it is an outgrowth of the ventral region, i.e., the ventrosternite, of the eighth and ninth segments. Some recent writers have reverted to this view. They claim that the gonapophyses are sternal outgrowths and not endites of the coxopodites. They would interpret, the close association of the two structures in Thysanura as secondary.

## THE OVIPOSITOR

### Thermobia Domestica

This is a common lepismatid, distinguished from *Lepisma saccharina* by the dark mottling on the dorsal side. *Lepisma* may be substituted as the structures are essentially alike. The female may be distinguished by the long median ovipositor on the ventral side. Note that styli are borne on the seventh, eighth and ninth segments. The seventh sternum, which overlaps the base of the ovipositor, resembles the anterior ones in consisting of a single sclerite without differentiated coxopodites. Behind the seventh sternum are two pairs of stylusbearing plates, the *gonocoxites,* i.e., the coxopodites of the eight and ninth segments. Remove the seventh sternum so as to expose the base of the ovipositor.

The ovipositor consists of two pairs of concave processes or *gonapophyses* fitted together by tongue and groove joints to form a tube. The two gonapophyses exposed in ventral view are the anterior ones. They are attached to the mesal angles of the bases of the gonocoxites of the eight segment. The posterior gonocoxite is longer than the anterior one and is divided into two sclerites, a small

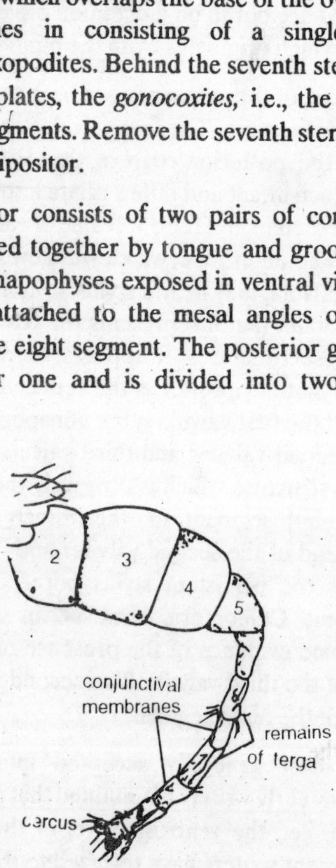


**Fig. 8.1. Abdomen of the housefly extended. The segments are numbered.**

triangular anterior sclerite and a larger. stylus-bearing sclerite. An oblique suture separates the two in such a way that the narrow anterior end of the larger sclerite lies mesal to the smaller one. The base of the posterior gonapophysis is attached to the somewhat thickened anterior edge of the smaller sclerite.

Gustafson identified the smaller sclerite as a fragment of the sternum, but in many Thysanura it is not detached from the stylus-bearing sclerite and the relation of the gonapophysis to it is similar to the relation of the corresponding structure to the second valvifer in pterygote insects. If Gustafson's interpretation is correct the valvifers and gonapophyses here and a Pterygota may conceivably be ventrosternal and not appendicular in origin.

**Conocephalus**

Apart from the size and shape of the ovipositor, differences between the species lie chiefly in the extent and distribution of the sclerotization in the roof of the basal region of the united second valvulae, and in the form of the intervalvulae, which may or may not be completely fused with the sclerities of the second valvulae. The *subgenital plate* is the apparent eight sternum, but some authorities claim that the eight sternum is reduced and the subgenital plate in this family is a secondary sclerite formed by a fold which grows out from the membrane between the seventh and eighth sterna.

It forms the floor of the genital chamber, a shallow invagination which communicates with the channel of the ovipositor on the one hand and with the oviduct on the other.

There is a conspicuous elongated ovipositor identifying the *first valvulae* on the ventral side and the *third valvulae* on the dorsal side. These are closely attached to each other by tongue and groove joints and together form the outer wall of the hollow *shaft* of the ovipositor. Note the enlarged base of the first valvula, partially concealed by the subgenital plate. Its anterior border continues into the pleural membrane of the eighth segment to which the appendage belongs. Dorsally the base of the first valvula is attached to a small sclerite which lies behind the ninth tergum and between it and the second valvifer. The sclerite is the *first valvifer*, the rudiment of the gonocoxite of the eighth segment, which is displaced posteriorly. The anterior edge of the valvifer forms a thickened ridge. Ventrally this ridge passes beneath the ninth tergum and forms an articulation with the tergum at its posteroventral border. The dorsal end of the ridge forms an articulation with the second valvifer.

The *second valvifer* and the *third valvula* are not separated by a suture but form a continuous process. The enlarged base, indefinite in extent, is the valvifer. Near the dorsal edge of the base of the valvifer there is a short longitudinal groove marking the position of a strengthening ridge which provides for the articulation with the first valvifer mentioned earlier. Anterior to the articulation, the narrow dorsal region of the valvifer passes into the body cavity as a thin sclerotized apodeme, the *superior apophysis*, for the attachment of muscles.

The valvulae can be easily spread apart and the bases of the second valvifers are connected dorsally by a T-shaped sclerite, the *posterior or superior intervalvula*. The *superior apophysis* can now be seen clearly extending anteriorly from the articulation with the intervalvula. If we spread the two valvifers apart and expose the common basal region of the second valvulae. It will be seen that these valvulae are completely fused at their base to form an arched roof over the egg passage. Remove the right valvifer carefully and note in doing so that there is an extensive membranous connection between the valvifer and the base of the valvulae. The dorsal wall of the base of the second valvulae is completely sclerotized except for the two membranous areas by which it is attached to the valvifers. This complete sclerotization of the dorsal wall is not found in all tettigoniids. The degree of sclerotization varies and in some forms there is only a

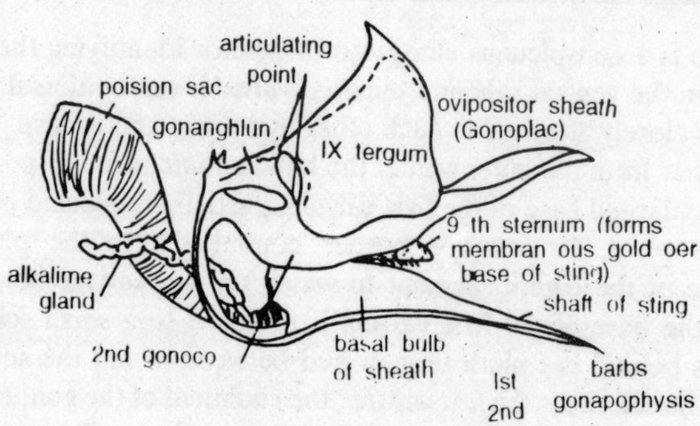


Fig. 8.2. Sting of the honeybee.

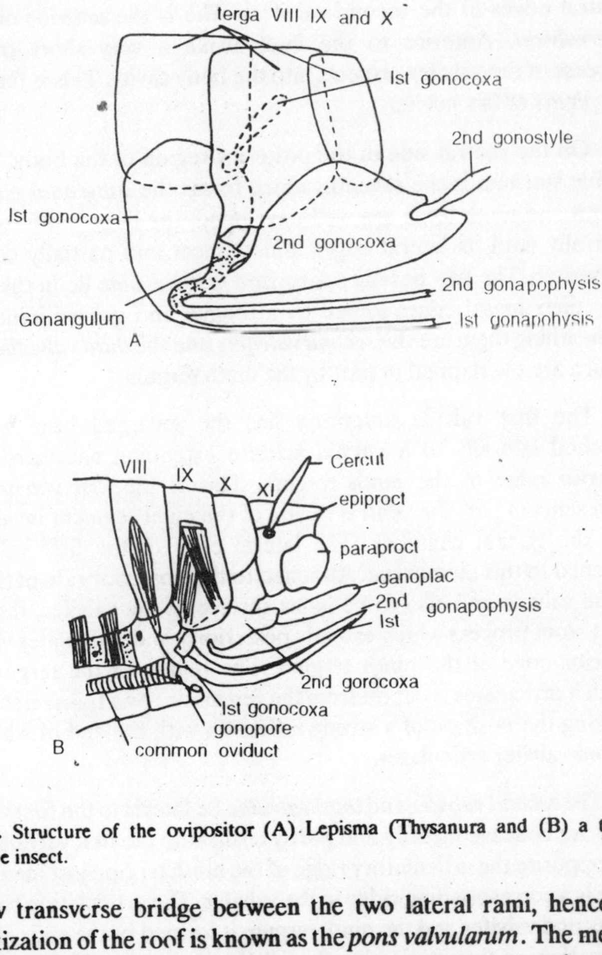


**Fig. 8.3.** Structure of the ovipositor (A) Lepisma (Thysanura and (B) a typical pterygote insect.

narrow transverse bridge between the two lateral rami, hence the sclerotization of the roof is known as the *pons valvularum*. The median region between the two membranous areas is a narrow sclerotized process closely united with the end of the median process of the intervalvula. Thus, the second valvulae articulate with the second valvifers through the mediation of the intervalvula. The lateral borders of the base of the valvulae are formed by narrow sclerotized rods or *rami*.

The ventral wall of the common base is membranous except for the rami which form its lateral borders. Proximally these curve inward and articulate with another small transverse sclerite which unites the ventral edges of the second valvifers. Thė is the *anterior* or *inferior intervalvula*. Anterior to the intervalvula a very short triangular process of the valvifer projects into the body cavity. This is the *inferior apophysis* of the valvifer.

On the ventral side of the posterior region of the body. The last visible sternum is the seventh, which forms the *subgenital plate*. The eighth and ninth terga are well-developed, the large ninth extending ventrally until its lateral edges almost meet and partially cover the ovipositor. The two heavily sclerotized *first valvulae* lie in the middle line, their mesal edges joined by a tongue and groove joint. Partly ensheathing them are the *second valvifers* and the *third valvulae*, which in turn are overlapped in part by the ninth tergum.

The first valvula anteriorly has the enlarged base which is attached laterally to a curved sclerite extending backward to the anterior edge of the ninth tergum. This is the *first valvifer*. This represents in part the ventral region of the eight segment invaginated into the genital chamber. The lateral edge of the first valvifer is attached to this membrane. Attached to the posterior side of the base of the valvula and also to be inner surface of the valvifer, there is a short stout process which extends posteriorly and articulates with the anterior edge of the ninth tergum. The region of the tergum with which it articulates is cut off from the remainder by a transverse suture marking the position of a strong inflection with the end of which the second valvifer articulates.

The *second valvifer* and *third valvulae* lie lateral to the first valvula. They are concave mesally and partly ensheathe the first valvula. Note that opposite the articulatory ridge of the ninth tergum just mentioned there is a corresponding ridge in the valvifer. The articulation between the second valvifer and the ninth tergum is formed by the ends of these ridges. Behind the articulation the valvifer is connected by membrane to the ninth tergum. The third valvula is distinctly marked off from the valvifer, unlike the condition found in the Orthoptera. The shaft of the ovipositor is formed from the first and second valvulae only. The second valvifer and third valvula form a sheath for this shaft. The valvula is provided with sensory hairs and is probably used to test the twigs in which the eggs are laid.

If we remove the first valvulae from the specimens the *second valvula* is seen. Observe their origin from the mesal side of the proximal region of the second valvifers and note the basic resemblance between the second valvifer, third valvula and second valvula in this insect and the coxopodite, stylus and gonapophysis of the ninth segment in *Thermobia*. It is somewhat doubtful, however, whether the third valvula can be homologized with the stylus of *Thermobia*. The two second valvulae are completely united except at their proximal ends which are connected by membrane. They thus form the roof of the egg passage.

## OTHER APPENDAGES

### Cerci

Paired Cerci occur in Diplura, Thysanurans, Ephemeroptera, Zygoptera (a suborder of the Odonata), the orghopteroid orders, and Mecoptera. It is generally agreed that the cerci are true appendages of the eleventh segment, although frequently all traces of the latter have disappeared. Typically, they are elongate multisegmented structures which function as sense organs. They may, however, be considerably modified. In nearly all Dermaptera the cerci form unjointed forceps.

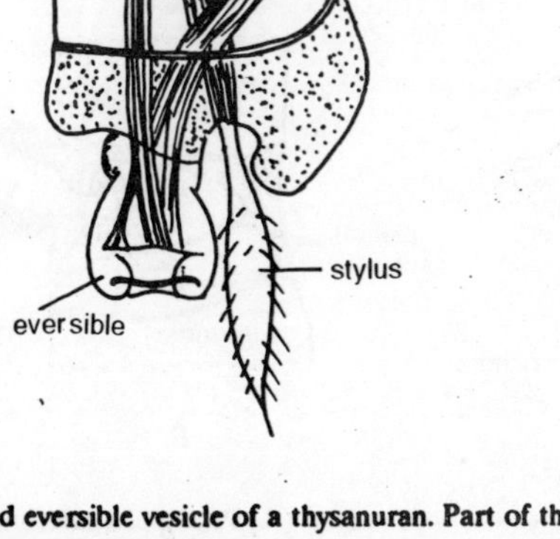


**Fig. 8.4. Stylus and eversible vesicle of a thysanuran. Part of the wall of the plate has been removed to show the musculature.**

The cerci of nymphs of Zygoptera are modified to form the *Lateral Caudal Lamellae.* The latter, along with the *Median Caudal Lamella* (developed from the epiproct) are accessory respiratory strucutures. In adult male Zygoptera the cerci form *Claspers* for grasping the female during copulation.. In most male Embioptera the basal segment of the left cercus forms a hook with which the insect can clasp his mate.

### Styli and Eversible Vesicles

*Styli occur* on most abdominal segments of thysanurans and Diplura, and on the ninth sternum of some male orthopteroid insects. In some thysanurans the styli are articulated with a distinct coxal plate but generally the original coxal segment is fused with the sternum. In thysanurans, at least, the styli serve to raise the abdomen off the ground during locomotion. The *Eversible Vesicles* are short cylindrical structures found on some pregenital segments of apterygotes. They are closely associated with the styli when present, but their homology is unclear. They are believed to have the ability to take up water from the environment.

### Prolegs

Segmentally arranged, leglike structures are present on the abdomen of many larvae. They are known as *Prolegs Pseudopods,* or *Larvapods.* Their structure is variable, though in a "typical" form (for

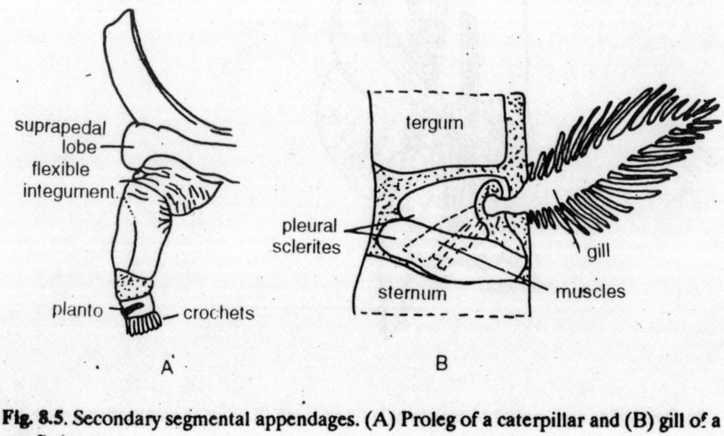


**Fig. 8.5.** Secondary segmental appendages. (A) Proleg of a caterpillar and (B) gill of a mayfly larva.

example that of a caterpillar), three regions can be distinguished: a basal membranous articulation, followed by a longer section having a sclerotized plate on the outer wall, and an apical protractile lobe. The *Planta,* which bears claws, *Crochets,* peripherally. The plants is protracted by means of blood pressure. Immediately above the leg there is a swollen area in the body wall, the *Suprapedal Lobe.*

**Gills**

A large number of aquatic larvae possess segmentally arranged *Gills* on a variable number of abdominal segments. These are flattened, filamentous structures, frequently articulated at the base.

**Nonsegmental Appendages**

In many insects nonsegmental structures are present. These are typically a mediodorsla projection on the last abdominal larvae and the *Caudal Filament* of Thysanura and Ephemeroptera. Occasionally these structures are paired (for example, the *Urogomphi* of some larval Coleoptera) and are easily mistaken for cerci. The *Anal Papillae* of certain dipteran larvae also fit in this category.

# 9

# The Integument

The outer covering of the body in insects is called the *integument.* It is the organ most suitable to begin with in a microscopic study of the insect body. It is physically the largest of all organ system and exhibits the greatest diversity in structure and in function. In each of its manifestations, three components can usually be identified; an outer, noncellular *cuticle*. a single layer of cells, or *epidermis*; and an inner sheet of connective tissue, or *basement membrane*.

The most easily seen portion of the integument is the hardened and jointed cuticle which covers the outside of the body and forms an external skeleton, or *exoskeleton.* As will be explained later, the exoskeleton is extremely resistant to decomposition. When protected from pests, dried specimens of insects may outlast the pins and cabinets of the museum and usually outlast the collectors and curators. When preserved as fossils, pieces of the exoskeleton may remain intact. The chemical component chitin, for example, has been detected in the fossil of an extinct eurypterid arthropod from the Silurian Period, over 405 million years ago. The special arrangements of the hairs and other features of the exoskeleton provide nearly all the diagnostic characters for the recognition of the species of insects. This means that easily a million or more distinctive designs of exoskeletons exist. Taxonomists are indeed fortunate that the exoskeleton is so simply preserved and that so much information concerning the insect's habits can be gained from it.

It addition to functioning in the support of the internal organs, the exoskeleton is involved in such important activities as feeding locomotion, and reproduction. The physically properties as well as the intricate shapes of the mouthparts permit the exploitation of a wide variety of foods. Similarly, the walking and flight mechanisms depend in large degree on the mechanical features of the exoskeleton. In reproduction, parts of the skeletal genitalia provide the means by which the male may transmit sperm to the female and the female, in turn, may deposit the eggs in special places. The exoskeleton determines the form and maximum size of the insect. At intervals during growth, portions of the cuticle are digested and a new larger cuticle is deposited. This process is termed the *moll*, and the actual shedding of the old cuticle is called the *ecdysis*. At times of molting much of the old exoskeleton, is reclaimed and the chemical components are used again. During starvation some of the cuticle is also reclaimed. For this reason, the exoskeleton is actually a kind of food reserve. The exoskeleton also serves many of the same functions as the skin of other animals. It is the main barrier to the loss of water, a problem which is particularly acute in small terrestrial organisms with a relatively large evaporative surface area in proportion to their volume. The entrance of disease organisms and the penetration of chemicals such as insecticides are impeded by the exoskeleton. Assaults by predators are not only resisted by the tough outer covering, but also avoided by cryptic or warning coloration. The sense organs of the nervous system are intimately connected with specialized areas of the exoskeleton because all the stimuli from the environment must be transmitted through the exoskeleton to the receptors beneath.

The integument includes not only the conspicuous exoskeleton but also extensive internal portions. The tracheal system, functioning in respiration, is created by invaginations of the integument. The ultimate subdivisions of these tubular ducts number literally in the millions and supply oxygen to virtually every cell of the body. The elimination of carbon dioxide is accomplished through the tracheae and also through the general body surface. In the alimentary canal, the foregut and hind gut are formed by invaginations of the integument during development. The integument also lines the ducts leading from the internal reproductive organs, providing not only passageways and storage vessels for the gametes, but also glandular secretions which function in their protection and transmission. Finally, a rich variety of other glands, including the salivary glands, is derived from cells of the epidermis. Their products function in nearly every phase of the insect's life history by supplying enzymes for digestion, materials for shelter and support, and other chemical compounds for communication, defense and offense. Among these integumental secretions, human beings have found some of great value, e.g., silk, shellac, and beeswax.

## THE STRUCTURE OF THE INTEGUMENT

The following account of the structure of the integument is intended to provide a basis for the discussion upon the control of the formation of its many layers. The detailed structure of integument is being explained under the following sub heads :

### The Basement Membrane

The epidermis is separated from the hemocoel by a layer of reutral muopolysaccharide secreted by a type of hemocyte. With the electron microscope, it appears as an amorphous granular layer up to 1/2 $\mu$m thick. Other cells may take part in the formation of the basement membrane in other parts of the body. Most is known about the basement membrane round the nervous system where it contains

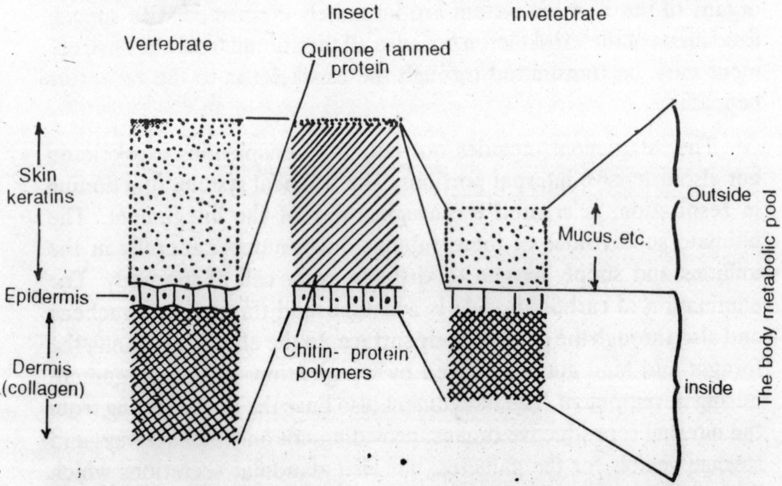


**Fig. 9.1.** The limits of epidermal control in vertebrate, insect, and in most nonarthropod, invertebrate integuments. In nonarthropod groups, inert "dead" material begins immediately outside the epidermis; the region within the body metabolic pool is defined by the cells. In insects and other arthropods, the limits of the metabolic pool are set by the exocuticle.

collagen fibrils. Nerves and tracheoles run between the basement membrane and the cells before ramifying further. The tracheoles may give the appearance of penetrating the epidermal cells but they always carry their own plasma membrane with them. There are many peripheral nerves but connections with the epidermal cells have not been described. If the basement membrane of the epidermal cells is comparable to the natural lamella in its permeability the cells may exist in an environment very different in composition from the blood.

**The Epidermis**

The epidermis forms a continuous sheet of polygonal cells below the cuticle. In *Locusta* the density of cells increases during each stadium until the molt, but at comparable times the density is similar in different stadia. Epidermal cells are often polyploid. Those in *Calpods* contain numerous nucleoli. In many Hymenoptera and probably in all Diptera Cyclorrhapha, the epidermal cells do not divide but grow in size as the larva grows. In *Calliphora erythrocephala* the epidermal cells are about eleven times as wide in the third as in the first instar. On the other hand in many Diptera Nematocera, the epidermal cells divide as the larva grows, although there is also an increase in the average size of the cells. The form and contents of the cells vary greatly with their activity. At some stages during cuticle deposition the outer plasma membrane is folded, increasing the surface area for the absorption of compounds from the blood. There are numerous connections between adjacent cells having the form of the septate desmosomes described by Wood (1959) in *Hydra*. The plasma membrane next to the cuticle is microvillate during cuticle deposition and lobed with much smooth endoplasmic reticulum during the secretion of wax. The axial filaments of the pore canals pass through the plasma membrane into the cell.

Many epidermal cells contain characteristic orange to red pigment granules. In osmium-fixed tissue they appear under the electron microscope as concentric shells of dense and less dense material. Some newly emerged insects – for example, *Rhodnius* and *Triatoma* – appear bright orange-red before the colour of these granules is obscured by pigment patterns in the cuticle. In other insects (*Calpodes*) the colour and number of granules appear to vary genetically. The granules disappear from cicatrices in wounded areas. These granules are probably the same as those described in the epidermis of gryllids and phasmids by Fuzeau-Braesch (1957) which are composed of a type of insectorubin capable of undergoing oxidation-reduction changes. They are unimportant in contributing to the gross coloration in gryllids but in phasmids they undergo a hormonally controlled migration with a diurnal rhythm. These granules have also been studied in acridiids by Nickerson (1956).

**The Endocuticle**

There have been no chemical assays of endocuticle alone, but since it constitutes the bulk of the integument the analyses of whole cuticles are probably nearly the same. One of the most recent analyses is that of DeHass *et al.* (1957) on *Anabrus simplex*. For other analyses and the general chemistry see Hackman. The inner face of the endocuticle next to the cells is frequently different from the more peripheral endocuticle; Schmidt (1956) called it the subcuticular layer and thought of it as a glue holding the cuticle to the cells, but is is more probable that it is newly secreted endocuticle in which the fibers are not ordered and have different chemical reactions. This layer shows up characteristically after oxidation and staining with paraldehyde fuchsin. It appears granular under the electron microscope but the granules already have the alignment characteristic of the microfibers of the endocuticle. In some cuticles the region next to the cell may be filled with lipid-water liquid crystals which later find their way through the pore canal to the surface.

The endocuticle is a layer from 10 to 200 $\mu$m thick composed of numerous lamellae. The electron microscope resolves these lamellae as patterns of microfibers arranged in sheets corresponding to a dense part of a lamella but which curve out at right angles between the sheets. So far, the microfibers have always been observed to have the same orientation in adjacent sheets, but there is no reason why the orientation should not change from layer to layer, giving structures with extra strength in particular directions. Although there are many parallels between plant cuticles with cellulose, and arthropod cuticles with chitin-protein polymers, the arrangement of microfibers is fundamentally different. Cellulose occurs in flat or woven sheets almost restricted to two dimensions. Chitin in the endocuticle occurs in this more complex three-dimensional pattern. The chitin of peritrophic membranes, however, is much more like the cellulose of plant cell walls. In insects and others arthropods peritrophic membranes are like loosely woven fish nets. Perhaps this difference from the endocuticle is due to the low chitin content. Chemically, the peritrophic membrane of several insects have been shown to more like a vertebrate gastric mucin.

The lamellae have a characteristic pattern of fibers. In sections normal to the surface, each lamella seems to be composed of fibers radiating out in a fan shaped pattern. The pattern is like a series of C's, the amplitude of the curve varying with the plane of section. Bouligand (1965) has shown how this appearance is an optical illusion crated by overlapping planes of fibers. The C pattern results from fibers in planes or laminae parallel to the surface. The fibers in each lamina are parallel to one another but the orientation changes slightly in the same direction from one lamins to the next. Each 180° turn gives rise to a lamella, each 360° turn to two lamellae differing from one another only if the component fibers are polarized. A lamella as such only exists with respect to the plane from which it is viewed, and the C pattern only appears in sections which are slightly oblique. The pattern is symmetrical except for the direction of rotation of the planes of successive laminae. It remains to be seen whether the direction of rotation is constant. The model is particularly satisfactory in accounting for helical pore canals which presumably follow gaps in successive layers in a helical fashion. Bouligand's model also makes it easy to comprehend transitions in structure between lamellar cuticle and cuticle with layers oriented primarily in one direction. There are two problems in lamellar structure. How is the orientation of fibers in each lamina determined, and what determines the frequency of microcycles of rotation of successive laminae to make lamellae of different dimensions? The lamellae system varies in the degree of cross linking and hence the stability of the lamellar microfiber system. Lack of cross links would be expected in cuticles needing plasticty of elasticity but not rigidity. In this connection it is of interest that the microfibers are most easily resolved in the plastic cuticles of caterpillars (*Calpodes* and *Galleria*). They are least well seen in the exocuticle and the hard endocuticle of *Tenebrio*. The orientation would also be advantageous to give uniform elasticity in cuticles heavily reinforced with resilin, because of the possibility of distortion in any direction. The elastic cuticle round the bases of hairs has the fiber pattern typical of a lamella.

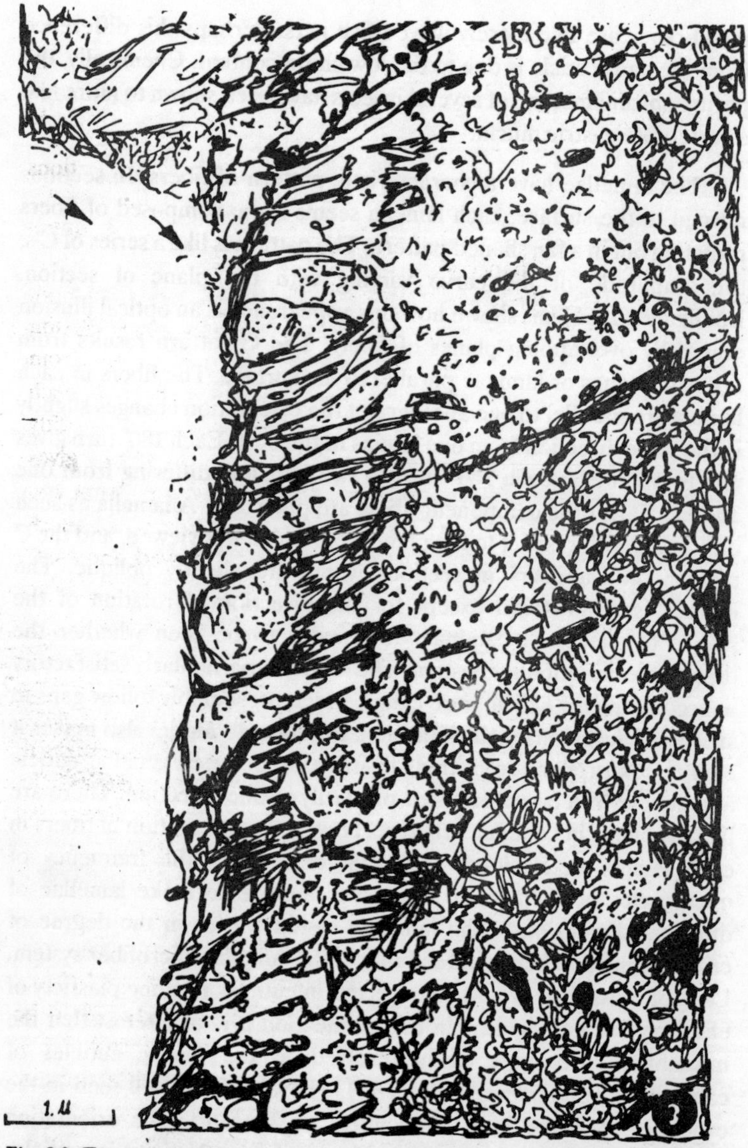



**Fig. 9.2.** Transverse section of the cuticle and epidermal edge in a *Calpodes* larva during the secretion of wax. The cells are filled with smooth endoplasmic reticulum, and the plasma membrane is lobed. Some middle phase lipid-water liquid crystals (arrows) penetrate the lamellae of the endocuticle.

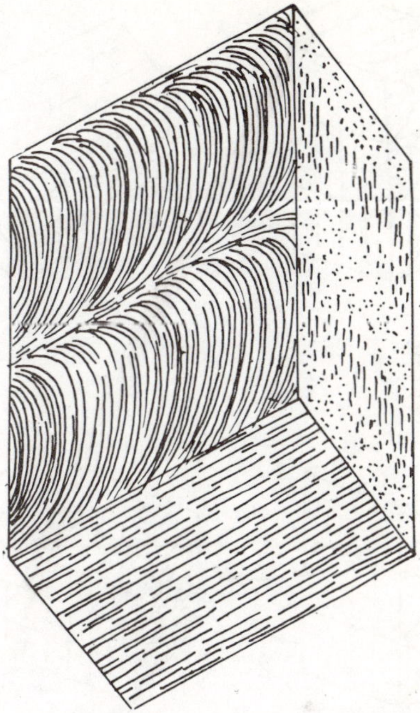

**Fig. 9.3.** Diagram of the apparent arrangement of microfibers to form lamellae in the endocuticle. The diagram shows two lamellae. The top of the cube is in the plane of the surface of the cuticle. This pattern has been interpreted by Bouligand to be the result of overlapping images of helicoidally arranged stacks of microfibers.

Weis-Fogh described a structural protein, resilin, which occurs in a pure state particularly in some wing hinges and muscle insertions. It shows long-range rubberlike elasticity with no permanent deformation. Unstrained, it is optically isotropic but becomes birefringent on deformation. It is probably built from randomly coiled peptide chains linked together is a stable three-dimensional network by two aromatic $\alpha$-amino acids containing a phenolic group, one of which is a diaminodicarboxylic acid and the other a traiminotricarboxylic acid.

### The Exocuticle

The exocuticle may be defined as the region of the endocuticle adjacent to the epicuticle which is so stabilized that it is not attached by the molting fluid and is left behind with the exuvium at molting. In

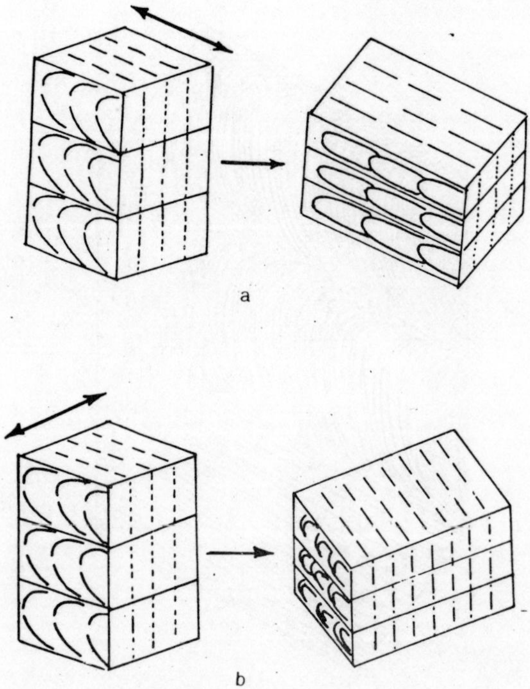

**Fig. 9.4.** Diagrams to show the effects of extension in two directions in the plane of the cuticle upon the orientation of microfibers. (a) Extension in the main axis of alignment of the microfibers would lead to the formation of laminae (as defined in the text) with the fibers predominantly in one direction in the plane of the cuticle; (b) extension at right angles to (a) would lead to laminae with microfibers still partly normal to the surface. In the light of Bouligand's work the above interpretation is incorrect.

this sense it includes all the darkly coloured, quinone-tanned cuticle. It may very in thickness from an thin outer layer, indistinguishable from the epicuticle (as in soft-bodied larvae to almost the whole thickness of the cuticle (as in adult beetles). I some larvae (Lepidoptera and Hymenoptera) the exocuticle contains calcium oxalate crystals. It is uncertain whether the first formed endocuticle destined to be stabilized as exocuticle is qualitatively different from that formed later, which remains labile. The extent of quinone tanning could be determined either by the suitability of the substrate or by the amount and time over which tanning agents are extruded. The term mesocuticle has been proposed for lightly coloured but stabilized

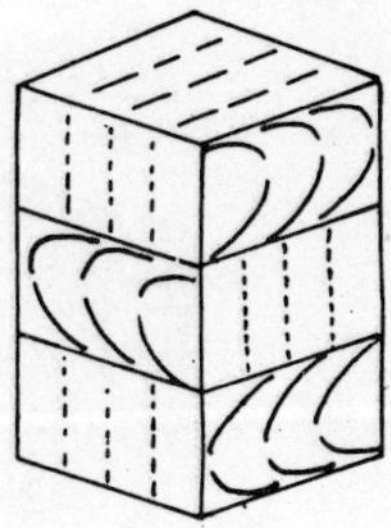


a

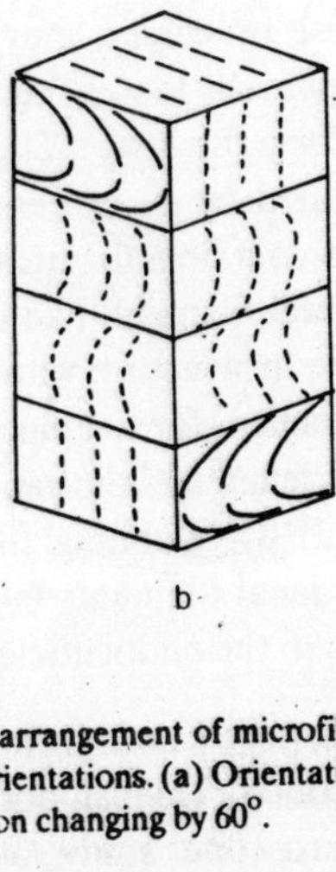


b

**Fig. 9.5.** Diagram of the arrangement of microfibers if successive lamellae were to be secreted with different orientations. (a) Orientation changing by 90° from one lamellae to the next; (b) orientation changing by 60°.

regions between the exo-and endocuticle. The mesocuticle differs in staining properties and presumably in chemistry from the cuticle on each side of it.

## Pore Canals

A number of processes occur at the surface of the insect cuticle, viz., secretion and repair of wax layers, tanning of the endocuticle to form exocuticle, and the repair of surface cuticle after abrasion. For

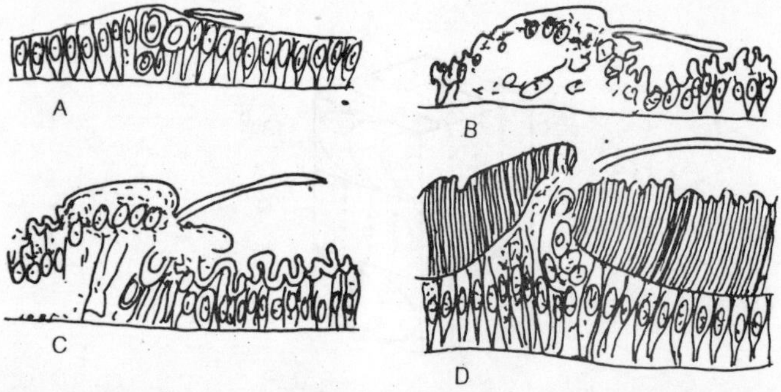


**Fig. 9.6.** Longitudinal sections showing formation of new cuticle during moulting in 4th stage larva of Rhodnius. A, 10 days after feeding; epicuticle just laid down. B, 11 days; epicuticle thrown into folds. C, 12 days; endocuticle being laid down. D, 15 days; just before moulting.

the cells to participate in these processes, they must maintain some route across, the cuticle, except in the loosely knit endocuticle of some caterpillars. The pore canals form this route. They are extracellular in the sense that they are not lined by an extension of the epidermal plasma membrane but they do contain a filament from the cell. Some pore canals seem to have several filament, like the cilia from the rat oviduct which are intermediate in structure with microvilli. The pore canal filament may be a mechanism for keeping a hole in the newly secreted cuticle until it has hardened and the canal is permanent. The filament might accomplish this by inhibiting fiber formation in its immediate vicinity. The pore canal filaments might also function as anchors to stick the epithelium to the endocuticle.

In light microscope studies, pore canals from several insects have been described as helically coiled. Helical pore canals were also reported in an electron microscope study of thick sections of cockroach cuticle and in *Rhodnius* the crescentic holes seen in cross section were taken to confirm their helical nature. However, another shape is more probable, and the hypothesis that pore canals are usually helical should be reconsidered. In many sections, some pore canals have a circular outline. The basic shape is probably that of a cylinder bent either in a helix or in some other way. If the pore canals are indeed helical, the shape and the orientations of the holes seen in cross section will be influenced by two main considerations: (1) the phase relations of different helices, and (2) the plane of the section.

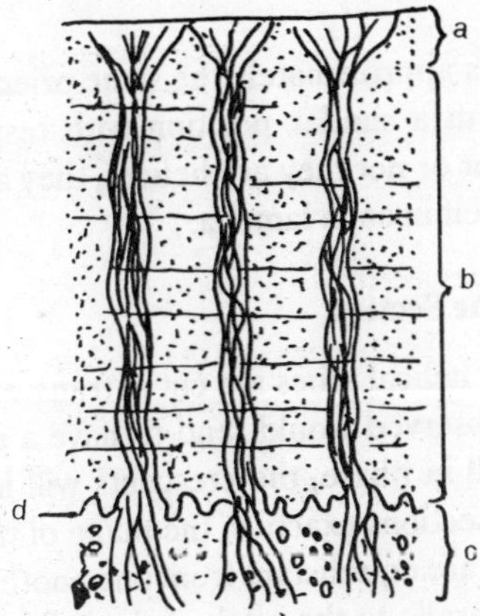


Fig. 9.7. Semi-schematic figure showing the canals in the cuticle extending from the plasma membrane (d), through the exo- and endocuticle (b), to the inner limit of the epicuticle, (a) : and the wax canal filaments, collected within the pore canals, and extending from the cytoplasm (c) to the outer surface of the epicuticle.

The shape of the tube and the pitch and regularity of the helix do not affect the following conclusion.

### (a) The Phase Relations of Different Helices

The orientation of the crescentic holes see in section is influenced by their position relative to each lamella. In any one plane the outline

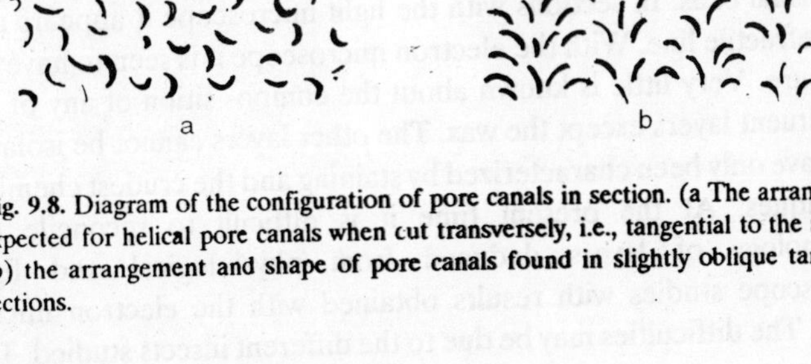


Fig. 9.8. Diagram of the configuration of pore canals in section. (a The arrangement expected for helical pore canals when cut transversely, i.e., tangential to the surface; (b) the arrangement and shape of pore canals found in slightly oblique tangential sections.

of a pore canal has approximately the same orientation as that of any other pore canal in a similar position with respect to the lamella. Therefore, whether or not they are helices, they are in phase, and the pattern repeats itself in each lamella.

### (b) The Plane of the Section

If a number of helical tubes are cut in transverse section, then the crescentic holes observed would tend to have a random orientation. Even if they are all in phase, the crescents will have all orientations except in the rare sections exacty in the plane of the lamella, or, when all the pore canals, are equidistant from one another, in sections cut in the set of planes paralle to the pitch of the helix. Many sections have been cut for the electron microscope in all planes, but the orientation expected for helices has not been seen. The most probably explanation is that the pore canals follow the arrangement of fibers in the lamellae, the hole taking a curved course repeated in the same plane in each lamella. The shape and orientation of the pore canals seen in some sections can only be explained in this way. In some sections the holes are symmetrical ellipses, the long axis of each hole being paralle to its neighbours.

Helical canals, if they occur, can readily be explained by this observation that a pore canal follows the fiber pattern of the lamellae. A helical pore canal would result if the fiber pattern were to rotate about an axis normal to the surface, from one lamellae to the next. This pattern has not been seen in the insects studied, but it may occur.

## The Epicuticle

The epicuticle is a composite structure involving secretions of the epidermis and also the detmal glands which are derived from epidermal cells. In sections with the light microscope it appears as a thin refractile line. With the electron microscope it is seen to have the structure. Very little is known about the compo- sition of any of the constituent layers except the wax. The other layers cannot be isolated and have only been characterized by staining and the crudest chemical techniques. At the present time it is difficult to reconcile the terminology of layers deduced from physiological and light microscope studies with results obtained with the electron microscope. The difficulties may be due to the different insects studied. The

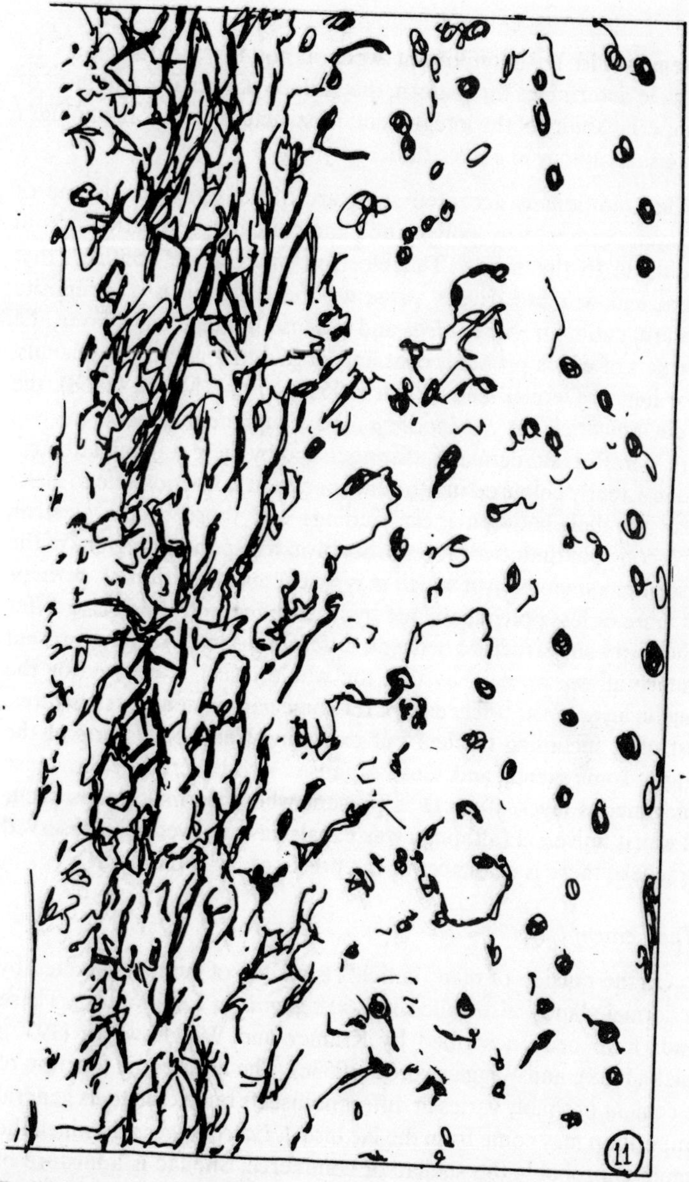

**Fig. 9.9.** Oblique section through the epicuticle of a *Tenebrio* larva. The cuticulin is cut obliquely and shows the black dots of the lipid-filled wax canals even where it is most thin. Internal to the cuticulin many middle-phase lipid-water liquid crystals ramify in the homogeneous inner epicuticle.

problem would be unimportant were it not for the fact that the epicuticle determines the pattern, the surface properties, and most of the impermeability of the integument; its structure therefore impinges upon insect biology in a most direct way.

The traditionally accepted structure of the epicuticle is one of four layers; cement wax, polyphenol, and cuticulin, respectively, from the outside to the inside. The electron microscope confirms that cement and wax are usually present. The cuticulin is a composite structure, cuticulin *sensu strictu* and a dense homogeneous layer, the outer part of which probably contains, or has contained, polyphenols. According to Dennell and Malek (1955a, b) and Malek (1958), the protein epicuticle in *Schistocerca* and *Periplaneta* consists of two layers, paraffin and cuticulin, distinguished by their staining although not consistently coloured indifferent species. It is not possible to make a firm parallel, between these findings and those of the electron microscope but their two layers probably correspond to regions of the dense homogeneous layer which may be chemically different, perhaps with more or less polyphenol and hydrocarbons, and which also differ in chemistry and structure between different insects. As far as present evidence allows we can say that the whole insect is covered by the cuticulin layer *sensu strictu* except for some gaps in sensory structures. Most of it, including trachea but excluding tracheoles, parts of the lining of some glands and sense organs is also covered by the dense homogeneous layer. Cement is present where dermal glands occur and wax is universal (although wax canals have not yet been observed in tracheae there is evidence for the presence of a lipid lining).

### *1. The Cement Layer*

On the outside of many cuticles is a layer of cement secreted by the dermal glands, also called *Verson's glands* in Lepidoptera. These glands have been described by Kramer and Wigglesworth (1950), Malek (1958), and Wigglesworth (1953c). The nature and function of the cement probably varies in different insects but a clue to its general composition may come from the lac insect, *Laccifer lacca*, in which the cement is probably the shellac of commerce. Shellac is a mixture of laccose and lipids. The cement may also be partly tanned protein, but we shall probably not be far wrong if we think of it functionally as a mixture something like varnish. It varies greatly in thickness and extent. It is thickest in the exposed cuticles of beetles and absent in many adults covered with scales although it is present in some, e.g., *Thermobia*. In *Rhodnius* it covers most of the surface but on the dorsal abdominal tergites of the adults it is incomplete under the wings, forming flat flowerlike plugs to the mouths of the glands. In caterpillars like *Diataraxia* the cement occupies the valley between the tubercles.

The outer surface of many insects has a fuzzy indistinct outline as seen in section with the electron microscope. The layer in much less distinct than the secretion of the dermal glands and is perhaps preedysial in origin resulting from the products of digestion of the old cuticle. Like the cement it would be important in determining the surface properties.

There are several possible functions for the cement. In some insects it is certainly protective. The delicate patterning of the underlying layers is given a complete coat of hard transparent varnish. This is impossible in cuticles which expands during an instar, and in caterpillars and blood-sucking larvae it is a meshwork which must distort during growth. In these insects the cement might serve as a reservoir of lipids which could leak out to replace lost surface lipids. This role as a sponge to soak up mobile lipids is probably the main one in cockroaches, which are covered with very low melting point waxes. If there is a reserve of mobile wax in the cement it could play an important role in rapidly sealing over surface abrasions to prevent water loss. Only very thin layers of lipid are necessary to prevent water loss, but such films are very vulnerable to absorption and mechanical rupture.

Wigglesworth's experiments (1947) show that the wax layer is overlain by the cement; the protection of the wax may be its most important function. In all small insects the high contact angle presented by a waxy surface is an essential safeguard against being trapped below a film of water. The surface of some large insects, on the other hand, is hydrophilic. The larva of *Calpodes*, for example prefers to live in a private bath of water within a folded leaf. The wax surface may be masked by a hydrophilic cement. In *Thermobia* the cement is highly hydrophobic and also repels most organic solvents.

### 2. *The Wax Layers*

The chemistry of cuticular waxes is well reviewed in Gilmour (1960) and will not be elaborated here. They are usually mixtures of hydrocarbons with odd and even numbers of carbon atoms in the range $C_{25}-C_{31}$, and esters of even-numbered fatty acids and alcohols in the range $C_{24}-C_{34}$. A recent and fairly complete analysis of the wax of the Mormon Cricket, *Anabrus simplx*, is given in Table II. The hydrocarbons are not unlike those in plant waxes in which the composition is sufficiently constant for an analysis to be useful in the systematic separation of closely related species. It would be interesting to know if the composition of insect waxes is also species specific. Koidsumi (1957) has recorded that fatty acids, particularly caprylic or caproic acids, are important in the resistance of *Bombyx* and *Chilo* to attach by fungi.

**Table 9.1**
**The Composition of Mormon Cricket Wax, *Anabrush Simplex***

| *Chemical class* | *Percent of wax* | *Individual component* | *Percent of class* | *Percent of wax* |
|---|---|---|---|---|
| Hydrocarbons | 48–58 | $C_{12}$, $C_{13}$, $C_{14}$, $C_{15}$, $C_{16}$, | 12.5 | 6–7 |
| | | $C_{17}$, $C_{18}$, $C_{19}$, $C_{20}$, $C_{21}$, | Equally | |
| | | $C_{22}$, $C_{23}$, $C_{24}$, $C_{25}$, $C_{26}$, | distributed | |
| | | $C_{27}-C_{28}$ | 20.6 | 10–12 |
| | | $C_{20}-C_{31}$ | 55.3 | 27–32 |
| | | $C_{32}$ ? | 11.4 | 5–7 |
| Free acids | 15–18 | $C_{12}$ | 0.4 | 1 |
| | | $C_{14}$ | 1.7 | 0.4 |
| | | $C_{14}$ – 1 double bond | 0.5 | 0.1 |
| | | $C_{14}$ – 2 double bonds | 0.3 | 0.1 |
| | | $C_{14}$ – 3 double bonds | 0.6 | 0.1 |
| | | $C_{16}$ | 7.9 | 1–1.4 |
| | | $C_{16}$ – 1 double bond | 4.0 | 0.6 |
| | | $C_{16}$ – 2 double bonds | 0.3 | 0.1 |
| | | $C_{16}$ – 3 double bonds | 0.7 | 0.1 |
| | | $C_{18}$ | 2.4 | 0.4 |
| | | $C_{18}$ – 1 double bond | 48.4 | 7–9 |
| | | $C_{18}$ – 2 double bonds | 26.8 | 4–5 |
| | | $C_{18}$ – 3 double bonds | 6.0 | 1 |
| | | $C_{20}$ | Trace | Trace |

| | | | | |
|---|---|---|---|---|
| Esters | 9–11 | Saturated | 34 | 3.2–3.9 |
| | | Saturated and unsaturated | 28 | 2.7–3.3 |
| | | Saturated and/or hydroxy | 15 | 1.4–1.7 |
| | | Unsaturated and/or hydroxy | 23 | 2.2–2.7 |
| Free alcohols | 2–3 | Cholesterol | | 2–3 |
| Polymers | 12–14 | Acidic resins of varying molecular size | | |
| Unidentified | 2–4 | | | 2–4 |

Three regions may be recognized in the wax at the surface of insects. According to Beament (1945, 1961) there is a monolayer of lipid next to the cuticle. This is sometimes visible in electron micrographs. Most of the lipid is less well ordered and forms a much thicker layer, often permeating the cement. On the very outside in many insects, the wax forms a bloom comparable to that described on plant surfaces by Juniper (1960).

The evidence for the presence of a monolayer at the surface is reviewed by Beament (1961). The most telling in evidence is perhaps the similarity in transition of water permeability between artificial

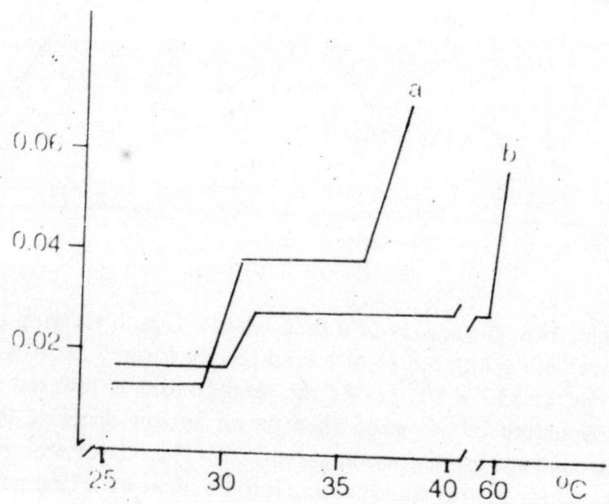


**Fig. 9.10.** The permeability of the cuticle to war. Graph showing changes in the permeability/temperature relationship with age of pupae of *Pieris*; (a) killed 24 hours after pupation; (b) killed 5 days after pupation. There are two discontinuities in permeability, the upper one varying with age.

monolayers on membranes and insect cuticles. In some insects *Rhodnius, Tenebrio, Pieris*; Beament, 1959) there are two critical temperatures for water loss. He argued from this that there might be two ordered layers forming the main barrier to water loss, one at the surface of the cement and one on the outer face of the cuticulin. Recent, still unpublished, work by Thompson (1964) on the changes in electrical resistance of lipid double layers with temperature, suggests an alternative explanation of Beament's observation. Thompson showed that the electrical resistance of an artificially prepared lipid double layer goes through two maxima and three minima in the temperature range from 19° to 36°C. It seems probable that the monolayer of lipid at the surface of the cuticle postulated by Beament could go through similar changes to give the two abrupt changes in water permeability which he found.

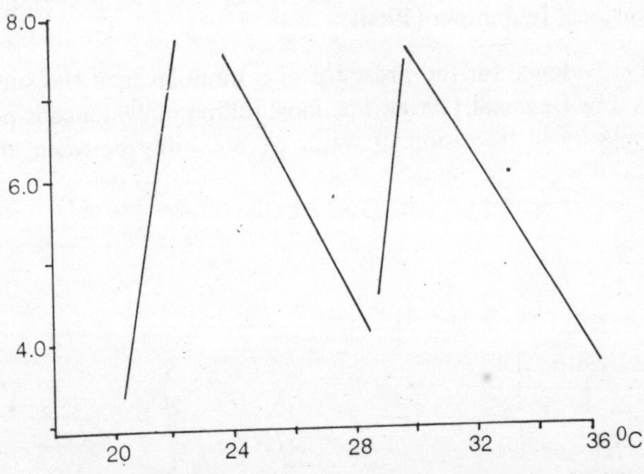


**Fig. 9.11.** The electrical properties of a lipid bilayer. Graph showing the changes in specific resistance with temperature of a lipid bilayer formed at 36° C. At 36°C the resistance per cm$^2$ is 33.7 x 10$^6$ Ω. As the temperature is lowered the resistance increases linearly united 29° C when there is an abrupt decrease followed by an increase and a second abrupt decease at about 22°C. The bilayer was created between two aqueous phases for the resistance across it to be measured. Compare the changes in resistance of a lipid bilayer with the permeability of an insect cuticle to water.

The monolayer is probably not static but is in a liquid phase being moved, lost, and replaced, C.T.Leewis (1962) used an autoradiographic technique to trace the spread of a radioiodine preparation of diiodooctadecane from the tarsi of adult *Phormia* over the body. found that oil films were established over the whole of the integument within a few minutes of exposure. Five to fifteen minutes was enough to coat even the wigs with a monomolecular film in a living fly.

The bulk of the lipid at the surface of insects, enough for a layer 1/10 to 1 $\mu$m thick, is probably in the less well-oriented layers permeating the cement. Juniper (1960) was able to study the waxy bloom at the surface of many plants by viewing direct carbon replicas in the electron microscope. There has not yet been any comparable study in insects, although Holdgate and Seal (1956) give a picture of the surface of a *Tenebrio* pupa taken with an electron microscope adapted for reflexion microscopy. In these examples, although the wax has a very varied and characteristic form, it never- theless arises from

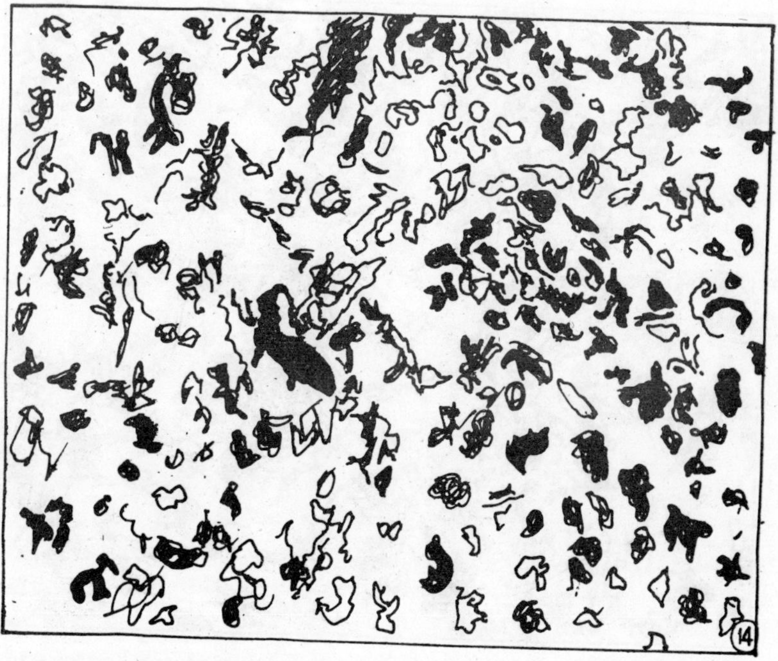

**Fig. 9.12.** Carbon replica of the early phases of formation of a wax bloom on the surface of a pupa of *Calpodes ethlius*. At a later stage, the surface is completely covered with extrusions.

a smooth surface of cuticle. The pattern of wax must therefore result from the distribution of wax canals through the epicuticle, the nature of the wax, and the conditions under which it is deposited. In other insects the surface of the cuticle is elaborately patterned to ensure a characteristic wax form. For example, the larva of *Calpodes ethlius* has crater-shaped tubercles with wax canals situated around the rims causing the wax to emerge in the form of cylinders. The molecules in the projections forming the blooms may be highly ordered. The wax cylinders from *Calpodes* larvae and the blooms on *Tenebrio* and *Calpodes* pupae are all strongly birefringent.

Although these wax blooms may look very different, they all have on feature in common. When projecting from the surface, and even more when broken up, they give wax in the form of thin sheets or

**Fig. 9.13.** Carbon replica of the wax bloom on the surface of a pupa of *Tenebrio* at an early stage. The bloom is mainly in the form of crystalline flakes but there are some pieces (arrowed) which appear almost as if they have been extruded like toothpaste from a tube.

plates having the maximum surface for volume of wax. The surface of the insects thus has a high contact angle for water which tends to roll up into spheres coated with wax fragments. The function of the specialized wax secretion in *Calpodes* is an enigma. It may beconcerned in the conservation of water which subsequently evaporates and cools the pupa. The wax filaments are broken up around the flimsy cocoon inside a folded leaf. At night, rain and dew collect in spheres within the leaf and evaporate in the heat of the day. Without wax, the water forms a thin film and drips from the coccoon.

### 3. Wax Precursors and the Liquid Crystalline Phases of Lipid-Water Systems

Elongate filamentous structures 60 to 130 A in diameter have been found in the cuticle of all insects examined with the electron microscope. They occur in abundance between the cells and the cuticle in the wax-secreting integument of the honey bee, in the pore canals of *Rhodnius, Galleria*, and *Tenebrio*, and penetrate the epicuticle in all species. The filaments appear more or less solid after osmium fixation and tubular after permanganate. They are similar in appearance to the liquid-crystalline phases of lipid-water systems.

Luzzati and Husson used X-ray scattering techniques to deduce the structure of liquid-crystalline phases of lipid-water systems, varying both the temperature and the proportion of water. They found that a number of soaps went through a sequence of phases – neat, cubic, complex hexagonal, rectangular, deformed middle, and middle – as the temperature was lowered and the proportion of water increased. The neat phase is an alternate sequence of planar layers of lipid and water. The cubic phase is isotropic and the structure is one of spheres closely packed in a cubic lattice. In the complex hexagonal phase, the lipids form a double-layered cylindrical shell, water being both outside and inside the cylinder. The rectangular and deformed middle phases are derived from the middle phase in which the lipids are in an array of single-layered cylinders of indefinite length the hydrocarbon chains filling the interior of the cylinders with the hydrophilic groups an water outside. The cylinders form a hexagonal array. Stoeckenius confirmed these findings with the electron microscope by direct observation of osmium-fixed soaps and phospholipids.

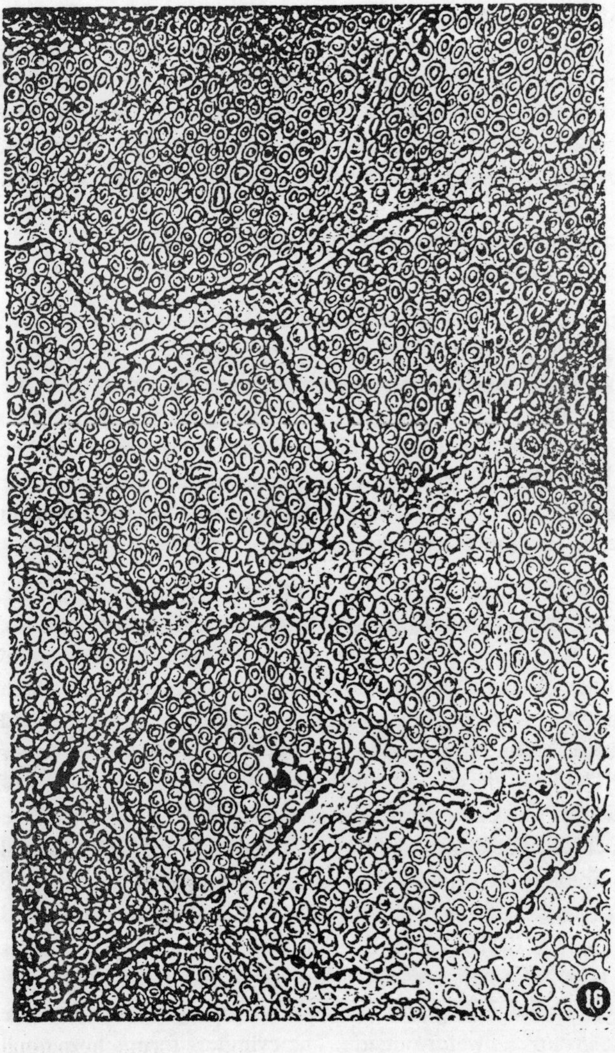

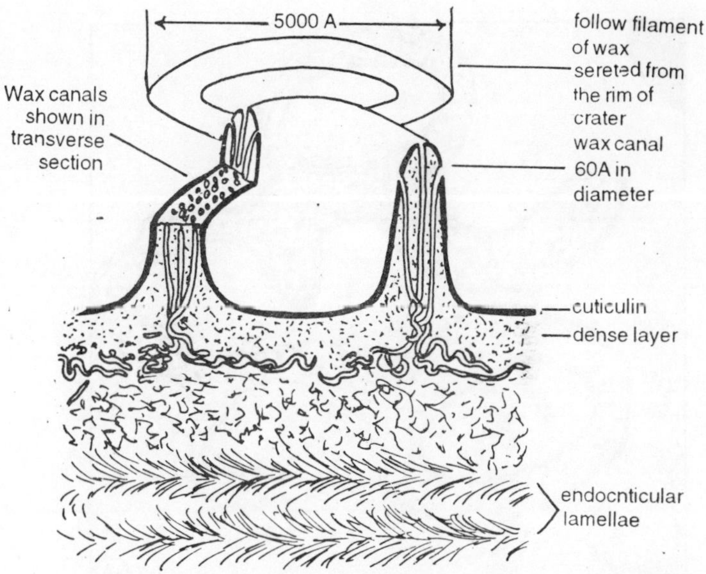


**Fig. 9.15.** The structure of the epicuticle in *Calpodes ethlius* larvae in the region where extra wax is secreted. The epicuticle is traveled by filamentous structures in hexagonal array believed to be lipid- water liquid crystals in the middle phase. These emerge around the rim and crater-shaped tubercles. The wax is secreted in the form of a cylinder from the rim of each crater.

The waxes of insects have longer hydrocarbon chains than the lipids studied above, and the liquid crystals are correspondingly larger. Even so in *Calpodes* the filaments are little more than 60 A in diameter, which would correspond to two hydrocarbon chains 20 to 30 A long arranged radially in the cylinders of the middle phases. In the epicuticle forming the wall of each crater-shaped tubercle there are 150 to 300 cylinders in a perfect hexagonal array. In *Galleria* and *Tenebrio* the filaments are a little larger but probably still corresponds

**Fig. 9.14.** Carbon replica of the surface of the cuticle which secretes extra wax in the larva of *Calpodes ethlius*. The surface is covered with tiny crater-shaped tubercles. Wax canals pass through the walls of each crater in hexagonal array. From the rim, a cylindrical filament of wax emerges. The polygonal grouping of the craters follows the pattern of the epidermal cells which secreted the epicuticle.

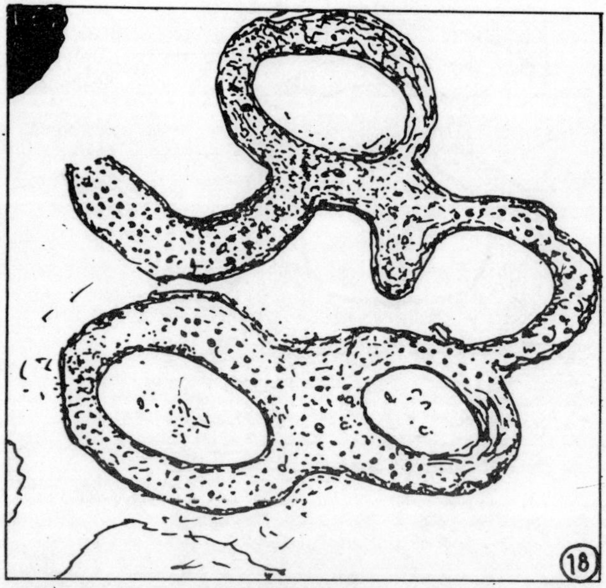

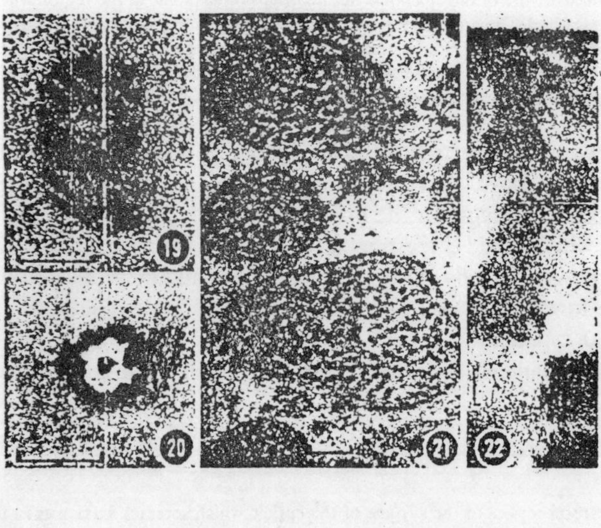

to the cylinders of the middle phase. In honey bee cuticle, however, there are cylinders almost 200 A in diameter which are probably in the complex hexagonal phase.

On the basis of structure along there is, therefore, a strong argument for supposing that the filamentous structures seen in cuticle are lipid-water liquid crystals. There is also the strongest circumstantial evidence. The presumed lipid-water liquid crystals are most abundant in regions where lipids would be expected, such as the cuticle where wax is secreted in the honey bee and, in *Calpodes*, the wax is secreted in a shape which follows exactly the pattern of the openings of the wax canals through the epicuticle to the exterior. It was for this reason that the liquid crystals were first called wax canal filaments, and the space they occupy in the epicuticle was referred to as a wax canal since it provides the route for the transport of wax to the surface.

There is no direct evidence (but it seems most probable) that the layer forming the cylinder of the liquid crystal is continuous with the surface monolayer. The figure suggests that the motive force driving lipids to the surface may be surface tension. On the free surface of the cuticulin, liquid lipids would be at a lower energy level than when constrained in the narrow cylindrical micelles; they would also flow out if lipids were to be continually drained from the surface, either by adsorption to the cement or by crystallization to form the blooms.

---

**Figs. 9.16—9.20.** Fig. 9.16, transverse section through several of the crater-shaped tubercles which secrete wax in *Calpodes ethlius* larvae. The hexagonal array of dots in the walls of each crater are believed to be sections of lipid-water liquid crystals in the middle phase. They are about 60 A in diameter. Fig. 17, tangenital section of *Tenebrio* larval cuticle about 3/4 of the way through the cuticle from the cells to show one pore canal. The pore canal contains numerous filaments believed to be lipid-water liquid crystals in the middle phase. Fig. 9.18, tangenital section of *Tenebrio* larval cuticle just below the epicuticle showing a single pore canal with an axial filament and lipid-water filaments arranged in a ring at the periphery. Fig. 19, transverse section of bundles of lipid-water liquid crystals enclosed by the plasma membrane of an epidermal cell below the wax-secreting cuticle in a honey bee. The filaments appear superficially to be intracellular but they are in finger like infoldings of the plasma membrane. They may form a layer below the cuticle before passing through the pore canals. Fig. 9.20, transverse section of the wax-secreting cuticle of the honey bee near the surface showing numerous filamentous structure about 200 A in diameter which may be lipid-water liquid crystals in the complex hexagonal phase.

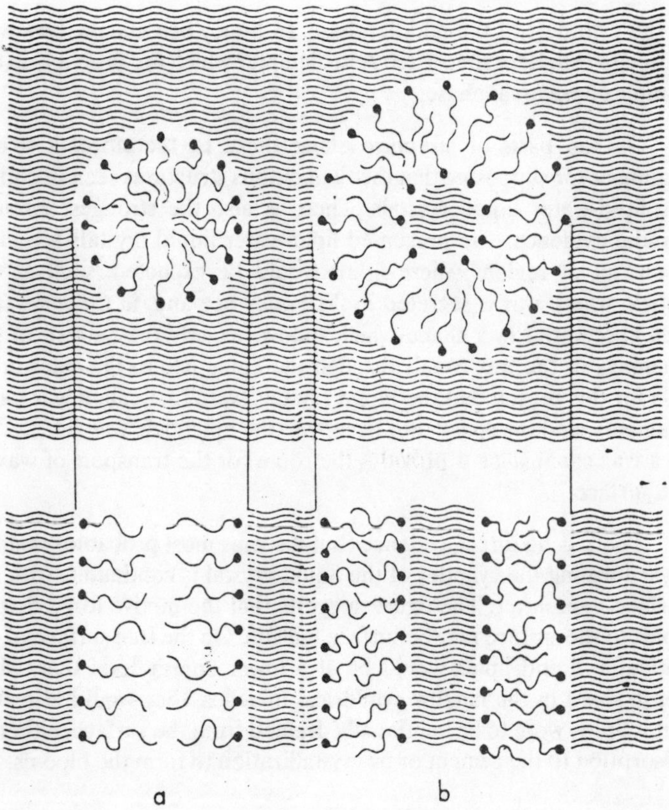


**Fig. 9.21.** Diagram of the structure of two phases of lipid-water liquid crystal. Above, transverse; below, longitudinal section. (a) The middle phase, in which the lipids form cylinders of indefinite length. Hydrocarbon chains fill the interior of the cylinders; hydrophilic groups are at the surface and water is outside; (b) the complex hexagonal phase, in which the cylinders are double-walled, with hydrophilic groups and water both inside and outside. Both phases tend to be in hexagonal arrays.

In addition to providing a route to the surface for waxes, the wax canals are probably also the route by which oil-soluble compounds enter an insect. The topical application of compounds with juvenile hormone activity is one of the most certain methods of inducting a systemic effect in *Rhodnius* (Wigglesworth, 1963a). The wax canals are presumably the port of entry for oil- soluble insecticides. Liquid-crystalline phases can incorporate large amounts of lipid-soluble

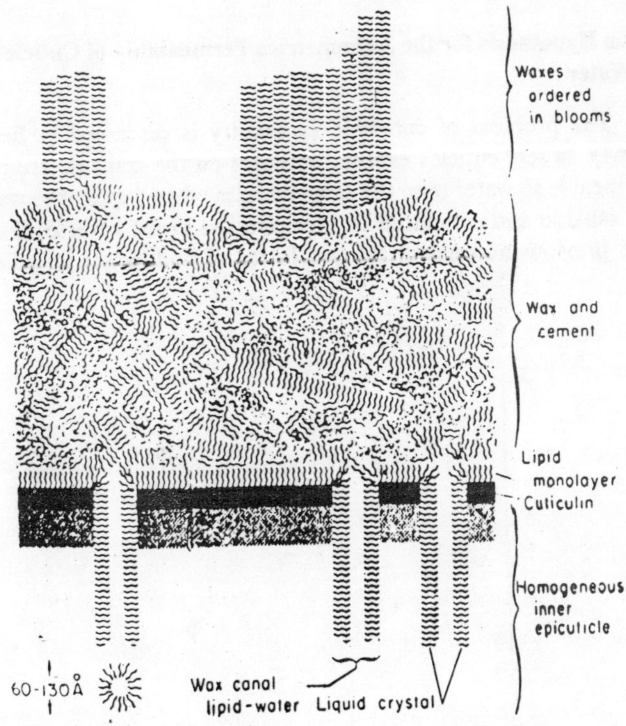


**Fig. 9.22.** The structure of the surface of an insect. Not all insects have al these structures. The cement is frequently lacking and wax blooms would then form directly over the wax canals. In contrast to the crystalline wax blooms, the inner waxes may be liquid and mobile. There is no direct evidence that the lipids in the wax canals are continuous with the surface monolayer, but it seems most probable. The lipids in the wax canals may not always be in the middle phase as drawn here. Complex hexagonal and cubic phases have also been seen. The homogeneous inner epicuticle is frequently denser next to the cuticulin as shown, perhaps owing to its content of osmiophilic polyphenols.

substances. Liquid crystals in the endocuticle and even adjacent to the epidermis might be expected to come into rapid equilibrium with lipid contaminants of the surface monolayer if the lipids do form a continuous liquid system.

**4. An Hypothesis for the Asymmetrica Permeability of Cuticle to Water**

The problem of cuticular asymmetry is discussed in Beament (1961). In sect cuticles exposed to water on the cell side are far less permeable to water than the same cuticles when exposed to water on the outside and submitted to transpiration in the reverse direction. The previous account of the two forms of lipid-water liquid crystal

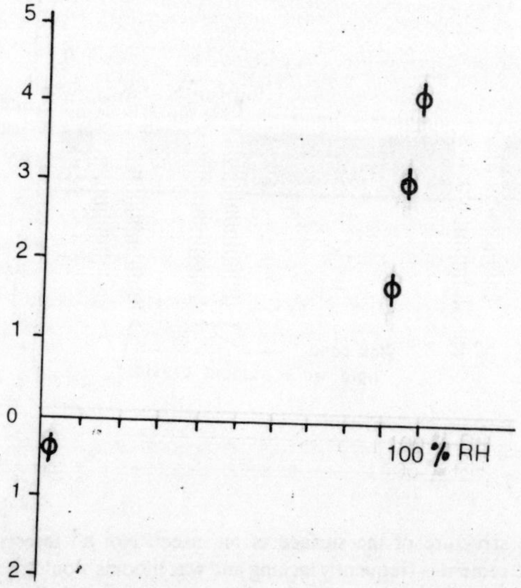


Fig. 9.23. The gain and loss of water by live *Tenebrio* larvae at different humidities at 30°C. Each point is the mean of about 25 animals. The cuticle behaves in a markedly a symmetrical manner. Even though the pressure drop into the animal cannot be as great as the pressure drop from body fluids to 0% RH externally, the passage of water into the animals in the direction epicuticle-endocuticle is much greater than the passage of water out in the direction endocuticle- epicuticle. Ordinate, rate or gain or loss of water in $mg/cm^2$/hour; abscissa, relative humidity.

suggests an hypothesis to account for this asymmetry. If both phases can exist in the epicuticle at the surface of the cuticulin we should expect high permeability to water with the complex hexagonal and very low water permeability with the middle phase. If the presence of water at the surface can induce the change, the observed asymmetry would be accounted for. An objection might be that the dimensions of the spaces occupied by the liquid crystals in the epicuticle are constant, and would not permit such a change in the contained water and lipid. The change from complex hexagonal to the middle phase need only take place over a very short length of the cylinder to reduce permeability markedly. It is difficult to determine what the dimensions of the wax canals may be since we only infer their presence from the lipid they contain. Fortunately, this hypothesis can be tested by electron microscope observations, although with some difficulty.

### 5. Cuticulin and the Protein Epicuticle

In early studies on the epicuticle Wigglesworth observed the formation of oil droplets when he treated it with Schultze's reagent. He deduced that it must be a lipoprotein and called it cuticulin. Since it turns out that this structure has several components, the term now lacks precise meaning. It has also been used in several different senses by other authors. It is proposed here to use cuticulin into mean the dense outermost layer of the protein epicuticle which is universally present and has about the thickness of a cell membrane (or a little more). It composition is unknown but it has been found to cover the entire surface of insects from tracheoles to gland ducts, hairs and scales. The only thing known of its structure is that it appears as thin line in electron micrographs. If the lining of tracheoles is the same, as it appears to be, then it may originate as a lipid double layer. In newly formed tracheoles the lining is about 85 Å thick and can be resolved as twin dense lines. In later tracheoles only a single dense line is resolvable. It is perhaps of interest that very thin membranes form spontaneously at the surface of injured cells in a number of animals. In these examples it seems that a lipid double layer could be protected by proteins in solution which are later stabilized by quinones produced in wounding. This would be a useful first hypothesis to test for the structure and formation of the cuticulin layer.

The cuticulin is perhaps the most important layer in the cuticle. Its functions are listed below.

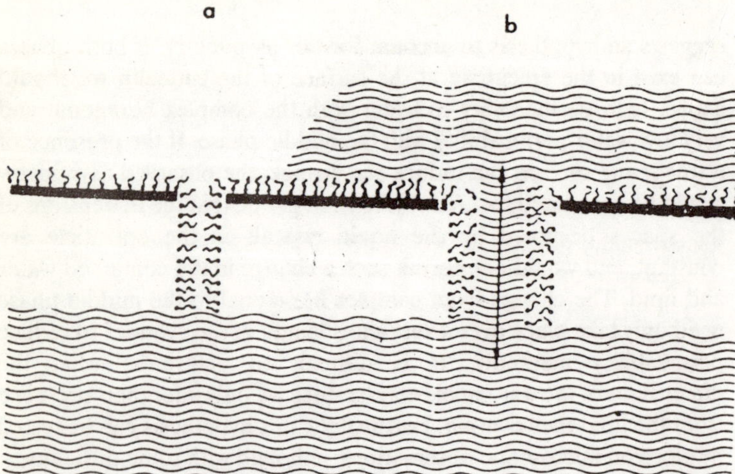


Fig. 9.24. Diagram to explain a possible effect of a phase change of the lipids within the wax canals upon the permeability of cuticle to water. A wax canal occupied by a lipid-water liquid crystal (a) in the middle phase and (b) in the complex hexagonal phase. If water or a high humidity in the environment can induce this phase change, then the apparent asymmetry in permeability to water would be accounted for.

### (a) Permeability Barrier

During its existence it has to be permeable to the factor activating the molting gel and the products of hydrolysis of the old cuticle, but impermeable to the molting fluid and resistant to digestion by it. It is permeable to waxe and under some circumstances, as when an insect is immersed in oil, it is permeable to water. In those insects which actively take up water it also has to be permeable to water passing into the insect.

### (b) The Limitation of Growth

The those insects such as blood-sucking bugs and caterpillars which grow within a stadium by expansion of the integument, the size achieved at the end of the stadium is limited by the dimensions of the epicuticle laid down on the surface of the epidermis before ecdysis. It is for this reason that in these insects the newly molted surface is deeply pleated.

### *(c)* The Surface Pattern

The elaborate sculpturing of the surface, the formation of hairs, scales, and microtrichia, and the elegant patterning of tracheal and tracheoles all take place at an epidermal cell surface at the time of formation of the cuticulin layer. It is questionable which is the prime mover, the plasma membrane assuming a shape for the deposition of the cuticulin, or the cuticulin responding to forces which carry the plasma membrane along with it.

### *(d)* Surface Properties

If Beament is corect in supposing that there is a lipid monolayer in the epicuticle, the cuticulin provides the surface to support it. If the waxes are polar, then the surface of the menolayer could be lipophilic or lipophobic depending upon the properties of the cuticulin and the orientation of the lipid molecules.

Immediately below the cuticulin in most cuticles is a much wider layer of homogeneous, rather dense cuticle which is the refractile line called the epicuticle in sections with light microscopy. Nothing is known of its composition apart from Wigglesworth's early work. Just before the secretion of wax Wigglesworth observed the progressive spread of argentaffin- positive material from the tips of the pore canals until it eventually covered the surface. He supposed that there was a discrete layer of polyphenols on the outside of the protein epicuticle, stabilizing both it and the underlying endocuticle. Although there is no discrete layer to be seen with the electron microscope, the region just below the cuticulin is more electron dense in osmium-fixed preparations, which would be expected if it contained polyphenols tending to bind osmium. The pore canals stop below this homogeneous dense layer in *Galleria* and *Tenebrio* but it is penetrated by the wax canals which in *Calpodes* are much wider on the inner edge than the diameter of a middle phase lipid-water liquid crystal.

## The Ecdysial Membrane

Passoneau and Williams (1963) first described a membrane between the pupa and adult cuticles of developing *Ceropia* which they called the *"ecdysial membrane."* Since then, the ecdysial membrane has been studied in *Schistocerca* by Malek (1956, 1958, 1959), in *Rhodnius* by Locke (1958a), in Lepidoptera by Richards (1955, 1958a)

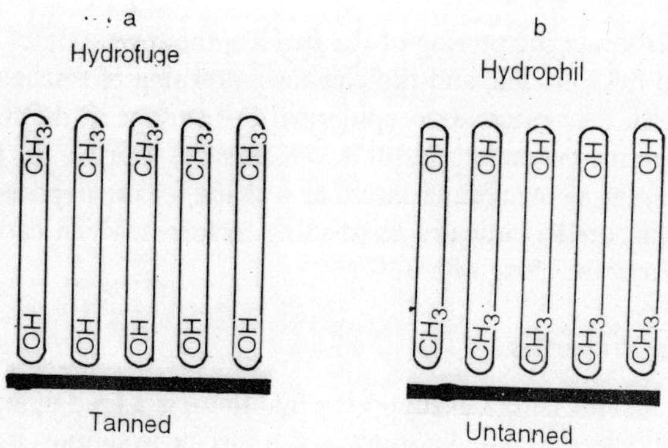


**Fig. 9.25.** Diagram showing how the condition of the cuticulin might affect the surface properties of cuticle through the orientation of the surface lipids.(a) Hydrofuge outer surface produced by mounting a normal monolayer of long-chain alcohol on a tanned cuticulin substrate; (b) hydrofoil outer surface produced by mounting a reverse monolayer on an untanned cuticulin substrate.

and Lower (1957a,b), and in *Sialis* by Selman (1960). In his earlier work on molting in *Rhodnius*, Wigglesworth figures a layer probably corresponding to the ecdysial membrane, but he included it with the debris of endocuticle attacked by molting fluid. In *Schistocerca*, the ecdysial membrane is formed from the inner few lamellae of the endocuticle which are resistant to the action of the molting fluid as well as to concentrated mineral acids. These laminae separate from the old cuticle when it is dissolved by the molting fluid and give rise to a discrete membrane which persists throughout the premolt period eventually being cast as the inner lining of the larval exuvium.

In *Calpodes* larvae the epidermal cells withdraw from the endocuticle and secrete numerous dense osmiophilic droplets, some of which run together to form a membrane on the inner face of the endocuticle. At a slightly later stage both droplets and complete membrane can be seen. Similar events take place during the formation of the tracheal cuticle in *Rhodnius*. There is a distinct stage after the deposition of the cuticulin but before the activation of the molting fluid when the membrane is a reticulum of coalescing droplets. The membrane, or the remains of it, is shed with the exuvium at ecdysis.

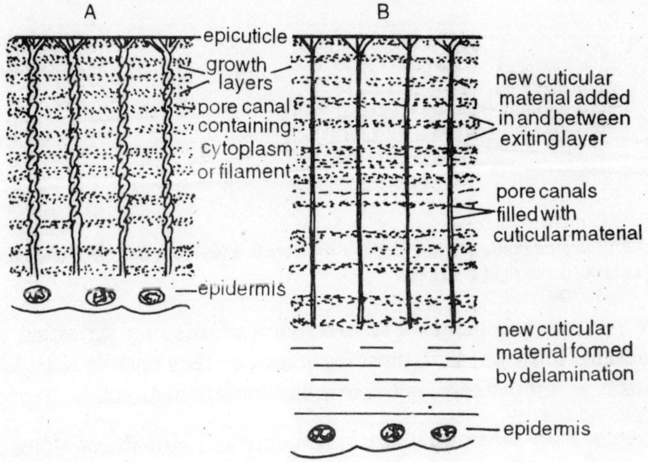


**Fig. 9.26.** A. Diagrammatic section through the integument of *Sarcophaga* larva soon after molting. B. The integument of the mature larva.

## Modifications of Cuticle

In order to give flexibility to the cuticle only part of it contain fully differentiated exocuticle. These parts are called sclerites and they are joined by regions in which the cuticle remains membranous. The extent of the membrane and the method of articulation of the two adjacent sclerites determines the degree of movement which can occur at the joint. Sometimes, as between abdominal segments, the membrane is extensive and there is no point of contact between adjacent sclerites so that movement is unrestricted. More usually the sclerites make contact with each other to form true articulations and the joints are called monocondylic or dicondylic depending on whether there are one or two points of articulation. Monocondylic articulations, such as that of the antenna with the head, permit considerable freedom of movement, whereas dicondylic articulations, which occur at many of the leg joints, give more limited, but much

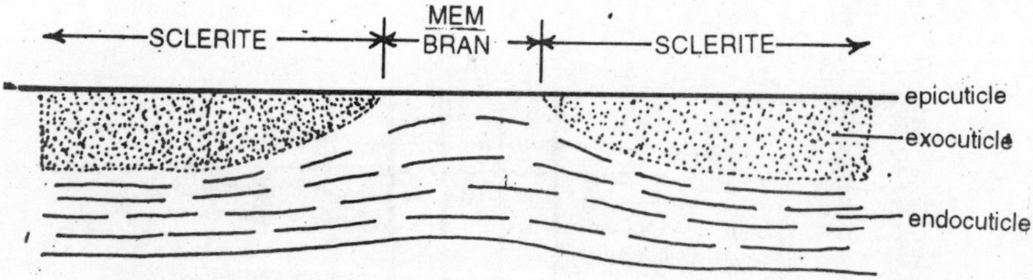


**Fig. 9.27.** Diagrammatic section through the cuticle showing a flexible membranous region between two rigid sclerites.

more precise movements. The articular surfaces may lie within the membrane, intrinsic, as in most leg joints, or they may lie outside it, extrinsic, as, for instance, in the mandibular articulations.

Apart from membranous area, exocuticle is also absent along the ecdysial lines of larval hemimetabolous insects. The cuticle along these lines consists only of undifferentiated procuticle and, usually,

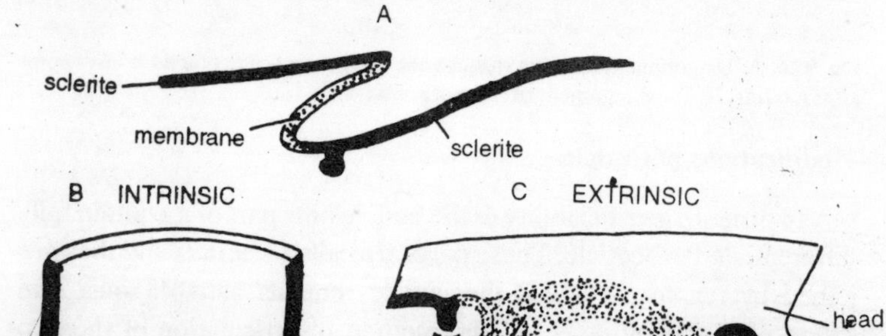


**Fig. 9.28.** Diagrams of different types of joints between sclerotised areas. A. Intersegmental. Extensive membrane with no articulation between sclerites. B. Dicondylic leg joint with intrinsic articulations, C. Dicondylic articulation of mandile with head capsule. Extrinsic aritculations. Edges of sclerotized regions shown black, membrane hatched.

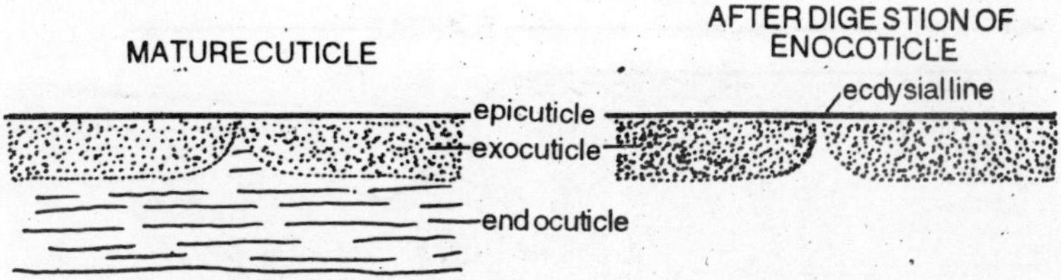


**Fig. 9.29.** Diagrammatic section of the cuticle transverse to an ecdysial line. Moulting fluid digests the endocuticle leaving the cuticle held together only by the epicuticle, producing a line of weakness.

epicuticle so that they constitute lines of weakness along which the cuticle splits at ecdysis.

In many larval holometabolous insects the greater part of the cuticle remains undifferentiated. This facilitates growth, since undifferentiated cuticle is more extensible than sclerotised cuticle, it also enables some larvae to crawl in a manner quite unlike the adults, but well-suited to some environments, and it is important for the conservation of materials. The bulk of the undifferentiated cuticle is digested and reabsorbed at moulting, while sclerotised parts are lost, hence an unsclerotised cuticle is more economic for a larval form

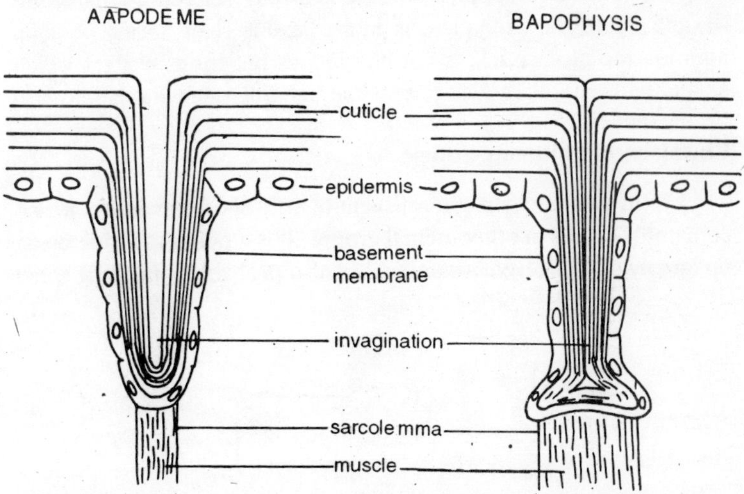


**Fig. 9.30.** Diagram of cuticular invaginations forming (A) an apodeme and (B) an apophysis.

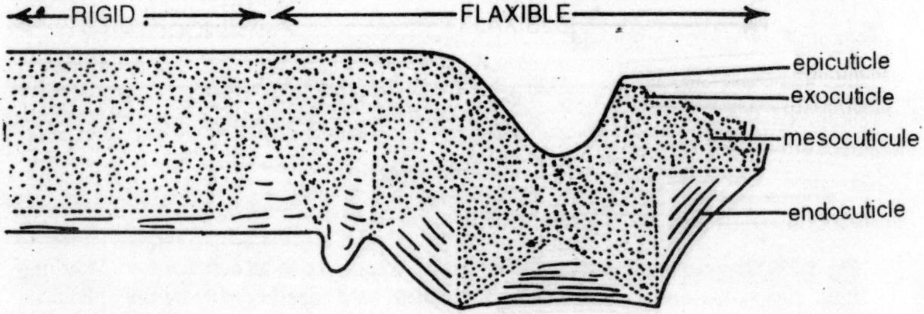


**Fig. 9.31.** Diagrammatic section of the flexible cuticle found in lepidopteran wing bases.

which moults several times. Collembola moult continuously even when they are adult and, in keeping with this, their cuticles contains very little exocuticle. In addition, they eat their exuviae so that they conserve as much as possible form the cuticle. Invaginations and inflexions of the cuticle form the endophragmal skeleton. Hollow invaginations for muscle attachments are called apodemes, solid structures are apophyses, but, in addition to muscle attachment, the endophragmal skeleton tentorium in the head and invaginations which are sometimes present covering the thoracic ganglia. In the wing bases of Lepidoptera some apparently sclerotised parts are flexible. Here the exocuticle is more flexible than normal, but, in addition the mesocuticle is not continuous but forms wedges which extend into the soft endocuticle so that bending is facilitated.

## Chemistry of Chitinous Cuticle

Chitin is an important constituent of arthropod cuticle, but it also commonly occurs in other animal groups. It is a polysaccharide made up largely of N-acetyglucosamine, but also probably containing some

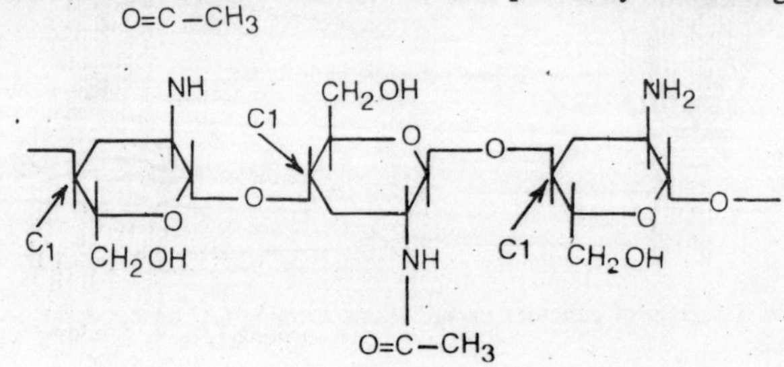


Acetylglucosamine Acetylglucosamine Glucosamine

glucosamine. The sugar residues are linked by 1 – 4 $\beta$ linkages so that they form a chain in which all the residues are orientated in the same direction (e.g., in the following diagram C I is always at the left-hand end of the sugar residue). Adjacent chitin chains are held together by hydrogen bonds to form micelles or micro-fibres, and hydrogen bonds probably also link the oxygen atoms of adjacent acetylglucosamine residues. Neighbouring chains run in opposite direction (i.e., with the C I atoms at opposite ends of the molecules) and the suggested linkages are as follows:

Chitin constitutes 25 – 50% of the dry weight of the cuticles; the remainder is largely protein. Chitin is always associated with protein and in the cuticle the two are linked by covalent bonds to form a glycoprotein. The protein of undifferentiated cuticle is water-soluble and is called 'arthopodin', but it is not homogeneous, consisting of at least three major components with differing amino acid composition. Sulphur-containing amino aids are absent. Part of the protein may later be tanned, stablilised by cross-linkages between the molecules, to from a hard, inflexible and usually darkened structure. Such tanned arthopodin is called sclerotin, and this produces the hardness of the sclerites. A cuticulin-like lipid is present throughout the refractile layers of the cuticle, in the pore canals and between the lamellae of the endocuticle.

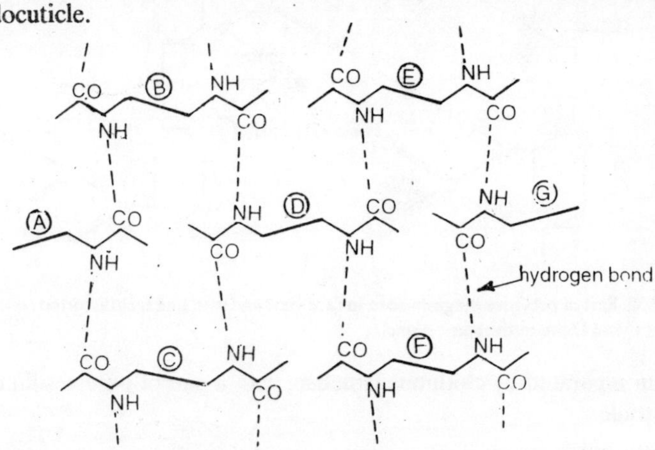


Fig. 9.32. Part of a chitin chain. It is suggested that every sixth or seventh residue is a glucosamine residue. Thick lines represent a chain of acetylglucosamine residues seen end on. Chain A, D and G run is one direction, chains B, C, E and F in the opposite direction.

**Resilin**

Some parts of the cuticle contain a colourless, rubber-like protein called resilin and such cuticle is called rubber-like cuticle. It is found in elastic hinges such as the wing-hinge ligament of *Schistocerca* (orthoptera) which lies between the pleural process and the second axillary sclerite. The ligament is sharply differentiated from adjacent sclerotised cuticle and consists of a ventral region of tough, dense chitin and fibrous protein, and a dorsal region containing layers of

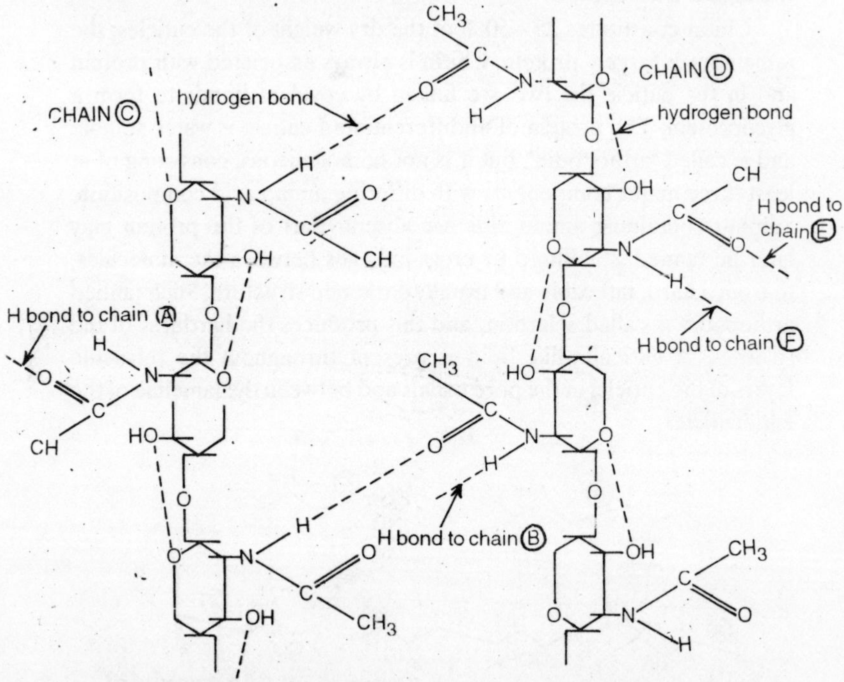


**Fig. 9.33.** Part of previous diagram seen in face view and with linkages distorted so that chains C and D are in the same plane.

resilin separated by chitinous lamellae, with a pad of pure resilin on the inside.

Resilin contains various amino acids including two previously unknown ones which provide the links between the protein chains. These linkage are produced continuously as the resilin is laid down and so the process contrasts with the linking of protein in sclerotisation which is limited to a brief period and occurs some time after the protein is laid down. The amino acid sequence in the protein is such as to prevent other cross-links occurring, since these would impair the rubber-like properties. Like rubber, resilin can be stretched under tension and stores the energy involved so that when the tension is released it returns immediately to its original length. In the locust between one quarter and one third of the recoil energy of the wing away from its equilibrium position is due to the elasticity of the wing-hinge ligament. Elsewhere resilin is found in the clypeo-

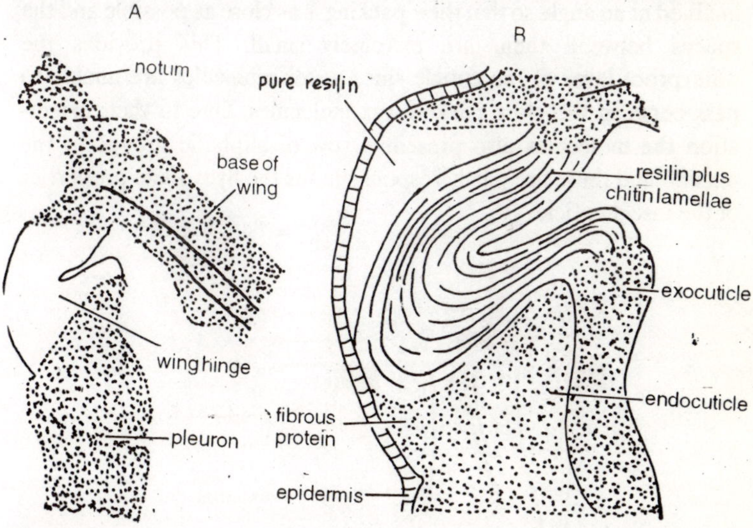


**Fig. 9.34.** A. Transverse section through the thoracic wall of *Schistocerca* showing the base of the wing and the wing hinge. B. Section through the wing hinge enlarged.

labral spring which keeps the labrum pressed against the mandibles, while in beetles, where there are no inspiratory muscles, inspiration is produced by resilin ligaments between the terga and sterna.

## Epicuticle

The detailed structure of the epicuticle varies from one insect to another. In general there is an inner cuticulin layer, then a layer of wax and frequently a layer of 'cement' over this. This cuticulin layer has two components; an outer very thin, but very dense layer (called cuticulin *sensu strictu*) and a thicker dense, relatively homogenous layer beneath. The wax of the epicuticle consists of long chain hydrocarbons and the esters of fatty acids and alcohols. It is proabable that the molecules adjacent to the cuticulin are strongly oriented as a result of their polar, hydrophilic groups being adsorbed on to the surface of the cuticulin so that they form a layer only one molecule thick and hence called a monolayer. The monolayer may be in a liquid phase which is continually being moved and replaced and experiments on *Phormia* (Diptera) show that such movements can occur since an oil film can spread over the whole surface of an insect within 15 minutes. The molecules of the monolayer are very close together and inclined at an angle so that their packing is as close as possible and the spaces between them are extremely small. This provides the waterproof layer of the cuticle since water molecules are unable to pass between the close-packed wax molecules. Due to their orientation the molecules also present a row of aliphatic groups to the outside and these are partly responsible for the hydrofuge properties of the insect cuticle.

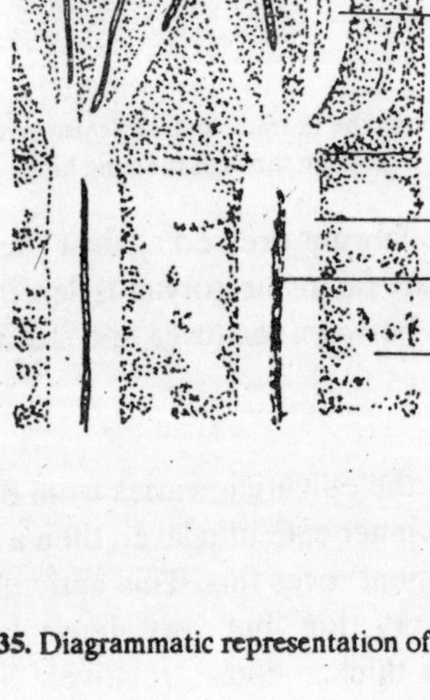


**Fig. 9.35.** Diagrammatic representation of a section through the epicuticle.

Outside the monolayer the wax molecules are randomly orientated and may permeate the cement, while on the outside of the cement in many insects is a bloom in which the molecules are ordered and form thin plates. The pattern of blooming is characteristic, depending on the distribution of wax canals and the form of the cuticle beneath.

The cement is a very thin layer outside most of the wax, perhaps consisting of tanned protein with lipids or of a shellac-like substance. It may serve to protect the underlying wax, but where the cuticle expands, as it does in the larvae of holometabolous insects, it might be in the form of an open meshwork which provides a reservoir of lipids to replace lost surface lipids. It is absent from many adult insects which have scales.

### Moulting and Cuticle Formation

Growth is limited by the cuticle which only undergoes a limited amount of stretching so that for any marked increase in size to occur the cuticle must be shed and replaced. Casting the cuticle is commonly known as moulting, but it involves two distinct processes which may be widely separated in time so that it is convenient to distinguish between them. The first process is the separation of the old cuticle from the underlying epidermal cells and this is called apolysis by Jenkin and Hinton (1966) although some authors use '*moulting*' in a restricted sense to include only this process. The second process is the shedding of the remnants of the old cuticle, and this is known as ecdysis.

### Changes in the Epidermis

The onset of moulting is usually first indicated by changes in the epidermal cells which divide mitotically and so become close-packed and columnar in form. In *Rhodnius*, at least, cell destruction accompanies call division and continues for a time after cell division stops, being dependent on the level of juvenile hormone in the blood. There is, nevertheless, a net increase in the number of cells per unit area so that when these are flattened the overall area of the epidermis, and hence of the cuticle, is increased. An increase in size does not always involve cell division and in the larvae of cyclorrhaphous

Diptera growth takes place solely as a result of the cells increasing in size. In *Lucilia* (Diptera) larva the cells increase in area by a factor of 20 between the first and third instars.

**Separation of the Cuticle from the Epidermis (Apolysis)**

Perhaps as a result of the changes in cell shape, a tension is generated at the surface of the epidermal cells which result in their separating from the cuticle. On the other hand in *Podura* (Collembola) the outer plasma membrane of the epidermal cells forms small out-pushing which separate off as vesicles and form a foam which lifts off the cuticle.

**Digestion of the Old Endocuticle**

As the cuticle separates from the epidermis moulting fluid is secreted into the space between the two. In *Podura* the fluids is

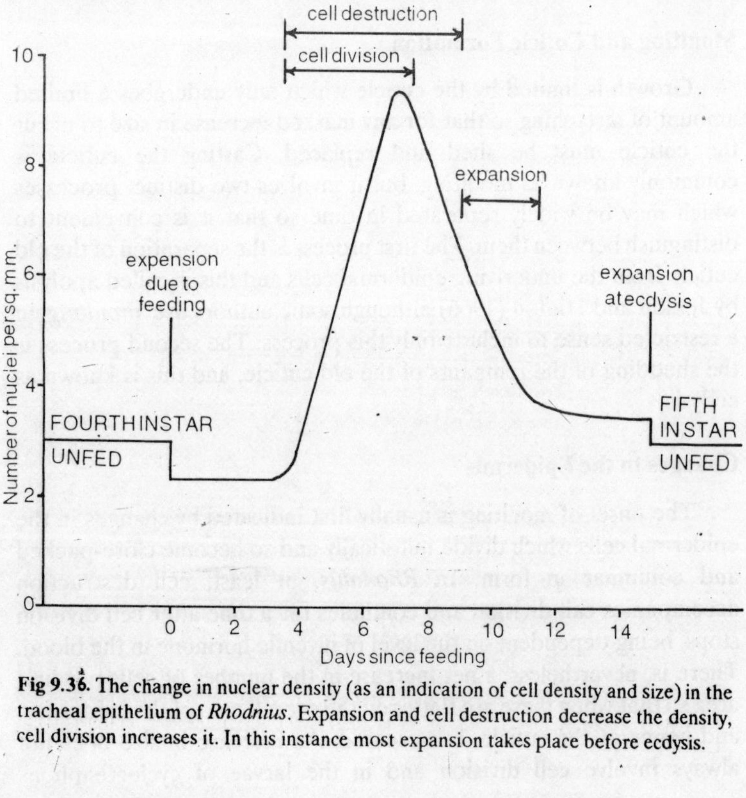


**Fig 9.36.** The change in nuclear density (as an indication of cell density and size) in the tracheal epithelium of *Rhodnius*. Expansion and cell destruction decrease the density, cell division increases it. In this instance most expansion takes place before ecdysis.

elaborated as granules on the endoplasmic reticulum of the epidermal cells at points where this comes to penetrate the outer plasma membrane. The granules are formed in groups and these are released into the subcuticular space as granule bodies. Secretion continues until the cuticulin layer of the new cuticle is complete. In the pupa of *Hyalephora* (Lepidoptera) the moulting fluid at this time is in the form of a gel. Moulting fluid contains enzymes, a proteinase and a chitinase, which will digest endocuticle, but when they are first secreted the enzymes are in an inactive state. The secretion of cuticulin follows the secretion of the moulting fluid, but precedes its activation. This is of great importance since otherwise the moulting fluid would digest the newly formed layers of the procuticle as well as the old endocuticle.

The substance of the cuticulin layer is probably produced in the oenocytes and then transferred to the epidermal cells which secrete it to the outside. In *Periplaneta* (Dictyoptera a very resistant paraffin layer is secreted first, while in *Rhodnius* polyphenols pass via the pore canals on to the outer surface of the cuticulin. These polyphenols do not form a discrete layer, but probably impregnate the cuticulin.

At first the cuticulin layer is smooth, but then it increases in area and this produces complex surface patterns such as the taenidia in tracheae and the transverse ripples of adult *Rhodnius*. Once it is complete the epidermal cells begin to lay down the procuticle underneath it. The mechanism by which the enzymes in the moulting fluid are activated is not known, but in *Hyalophora* it may be related to sclerotisation. They proceed to digest all the endocuticle of the old cuticle except for a thin layer which is modified in some way and persists as the ecdysial membrane but they have no effect on the exocuticle or on the muscle and nerve connections to the old cuticle. These persist so that the insect is still able to move and receive stimulate from the environment, but these connections are finally broken at ecdysis by the muscular activity of the insect. The products of digestion of the cuticle are absorbed, it is said, through the integument, and up to 90% of the materials present in the cuticle may be conserved in this way.

As a result of the activity of the moulting fluid, lines of weakness appear in the cuticle along the ecdysial lines. These vary in position, but in the locust there is a Y-shaped line on the head, the ecdysial cleavage line.

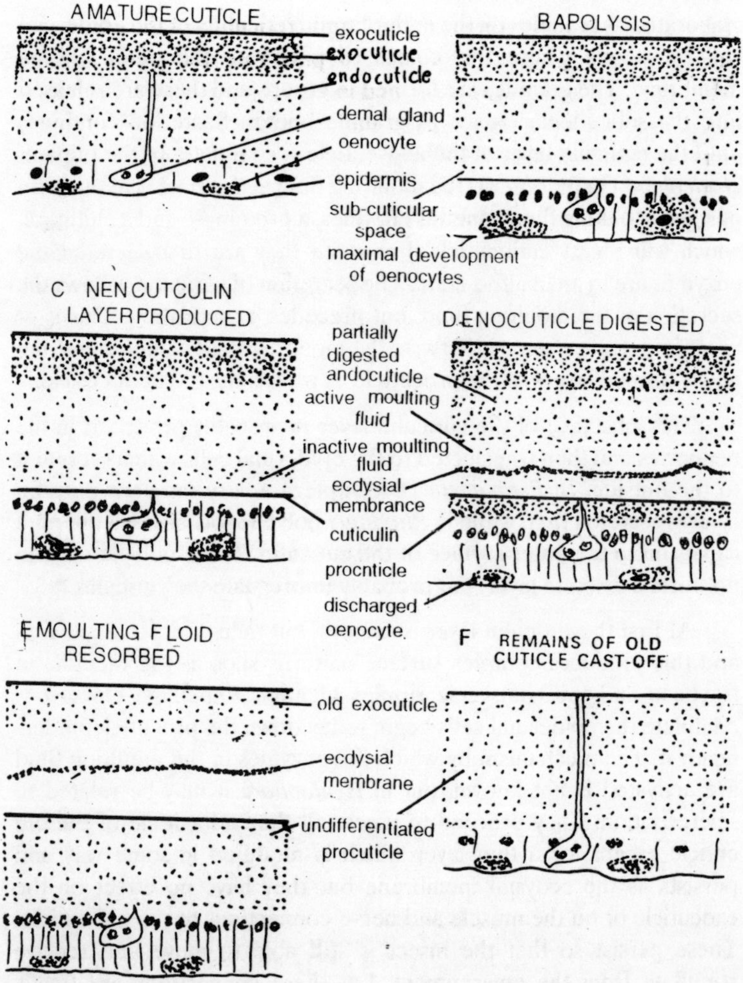


**Fig. 9.37. Diagrammatic representation of the changes occurring in the integument during the moulting cycle.**

## Wax Secretion

Shortly before ecdysis wax is secreted on the surface of the new cuticle, and the layer adjacent to the cuticulin forms the orientated monolayer. In the cockroach, at least, the orientation of the molecules in the monolayer in dependent on the water-saturated layer of cuticulin beneath. The cuticulin adsorbs the hydrophile polar groups so that the hydrofuge aliphatic groups are towards the outside. Tanned cuticulin has a very high affinity for the polar groups so that tanning is also important in the production and maintenance of the hydrofuge monolayer. Wax secretion is well-advanced at ecdysis, but even so, in *Tenebrio*, the rate of water loss during the first 24 hours after ecdysis is four to six times as high as normal due to the incompleteness of wax production which continues to a greater or lesser extent through the intermoult period.

It is possible that the wax is transported to the surface as lipid-water liquid crystals. The polar groups of the wax molecules lie at the hydrocarbon water interface and various arrangements are possible. The wax filaments of the pore canals and wax canals may be liquid crystals and these would be pulled outwards as the surface wax, with which they are continuous, was absorbed in the meshwork of the cement or crystallised into the solid surface bloom. At the same time further molecules of polar lipids are added to the inner ends of the lipid crystals by the epidermal cells. Non-polar lipids, which form some of the surface bloom, could diffuse through the cylinders of polar lipids and even where, as in the larva of *Calpodes*, there are no pore canals the texture of the endocuticle is sufficiently coarse to allow lipids to pass from the epidermal cells to the surface. Soon after ecdysis a layer of cement produced by the dermal glands is formed over the surface of the wax.

## Ecdysis

When the moulting fluid and the products of digestion of the moulting fluid are resorbed the old cuticle consists of little more than epicuticle and exocuticle and it is quite separate from the new cuticle. Usually ecdysis follows as soon as digestion is complete, but sometimes the old cuticle may be retained from sometime and the insect enclosed in this way is referred to as a pharate instar. In addition to these structural proteins, enzymes also exist in the cuticle, including diphenol oxidase, which catalyzes the oxidation of dihydric phenols used in the tanning process. This enzyme appears to be located in or just beneath the epicuticle. A variety of pigments have been found in the cuticle (or in the epidermis) which may give an insect its characteristic colour.

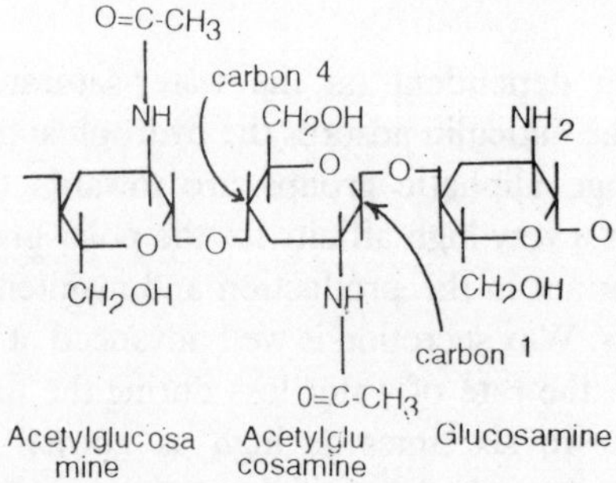


**Figure 9.38.** The chemical structure of chitin.

Certain processes occur at the surface of the cuticle after it has been formed, for example, secretion and repair of the wax layer and tanning of the outer practical. Thus a route of communicàtion must remain open between the epidermis and cuticular surface. This route takes the form of pore canals which are formed às the new procuticle is deposited, and which may or may not contain a cytoplasmic process. Most often, the canals do not contain an extension of the epidermal cell but have at least one "filament" produced by these cell. Locke (1974) suggests that the filament(s) might keep a channel open in the newly formed cuticle until the latter hardens. and anchor the cells to the cuticle. In some insects the pore canals become and anchor the cells to the cuticle. In some insects the pore canals become filled with cuticular material once epicuticle formation (including tanning) is complete. The pore canals terminate immediately below the epicuticle. Running from the tips of the pore canals to the outer surface of the epicuticle are lipid-filled channels known as wax canals.

The epicuticle is a composite structure produced partly by epidermal cells and partly by specialized glands. It ranges in thickness from a fraction of a micrometer to several micrometers. Mainly on the basis of its staining reactions earlier authors concluded that the epicuticle was a four-layered structure, and electron microscopy has tended to confirm this. However, one or more of these layers may be absent in some species and even over certain arras of cuticle in an individual under particular conditions. The layer are, from outside to inside, cement, wax, polyphenol and cuticulin. Nowadays it is considered that the polyphenol layer is the part of the cuticulin which has been tanned. The nature of the cement is probably variable, though it is likely to be approximately similar to shellac. The latter is a mixture of laccose and lipids. The cement is undoubtedly a hard, protective layer in some insects. In others it appears to be more important as a sponge which soaks up excess wax. The latter could quickly replace that lost, for example, by surface abrasion. The waxes are long-chain fatty acid esters. Within the wax layer three regions can be distinguished. Adjacent to the cuticulin is a monolayer of tightly packed molecules in liquid form. The molecules are oriented with their monpolar (water-repellent) groups facing outward, giving the epicuticular surface its high contact angle with water and its resistance to water loss. Most wax is in the middle layer, which is less ordered and permeates the cement. The outer layer, which comprises crystalline wax blooms, is not present in all insects. The cuticulin sensu *stricto* (that is, the untanned inner portion) extends over the entire body surface and ecotodemal invaginations, including the most minute tracheoles, but is absent from specific areas of sense organs and from the tips of certain glands cells. It may be considered the most important layer of the epicuticle for the following reasons. (1) It is a selectively permeable barrier. During breakdown of the old cuticle, it allows the "activating factor" for the moulting gel to move out and the products of cuticular hydrolysis to enter, yet it is impermeable to the enzymes in the molting fluid. It is permeable to waxes (since these are deposited only after the cuticulin layer has formed) and, in some insects, it permits the entry of water. (2) It is inelastic, and therefore, serves as a limiter of growth. (3) It provides the base on which the wax monolayer sits. The nature of the cuticulin will therefore determine weather the wax molecules are oriented with their polar or nonpolar groups facing outward and therefore, the surface properties of the cuticle. (4) It may play a role in determining the surface pattern of the cuticle. Despite its importance the composition of the cuticulin is unknown, though its staining properties indicate that it contains lipoprotein, which becomes tanned in the outer polyphenol layer.

**Cuticle Formation**

Formation of new cuticle may be viewed largely as a succession of syntheses by epidermal cells, with dermal glands and oenocytes adding their products at the appropriate moment. It must be realized however, that other, related processes such as dissolution of old cuticle are going on concurrently and that cuticle formation is partly a preecdysial and partly a postecdysial event; that is, much endocuticle formation, tanning of the outer procuticle, wax secretion, and other processes occur after the remains of the old cuticle are shed.

**Preecdysis**

In most species the onset of a moulting cycle is marked by an increase in the volume of the epidermal cells and/or by epidermal mitoses. These events are soon followed by apolysis. The detachment of the epidermis from the old cuticle. The epidermal cells, at this time, show signs of preparation for future synthetic activity. One or more nucleoli become prominent, the number of ribosomes increases, and the ribonucleic acid content of the cells is elevated. As apolysis occurs, the space between the epidermis and old cuticle becomes

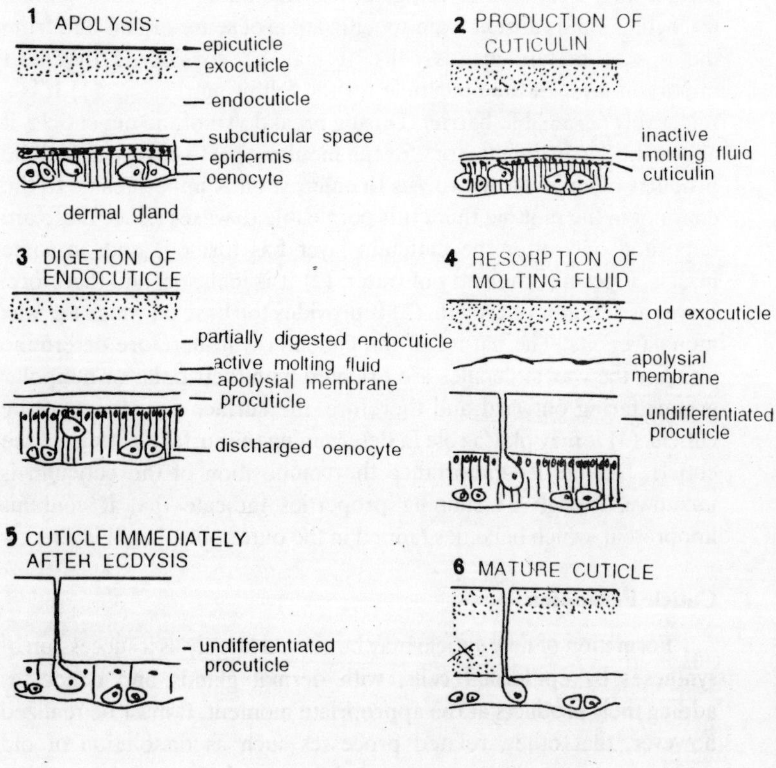


**Figure 9.39.** Summary of events in cuticle formation.

filled with moulting gel. Initially the gel is inactive, and it is only after formation of the cuticulin layer of the epicuticle that the epidermal cells release on "activation factor" so that digestion of the old endocuticle can begin. Molting fluid contains proteinases and a chitinase. Between 80 and 90 per cent of the old cuticle is digested, resorbed, and reused in the production of new cuticle. However, molting fluid does not affect the exocuticle, or muscle insertions and sensory structures in the integument. Thus, an insect is able to move an receive information from the environment more or less to the point of ecdysis.

The first layer of new cuticle deposited is the cuticulin component of the epicuticle. Minute papillae form on the distal side of the epidermal cells and fine droplets of material are secreted from them. Eventually the droplets coalesce to form a continuous layer. Oenocytes are maximally active at this time, and it is possible that they are involved in cuticulin formation, perhaps by synthesizing a precursor. Initially the cuticulin layer is smooth but soon becomes wrinkled. This permits expansion of the cuticle after molting. Tanning of the other cuticulin layer then occurs, following secretion of polyphenols. At about the same time production and deposition of the new procuticle begin. In contrast to the epicuticle, whose layers are produced sequentially from inside to outside, the new procuticle is produced with the newest layers on the inside. As the procuticle increases in thickness, the original papillae elongate to become the cytoplasmic processes within the newly formed pore canals. Deposition of the wax layer of the epicuticle begins a few hours prior to ecdysis. The wax is secreted by the epidermal cells, probably as lipid-water liquid crystals, and passes along the pore canals to the outside. Wax production continues after ecdysis and, in some insects, throughout the entire intermolt period.

**Ecdysis**

At the time of ecdysis, the old cuticle comprises only the original exocuticle and epicuticle. In many insects it is separated from the new cuticle by an air space and a thin ecdysial membrane that is formed from undigested inner lamellae of the endocuticle. Shortly before molting an insect begins to swallow air (or water, if aquatic), thereby increasing the hemolymph pressure by as much as 90 mm Hg. Hemolymph is then localized in the head and thorax following contraction of intersegmental abdominal muscles. In many insects these muscles becomes functional only at the time of ecdysis and histolyze after each molt. The local increase in pressure in the anterior part of the body causes the old cuticle to split, along the ecdysial line when this is present. An insect continues to swallow air or water after the molt in order to stretch the new cuticle prior to tanning.

**Postecdysis**

Several processes are continued or initiated after ecdysis. Wax secretion continues, and the major portion of the endocuticle is deposited at this time. Indeed, endocuticle production in some insects appears to be a more or less continuous process throughout the intermold period. It is also at this time that the cement layer of the epicuticle is laid down. The most striking postecdysial event, however, is the hardening of the outer procuticle (i.e., the differentiation of the exocuticle). Hardening is usually accompanied by darkening (melanization), though the two may be distinct processes; that is, some species have pale but very hard cuticles. Much information of the mechanism of hardening has come from study of tanning of the fly puparium and cockroach ootheca, though both are "special cases" and may not indicate precisely the nature of tanning in normal procuticle. In fly larvae the level of the amino acid tyrosine in the hemolymph reaches a maximum prior to puparium formation, then declines as tanning proceeds. The tyrosine is apparently accumulated by hemocytes where it is oxidized to "dopa" (dihyroxyphenylalanine) which is then decarboxylated forming dopamine. It has been suggested that dopamine is accumulated by the epidermal cells and converted to N-acetyldopamine which moves, presumably via pore canals, to the epicuticle where it is oxidized to a quinone under the influence of diphenol oxidase. The quinone diffuses back into the procuticle and links protein molecules together. According to Hackman (1974), the quinone may polymerize, a process which makes the molecule a larger, more effective tanning agent. The polymer will have more reactive sites and be capable of bridging larger distances, the result being that more proteins molecules can be linked. The quinone combines covalently with the terminal amino group or a sulfhydryl group of the protein to form an N-catechol protein, which in the presence of excess quinone is oxidized to an N- quinonoid protein. The latter is able to react with the terminal amino group or

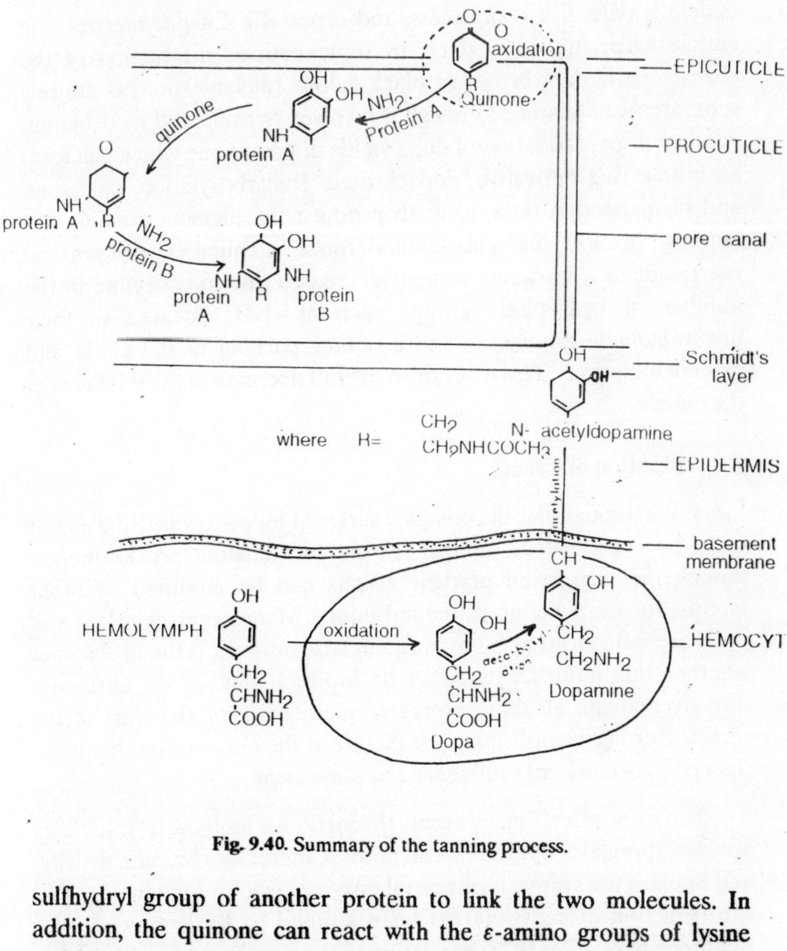


**Fig. 9.40.** Summary of the tanning process.

sulfhydryl group of another protein to link the two molecules. In addition, the quinone can react with the ε-amino groups of lysine residues in the protein molecule to effect further binding. The possibility of autotanning also exists, though the process has never been demonstrated. This involves the oxidation of the constituent aromatic amino acids (for example, tyrosine) in a polypeptidae chain to form quinone compounds which can subsequently link with amino groups in adjacent chains. This process could account for tanning that occurs in cuticles known to contain little free diphenol.

The colour of tanned cuticle depends on the amount of o-quinone that is present. When this molecule is present in small quantities the cuticle is pale; if it is in excess, and especially if it polymerizes, the cuticle when tanned is dark. In some cuticles production of the characteristic dark brown or black colour (melanization) is entirely separate from tanning, although it involves tyrosine and its oxidation product dopa. Oxidation of dopa yields dopaquinone which can form an indole ring derivative, dopachrome. Decarboxylation, oxidation, and plymerization then occur to produce the pigment melanin. As tanning occurs, the cuticle undergoes a number of physical rearrangments. Its water content decreases due to a decline in the number of hydrophilic groups such as $-NH_2$ because of their involvement in tanning and to a tighter packing of the chitin and protein molecules. This leads to an overall decrease in the thickness of the cuticle.

**Co-ordination of Events**

It is essential that the complex series of events comprising a molt cycle be coordinated. Central to this coordination are hormones, though the expression of their effects can be modified by other factors, for example, nutrition and injury. Many events within a milt cycle are influenced by hormones, but still unsolved is the problem of whether this influence is direct or indirect; that is, do hormones directly regulate all these events or merely trigger the start of the series? Further complicating the picture is the observation that more than one hormone may influence the same event.

Ecdysone affects many events throughout a molt cycle. It initiates apolysis stimulates epidermal cell mitosis, increases chitinase activity, and induces the synthesis of several enzymes involved in tanning. The site of action of ecdysone has been studied by Karlson *et al.* (see reference in Neville, 1975) in relation to tanning. Ecdysone stimulates synthesis of the messenger RNA specific for dopa decarboylase. Karlson's group suggested that ecdysone acted on the epidermal cells; however, it now seems likely that the hemocytes are a target organ. Though the modifying influence of juvenile hormone on ecdysone effects has been known for a considerable time, it has been shown only recently that it, too, acts on the genome. Its effect is to program the epidermis to secrete a cuticle characteristic of the juvenile stage.

Bursicon, like ecdysone, is known to affect many processes though all are related to tanning and melanization of the cuticle. Bursicon's primary effect appears to be to increase the permeability of the hemocyte wall to tyrosine and the epidermal cell wall to dopamine. The hormone may exert this effect via a cyclic AMP-mediated system. In addition bursicon may activate tyrosinase in hemocytes, which catalyzes the oxidation of tyrosine to dopa. Finally, ecdysis is regulated by an eclosion hormone secreted by the brain in silk moths and wax synthesis and endocuticle deposition require the presence of the corpus allatum corpus cardiacum complex in the lepidopteran *Calpodes ethlius*.

## FUNCTIONS OF THE INTEGUMENT

Most functions of the integument relate to the physical structure of the cuticle though the latter may serve as a source of metabolites during periods of starvation. These functions may be discussed under three headings, strength and hardness, permeability, and production of colour.

### Strength and Hardness

The few studies that have been carried out on the mechanical properties of insect cuticle indicate that it is of medium rigidity and low tensile strength. There is, however, wide variation from the general statement, for example, the cuticles of most endopterygote larvae are extremely plastic, whereas the mandibular cuticle of many biting insects may be extremely hard, enabling them to bite through metal. Further, there is an obvious difference in properties between sclerites and intersegmental membranes, and between typical nonelastic cuticle and that which contains a high proportion of resilin.

Though the above properties indicate that the cuticle is satisfactory as a "skin" preventing physical damage to internal organs, discussion of the suitability of the cuticle as a skeletal component must include an appreciation of overall body structure. Most components of insect (and other arthropod) bodies may be considered as cuticular cylinders or spheres. Such a tubular shell is about three times as strong as a solid rod of the same material having the same cross-sectional area as the shell (i.e., they both contain the same amount of skeletal material). The force required to distort the shell is proportional to the thickness of the shell and inversely proportional to the cross-sectional area of the whole body. Thus, in small organisms where the thickness of the shell is great relative to the cross-sectional area of the body, the use of a shell as an exoskeletal structure is quite feasible. In larger organisms the advantage of the extra strength due to a shell type of skeleton is greatly outweighed by the massive increase in thickness of the shell that would be necessitated and, perhaps, the physiological problems of producing the large amounts of material required for its construction.

### Permeability

For different insects there exists a wide range of materials that are potential permanents of the integument, and of factors that affect their rate of permeation. Sometimes specific regions of the integument are constructed to facilitate entry or exist of certain materials; more often the integument is structured to prevent entry or loss. As this time we shall consider only the permeability of the cuticle to water and insecticides, of which the latter may now be considered a "natural" hazard for most insects.

### Water

Water may be either lost or gained through the integument. In terrestrial insect, which exist in humidities that are almost always less than saturation, the problem is to prevent loss through evaporation. In freshwater forms the problem is to prevent entry due to omosis.

In many terrestrial insects the rate of evaporative water loss is probably less than 1% per hour of the total water content of the body. Most of this loss occurs via the respiratory system, despite the evolution of mechanical and physiological features to reduce such loss. Water loss through the integument *(sensu stricto)* is extremely slight due mainly to the highly impermeably epicuticle and in particular the wax components. Early experiments demonstrated that permeability of the integument is relatively independent of temperature up to a certain point (the transition temperature), above which it increases markedly. As a result of his studies on both artificial and natural systems, Beament (1961) has concluded that the initial impermeability is due to the highly ordered wax monolayer whose molecules sit on the tanned cuticulin at an angle of about 25° to the perpendicular axis, with their polar ends facing inward and nonpolar ends outward. In this arrangement the molecules are closely packed and held tightly together by Van der Waals forces. As temperature increases, the molecules gain in kinetic energy, and eventually the bonds between them rupture. Spaces appear and water loss increases significantly. The nature of the wax and its transition temperature can be correlated with the normal niche of the insect. Insects from humid environments or that have access to moisture in their diet, for example, aphids, caterpillars, and bloodsucking insects, have "soft" waxes with low transition temperatures. Forms from dry environments or stages with water-conservation problems, for example, eggs and pupae, are covered with *"hard"* wax, whose transition temperature is high (in most species above the thermal death point of the insect).

Some insects which are normally found in extremely dry habitats and may go for long periods without access to free water, for example. *Tenebrio molitor* and prepupae of fleas are able to take up water from an atmosphere in which the humidity is relatively high. Until recently it was believed that uptake occurred across the body surface perhaps via the pore canals. However, it has now been demonstrated that uptake occurs across the wall of the rectum.

In many freshwater insects, for example, adult Heteroptera and Coleoptera, the cuticle is highly impermeable because of its wax monolayer and water again is probably 4% or less of the body weight per day. In most aquatic insects, however, the wax layer is absent. Thus, gains of up to 30% of the body weight per day are experienced, the excess water being removed via the excretory system.

## Insecticides

Economic motives have stimulated an enormous interest in the permeability of the integument to chemicals, especially insecticides and their solvents. Though, for the most part, the cuticle acts as a physical barrier to decrease the rate of entry of such materials, there is evidence that in some insects it may bring about metabolic degradation of certain compounds, and consequently reduction of their potency. It follows that increased resistance to a particular compound may result from changes in either the structure or the metabolic properties of the integument. The primary barrier to the entrance of insecticides is for most insects the wax layer of the epicuticle. In addition, the cement layer probably provides some protection against penetration. The procuticle offers both laid and aqueous pathways along which an insecticide may travel, but the precis rate at which a compound moves depends on many variables, especially thickness of the cuticle, presence or absence of pore canals, and whether the latter are filled with cytoplasmic extensions or other material. It follows that the rate of penetration will very according to the location of an insecticide on the integument. Thin, membranous cuticle such as occurs in intersegmental regions or covers tactile or chemosensory hairs generally provides little resistance to penetration. The tracheal system is another site of entry. The extent to which tanning of the procuticle occurs is also related to penetration rate. As the chitin-protein micelles become more tightly packed and the cuticle partially dehydrated, permeability decreases.

In addition, but obviously related to the physical features of the cuticle, the physicochemical nature of an insecticide is an important factor in determining the rate of entry. Especially significant is the partition coefficient (the relative solubility in oil and in water) of an insecticide or its solvent. In order to penetrate the epicuticular wax the material must be relatively lipid soluble. However, in order to penetrate the relatively polar material of the procuticle and, eventually, to leave the integument to move toward its site of action the material must be partially water-soluble. Thus, correct formulation of an insecticidal solution is an important consideration.

It should be apparent from the above discussion that few generalizations can be made. At the present time, therefore, the suitability of an insecticide must be considered separately for each species. Because of the factors which affect the entry of insecticides, a great difference usually exists between *"real toxicity."* that is, toxicity at the site of action, and *"apparent toxicity,"* the amount of material that must be applied topically to bring about death of the insect. The chief feature which relates the two is obviously the "penetration velocity," that is, the rate at which material passes through the cuticle. When the rate is high, the real and apparent toxicity values will be nearly identical.

## CONTROL OF MOULTING \ND ASSOCIATED PROCESS

Moulting and cuticle formation are complex phenomena involving a number of discrete processes. Apart from wax formation and endocuticle production which may continue for some time there is a fairly distinct sequence of events:

1. changes in the epidermal cells
2. secretion of the moulting fluid
3. secretion of the outer layer of cuticulin
4. secretion of the homogeneous layer of cuticulin
5. activation of the moulting fluid
6. absorption of digested remains of old cuticle
7. start of secretion of new procuticle
8. ecdysis and expansion of the new cuticle
9. sclerotisation
10. start of wax secretion.

The hormone ecdyson plays a vital role in controlling these events, but evidence is accumulating which indicates that it does not simply trigger the epidermal cells so that the sequence follows automatically. Some of the processes are known to be controlled by other hormones or nervous stimuli, perhaps acting together with ecdyson, while the juvenile hormone exerts an overall modifying effect.

For instance, the period during which the cuticle of *Calliphora* is sufficiently plastic to be expanded is short and is restricted to the period during which the insect is swallowing air. This suggests that cuticular plasticity may be under nervous control. In *Rhodnius* the swelling of the abdomen at the time of feeding is controlled by neruosecretory axons to the abdominal wall. Sclerotisation is controlled by a hormone, called bursicon, which in *Sarcophaga* is related from the brain and the large compound thoracic ganglion, and it appears to be quite separate from other hormones produced in the brain. It is only produced at the time of moulting and its release is initiated by receptors in the cuticle operating when the fly is no longer confined to the medium in which it emerges. If the fly is forced to keep digging through soil after it has emerged from the puparium the release of this factor is delayed. Bursicon is probably not involved in quinone formation, which may be controlled by ecdyson itself, but it may be concerned with some later step in tanning. The secretion of wax by *Calpodes* larvae is controlled by a hor- mone released from the corpus allatum-corpus cardiacum complex, but produced elsewhere. This hormone, however, is only effective when together with ecdyson; endocuticle production is probably similarly controlled by ecdyson with another hormone.

## FUNCTIONS OF THE CUTICLE

The cuticle is one of the features of insects which is primarily responsible for their success. It plays an important part in supporting the insect, an essential requirement of terrestrial animals. Further, the presence of hard, jointed appendages makes accurate movements possible with a minimum of muscle, thus effecting an economy of muscle, and, by lifting the body off the ground, facilitates rapid movement. Flight depends on the possession of rigid wings and in insects the cuticle provides the rigidity. In the absence of the cuticle flight would not be possible.

Protection is also provided by the cuticle. Some insects, such as adult beetles, have hardened heavily sclerotised cuticles which make them difficult for predators to catch or parasites to parasitise. Protection from the physical environment is also afforded. Again the beetles, the upper cuticle of the abdomen, protected by the elytra, is very thin, but the cuticle of the ventral surface, exposed and subject to abrasion by the substratum, is very thick. The cuticular lining of the fore-and hind-guts also protects the epithelium from abrasion by the food while the wax layer is of great importance in the restriction of water loss. Finally, parts of the cuticle are modified to form sense organs and its physical structure is also often important in the production of colour.

**Pigments**

There is a considerable body of literature describing pigments in insects but often the location of the pigment is not given. Pigment granules occur in the hypodermal cells of many insects and in some insects the granules can be moved under hormonal influence to produce transient colour changes. Coloured hemolymphs may also be visible through the cuticle. The colour of an insect may be due to the present of more than one class of pigment and they may be allocated in different tissues, e.g., the green colour of pupae of the cabbage white butterfly, *Pieris rapae crucivora*, is due to the combination of a yellow cuticle (carotenoid pigment and an underlying bluish-green epidermis (tetrapyrrole pigment). Several general reviews on pigments in insects have appeared recently which should be consulted for the references to and for further details on some of the work discussed below. Rowell (1971) has reviewed work on colour variation in Acridoid grasshoppers. Optical effects, brought about by the structure of the cuticle, are responsible for the colour of some insects.

*(a) Sclerotin and melanin*. The cuticles of some insects possess a black melanic pattern as well as the normal amber colour of the hard cuticle. The chemistry of sclerotin, the coloured pigment generally formed when cuticles harden, is discussed in Section IV, but little is known about the structure of sclerotin (s) apart from their quinonoid and phenolic character. Melanin also has quinonoid and phenolic properties and an interesting problem in the field of cuticular pigments is that of the relation between melanin and sclerotin.

Melanin is a somewhat ill-defined term and many pigments which are dark in colour and insoluble in most solvents have been referred to as melanins without regard to their chemistry. Sclerotin was originally described as a melanin, but spectroscopically they are different. Melanins are high molecular weight compounds formed by oxidation of polyhydric phenols, e.g., dopa, but a complete structural formula cannot as yet be written for a melanin. Nicolaus (1968) has reviewed the chemistry and formation of melanins and he classifies such pigments as eumelanins (black, indole-type), phaeomelanins (yellow, red, or brown), and allomelanins (black catechol-type). Methods for the microdetection of melanins are given by Hackman and Goldberg

(1971b. Mason (1967) regards melanin as a homopolymer, a polymer built up from only one type of monomer, but Nicolaus (1968) Nicolaus *et al*, 1967 regard melanin as a polymer built up from a number of different but related monomeric units, perhaps linked in different ways. Most melanins of animal origin, including insects, are eumelanins (e.g., Hackman 1967; Nitolaus, 1968), but melanins from *Calliphora augur* puparia and *Xylotrapes gideon* cuticle give both indole and catechol degradation products and have nitrogen contents intermediate between those of eumelanins and allomelanins. Thus, these two insect melanins are either mixtures of indole- and catechol-type melanins or are polymers containing both indole and catechol units.

Melanins is often bound to protein to form melanoproteins. In fact, it is difficult to imagine that there could exist in a biological system quinones and proteins without some reaction taking place between them. Methods used to isolate melanins generally destroy the protein but cysteine survives and remains attached to the melanin nucleus. The link is presumably covalent and of the thioether type. There is some evidence to suggest that protein may also be linked to melanin by peptide bonds.

Although melanin formation is sometimes included in the general process of sclerotization, the evidence suggests that sclerotization and melanization are two independent processes. In a number of insects it has been shown that the two processes are separated in time and depending on the species, either hardening of the cuticle or the appearance of black pigment may occur first, e.g., *Calliphora* flies. *Ephestia Kühniella*, *Schistocerca gregaria*, and various Diptera species. In other insects melanin formation can be prevented while cuticular hardening continues normally, e.g., *Calliphora* larve and *Gryllus bimaculatus*. A study of enzyme activity and the utilization of radioactive phenolic compounds has shown the same process of cuticular tanning to take place in both the melanic and nonmelanic strains of *Schistocerca gregaria* and of *Lucilia cuprina*. In these two species malanization appears to proceed by a discrete metabolic pathway.

Melanin would appear to have a protective function in insects, e.g., adaptive colouration and heat control. Deposition of melanin takes place where a cuticle is damaged mechanically or by a parasite implanted tissues frequently darken, and there are many references to dark-pigmented encapsulations around parasites or other foreign bodies. Taylor (1969) has suggested that the polyhydric phenol-phenol oxidase system which produces melanin is directly involved in invertebrate immunity. Quinones are powerful oxidizing agents and they rapidly destroy enzymes, so there could be considerable advantage attached to the possession of a system which, in the event of damage, produces quinones able to inactivate rapidly enzymes and other proteins and to act as antibiotics. A melanin-producing system may therefore be one of the most primitive, nonspecific defense system of al living forms.

*(b) Carotenoids*. Carotenoids occur in many different species of insects, usually as conjugates with proteins. The colours of most carotenoids are yellow, orange or red and these are the colours they give to insects. Together with a blue tetrapyrrole pigment (see below), carotenoids produce the green colour of many insects. Carotenoids have been isolated from the "body wall" or integument of Orthoptera, Coleoptera, Lepidoptera and Hemiptera and from the wings of Orthoptera, the compounds isolated were $\beta$-carotene (red,) lutein (yellow), or astaxanthin (purple). Carotenoids are present in many other insect tissues. Cheeseman *et al*. (1967) have reported on carotenoproteins in invertebrates.

*(c) Tetrapyrrole Pigments*. Tetrapyrrole pigments (bile pigments) occur in the hemolymph of many insects, generally as protein conjugates and they are blue. Two pigments belonging to this class have been reported from integuments; biliverdin 1Xa in Orthoptera and biliverdin 1X in Lepidoptera, Pterobilin, previously thought to be mesobiliverdin, is now reported to be biliverdin 1Xy. The reported occurrence of biliverdin 1Xa and 1Xy if of considerable interest since they are generally associated with the degradation products of hemoglobin and so, *a priori*, would not have been expected to occur in these insects. Rudiger (1970) has reviewed the occurrence of tetrapyrrole pigments in invertebrates. Some brown and yellow pigments of Orthoptera are said to be formed by oxidation of biliverdin.

*(d) Pteridines*. Pterins (i.e., pigments with a pteridine nucleus) are a class of fluorescent pigments which was first discovered in the wings of butterflies, but are now known to occur in other insect tissues, e.g., eyes. Xanthopterin (yellow), leucopterin (white), isoxanthopterin

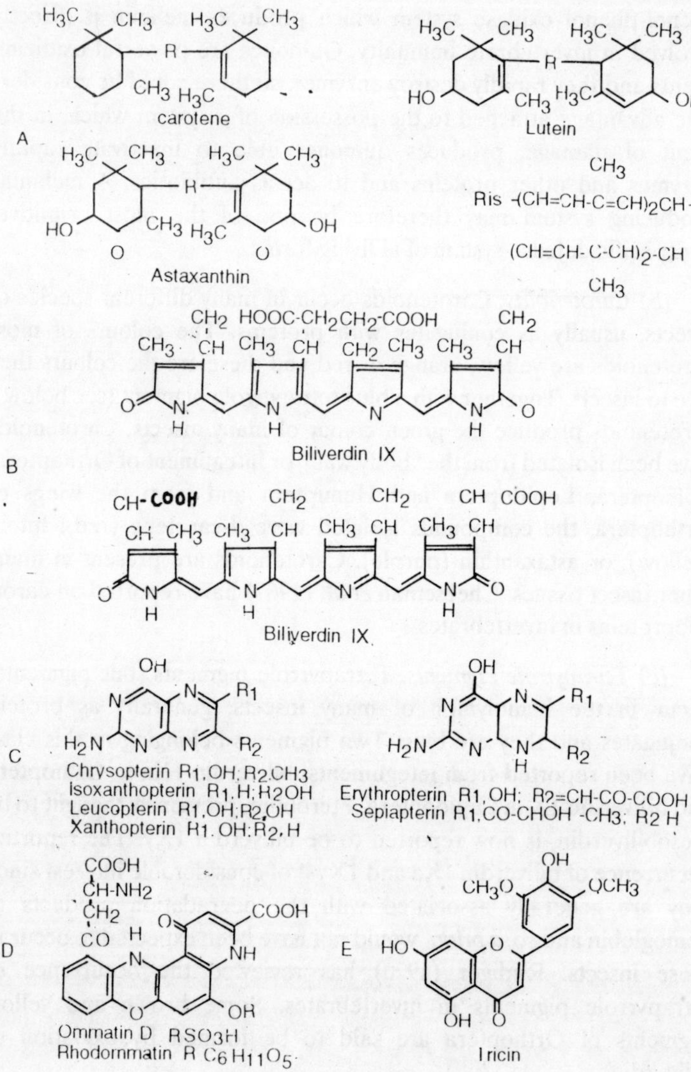


Fig. 9.41. Cuticular pigments.

(white), erythropterin (red), chysopterin (yellow), and sepiapterin (yellow) have been identified in the wings or integuments of Lepidoptera, Hemiptera, Hymenoptera, and Neuroptera. Of the pterins, xanthopterin is the most widespread in nature and its fluorescence and absorption spectrum make it readily identifiable in small amounts. Some of the white, yellow, orange, and red colours in insects are due to differential concentrations of the above pterins. In contrast to melanins which are usually embedded in the cuticle, pterins are located in the epidermal cells or in the interlamellar spaces in wings, scales, and hairs.

*(e) Ommochromes*. These pigments were first isolated from the ommatidia of insect eyes and red and brown ommochromes are present in the wings of Nymphalidae. (Lepidoptera) and in the epidermis of some Orthoptera and Lepidoptera. Rhodommatin, xanthommatin, and ommatin D are the pigments most commonly found in these tissues.

*(f) Flavones*. The wings of the marble white butterfly *Melanargia galathea* and the small heath butterfly *Coenonympha pamphilis* contain the flavone tricin both free and as a glycoside.

**Inorganic Constituents**

The amount of ash formed when insect cuticles are incinerated is generally low (less than 1 % of the dry weight, but in puparia and cast skins it may be higher). Inorganic materials, therefore, form only a very small part of the cuticle. Richards (1956) has shown that in the larval cuticle of *Sarcophaga bullata* there is a higher concentration of ash components in the epicuticle than in the procuticle. So puparia and cast skins may be expected to give a higher percentage of ash than would the intact cuticle. Both puparia and cast skins contains the remains of secretions and care must be taken to remove these contaminates. Richards (1956) has given quantitative values for the 25-elements in the ash of *S. bullatta* larval cuticles. Na, K, Mg, Ca, Fe, Si, Cl, $SO_4$, $CO_3$ and $PO_4$ account for almost all the ash, Mg and K are the preodminant cations, Fe is concentrated in the epicuticle, and the cations are balanced by the anions. Cuticles and cast skins from other species of insect also contain a large number of ions.

Insect cuticles are not calcified as are crustacean cuticles but some specialized forms of calcification are known. Weaver (1958), in a survey of 147 species of immature insects (11 orders), has reported the presence of calcium oxalate crystals in the exo- or procuticle of thirteen species (twelve Lepidoptera and one Hymenoptera). Certain aquatic dipterous larvae (living in calcium-rich waters) belonging to the families Psychodidae (e.g.,*pericoma* spp.) and Stratiomyidae (e.g. *Sargus* spp) have calcereous deposits in the form of warts or nodules on the outer surface of the cuticle. The inner surfaces of the cast skins of some insects are covered with a deposite of calcium slats. These deposits are not part of the cuticle proper but are formed by materials from the malpighian tubules being secreted into the space between the old and new cuticles prior to ecdysis. Thus an inner shell is formed within the puparium of the fly *Acidia heraclei*. In larvae of the fly *Rhagolet is cerasi* the calcium salts are exerted before separation of the old cuticle and they are deposited as granules in the procuticle of the old cuticle. With the transformation of the larval cuticle into the puparium the granules remain with the cuticular structure and may account for more than half of its weight. Calcification appears to have replaced sclerotization in the puparium of *Musca autumnalis*.

Calcium oxalate occurs in the oothecae of many, but not all, species of cockroach. Oothecae of *Periplaneta americana* contains 6.5% calcium (the only major cation present) but only 35 % of the calcium was present as the oxalate. The oothecae of *Blattella germanica* gave similar results. Parker and Rudall (1955) reported crystals of calcium citrate in oothecae of praying mantids, but Hackman and Goldberg (1960) failed to detect citrate in oothecae of the mantid *Orthodera iministralis*, the oothecae, inn composition, resembling closely that of cockroaches. Kato and Kubomura (1956) have reported the identification (histochemically) of calcium carbonate in egg packets of Japanese mantids. Calcium oxalate, as a crystalline powder, covers the cocoon of the Lepidopteran *Malacosoma neustria testacea*. The biological significance of inorganic ions in insect cutieles is not known, but Kikkawa *et al.* (1954, 1955) have suggested that certain metals (Fe, Cu, Co, Ni, Ti, Mo) are associated with melanoproteins and other natural pigments. Hackman and Goldberg (1960) and Kenchington and Flower (1969) have discussed the role calcium may play in the oothecae of cockroaches and mantids.

## Other Components

### 1. *Sugars*

Sugars other than glucosamine or *N*-acetyl-d-glucosamine (from chitin) have been identified in hydrolyszates of cuticles, e.g., in *Periplaneta americana* (arabinose, galactose, glucose, mannose, and xylose, together with ketoses and pentoses and in *Lucilia cuprina* larvae (galactosamine and mannose. The role of these sugars is not known.

### 2. *Mucopolysaccharides*

On histochemical evidence acidic mucopolysaccharides have been reported to be present in some cuticle but their significance is obscure. In crustacea mucopolysaccharides appear to be involved in calcification of tissues.

### 3. *Insoluble Membranes*

The epicuticle of some insects contains as a membrane which is very ressitant to chemical action; in dipterous larvae and puparia this membrane has considerable mechanical strength. For example, acidic hydrolysis of blow fly larval cuticles (or puparial cases leaves insoluble membranes which are ghosts of the original cuticle and they are insoluble in boiling aqueous alkali and in all common organic solvents. The membranes are nitrogen free and, as judged by the infrared spectrum, appear to be built up from long chains of $CH_2$ groups. They resemble membranes isolated by less drastic means from blow fly larval cuticles and from cockroach cuticles by Dennell and Malek (1955a) and said to contain mostly "paraffin structures." A similarly resistant layer occurs in *Calpodes ethlius*. Filshie (1970a,b) suggests that the resistant membrane is composed of the outer epicuticle together with the cuticulin layer, the latter as defined by Locke (1966). Hot acid treatment removes the cuticulin leaving only the outer epicuticle.

## Hardening of the Cuticle

The cuticle of the newly ecdysed insect is soft and pale in colour, but over a period of a few hours it darkens and becomes harder, a process known as sclerotization or tanning. Tanning occurs first in the outer layers of the cuticle and proceeds inwards. The water content of the cuticle decreases (i.e., it becomes less hydrophilic), the protein and chitin chains become more highly oriented and close packed and inorganic components are extruded from the exocuticle. Similar changes occur during formation of the cockroach ootheca and fly puprium and they have been studied extensively as examples of sclerotization. However, amino acid analyses of their protein components differ markedly from those of cuticles which harden after a milt. Some cuticles are harder and darker than others, while some have in addition black melanic patterns. Hardness of a cuticle is not related to its chitin content, but occasionally the term "heavily chitinized" is still used to describe a hard cuticle.

Sclerotization is considered to involve reaction between cuticular proteins and *o*-quinones, derived from *o*-dihydric phenols, *o*-Benzoquinones react with $-NH_2$ and $-Sh$ groups and so in the cuticulr protein they would react with a terminal amino group, a sulfhydryl group or the ε-amino group of lysine to form, in the first instance, an *N*-catechol protein which, in the presence of excess quinone, would be oxidized to the *N*- quinonoid protein. The quinonoid protein would then react with another $-NH_2$ or $-Sh$ group to a disubstituted derivative, first as the colourless catechol form which would be oxidized by excess quinone to the coloured quinonoid form. Cross-linked structures are formed in which the N of the $-NH_2$ groups and the S of the $-SH$ groups in the protein chains become attached directly by covalent bonds to the aromatic nuclei of the quinones, the protein chains being linked end to end and also at intermediate points. Since reaction takes place with hydrophilic groups, the cuticle becomes less hydrophilic. Sclerotized cuticles are highly resistant to enzymolysis and this may indicate extensive hydrogen bonding between components. Chitin is bound to protein and so would form part of the network, Dennell (1958c). Dennell and Malek (1955b), and Malek (1952), on histochemical evidence only, suggested that sclerotization involves a sterol- protein complex. Simple *o*-benzoquinones may not be efficient cross-linking agents. Polymerization and condensation would from larger molecules with many reactive positions; those able to bridge longer distances than can *o*-Benzoquinones and those which would impart a deeper colour to the cuticle. It is also possible that polymerized quinones or phenols are simply embedded in the cross-linked protein structure or are complexed to the protein in a manner analogous to the tanning of proteins with vegetable tannings. With time, covalent bonds would be formed and this could explain the slow continuous insolubilization of cuticulr proteins observed by Andersen (1971b). The minimum number of cross-links needed to stabilize the cuticle is not known but, in view of what is known about other structural proteins, there is no reason to believe that it is great.

Formation of the cockroach ootheca is the best understood example of sclerotization, although it appears to be a special case. The ootheca is formed from the secretions of the left and right collatcral glands, the left gland secreting the structural protein, a phenoloxidase and a phenolic glucoside while the right gland secretes a $\beta$-glucosidase. When the secretions mix, the glucosidase liberates the free phenol which is oxidized by the phenoloxidase to an *o*-quinone which cross-links the structural protein and so forming the hard,dark ootheca. Two phenolic glucosides have been identified in the left gland, the 4-*o*-$\beta$-glucosides of 3,4-dihydroxybenzoic acid and 3,4-dihydroxybenzyl alcohol. The phenol oxidase is of the laccase type; it oxidizes a number of *o*- and *p*-dihydric phenols with the notable exceptions of dopa and dopamitne. It has been suggested that calcium ions and calcium oxalate (which are in the ootheca) help to organize and stabilize the structural protein.

The puparium of flies is the sclerotized last larval cuticle. The tyrosine content of the larval hemolymph of blow flies reaches a maximum just before puparium formation and then decreases rapidly as the puparium hardens and darkens. If labeled tyrosine, dopa, dopamine, or *N*- acetyldopamine are injected into the larvae, most of the radioactivity becomes located in the puparia, thus showing that tyrosine or its derivatives are incorporated into the sclerotized structure. Karlson has proposed that the quinone precursor is *N*-acetyldopamine, that the phenol oxidase originates, from a proenzyme in the hemolymph with the activator in the cuticle, that the activator is a protease, and that the protein which reacts with the quinone may be the phenol oxidase itself. In *Calliphora erythrocephala* the activator of the preoenzyme is in the outermost layers of the cuticle and this would explain why sclerotization occurs first in the outer layers. A similar but not identical sclerotization pathway to that for blowflies (one involving tyrotsine *o*- phos-phate) has been suggested for *Drosophila*.

It has generally been assumed that the phenol oxidase needed to form quinones arises from a proenzyme in the hemolymph, the proenzyme moving into the cuticle shortly before sclerotization commences, but this view was questioned by Hackman and Goldberg (1967). Recent work suggests that hemolymph enzymes may not be involved in sclerotization. A phenol oxidase appears to be secreted as part of the cuticle and only the phenolic substrates diffuse outward through the cuticle prior to sclerotization. If the phenol oxidase forms part of the cuticle as it is laid down then, from the work of Hackman and Goldberg (1967) on fly larvae, the enzyme must initially be in an inactive form.

*o*-Diphenoloxidases occur in all hemolymphs, in the plasma, in hemocytes, or in both and usually as proenzymes. Cuticular extracts contain an activator for the proenzyme which in some species is said to be a protein and in others a lipid, but the proenzyme can often be transformed into an active enzyme by heat, a change in pH. surface-active agents, or heavy metals. The cuticular activator in *Calliphora* and *Drosophila* spp. is not species specific. In *Drosophila melanogaster* and *Lucilia cuprina* the larval hemolymph contains either more than one prodiphenol oxidase or a proenzyme which with different activators gives enzymes with different specificities, while Mitchell (1966) has shown that the proenzymes in *Drosophila* larve appear to be used for melanization in the adult. These observations may explain why hemolymph phenol oxidases are less specific than are cuticular phenol oxidases and, if used for sclerotization, why activity is still present in the hemolymph during and after sclerotin formation, Sin and Thomson (1971) have separated the hemolymph prophenol oxidases from *Calliphora* spp. into components similar to those isolated from homogenates of whole *Drosophila*.

Cuticular sclerotization occurring after ecdysis, in the few species which have been investigated, appears to be similar to that outlined for the puparium of flies. Work has been confined principally to *Periplaneta* but *N*-acetyldopamine is also considered to be the quinone precursor in *D. melanogaster, Schistocerca gregaria, Tenebrio molitor*, and adult *C. erythrocephela*. When dewaxed abdomens of *S. gregaria* albino nymphs are dipped into solutions of catechol or tyrosine the entire cuticle takes on the normal black pattern; 3,4-dihydroxybenzoic acid and dopa only cause an intensification in the amber colour of the cuticle. The absence of a differentiation along chemical lines (phenols containing or not containing nitrogen) indicates that sclerotization may be more specific than is generally thought.

Prior to cuticular hardering an adult *Calliphora* the cuticle becomes plastic, and Cottrell (1962b) suggests that plasticization may be a stage in the process of sclerotization. Many different labeled compounds derivable from tyrosine have been injected into insects and the percentage of the radioactivity incorporated into the cuticle measured. Tyrosine, dopa, dopamine, and *N*-acetyldopamine away give high rates of incorporation, very much higher than tyramine and its derivatives of *N*-acetylarterenol or *N*- acetylarterenone. Koeppe and Mills (1970) suggest that arterenol (noradrenaline) may be a sclerotizing substance in the cockroach cuticle, but Andersen (1971a) has shown in the locust cuticle that although this and related compounds are incorporated in low yields, it is unlikely that they are natural tanning agents since they do not give rise to the usual degradation products.

## THE PRODUCTION OF COLOUR

Beautiful colours or even the general darkening of the cuticle, are created by processes that may be quite apart from the events leading to hardening. In some species of wild bees, for example, certain areas of the cuticle may be perfectly transparent and reveal the colours of the tissue beneath. Yet they are as rigid as the adjacent black cuticle. Except for special mutant forms or some subterranean species all insects exhibit colours. Here colour will be discussed from three points of view: (1) the anatomical location of the colour-producing structure; (2) the manner in which the colour is generated; and (3) changes in colour during the life of the insect. With regard to the first topic, most colours are produced by the cuticle. When the cuticles is translucent or transparent, the underlying epidermal cells, hemolymph or other internal tissues may produce colour. Both the cuticle and the deeper tissues may contribute to the production of a single pattern of colour.

White light is composed of various wavelengths of light, commonly identified as red, orange, yellow, green, blue, indigo and violet. When an object or insect absorbs all but the red portion of the spectrum, we perceive a red colour, etc. The selective reflection of certain colours may be due to the physical structure of the object, producing *structural* colours; or it may be the result of selective absorption by certain chemical compounds, producing *pigmentary* colours; or a colour may be produced by a combination of both structure and chemical compounds together. When the cuticle contains the pigment or physical structure for colour, the colours often remain lifelike in museum specimens. On the other hand, when the colour resides in the cells or hemolymph beneath the cuticle, one may expect fading and discolouration as the internal organs decompose after death.

**Structural Colours**

To produce a structural colour, white light is altered in one of three ways; by the interference, diffraction, or the scattering of light waves. Interference colours are familiar as the iridescence of a thin film of old on water or a soap bubble. Both the upper and lower surfaces of the film reflect light. The thickness of the film will determine the distance these reflecting surfaces are separated. Since light reflected from the lower surface must travel a minute distance farther to reach the eye of the observer than the light from the upper surface, the former is said to be retarded. When waves are retarded by an odd number of half-wavelengths they will be out of phase with the waves of the same length reflected from the upper surface. In this case the two reflections cancel each other and only that part of the spectrum with the waves in phase will be seen. The colours vary because of the varying thickness of the film. Changes in the angle of observation will also change the thickness traversed by the retarded waves and likewise produce iridescence.

Interference colours are widespread among the insects. Least conspicuous, but extremely common, is the play if iridescent colours on the transparent wings of insects. The membranes of the wings have the same effect as thin films. More spectacular are the butterflies and moths whose scales are intricately modified to produce striking colours. The blue of the *Morpa* (Nymphalidae) arises from a series of vertical vanes on each scale. Each vane has a series of delicate ribs spaced at the correct distance for canceling or transmitting without reflection all wavelengths but the blue which is reflected. A series of superimposed laminae on each scale of the diurnal moth *Urania* (Uraniidae, Geometroidae) has a similar effect. Lamellate scales or lamellae within the cuticle itself may produce interference phenomena in certain metallic beetles. When underline by dark pigments, the interference colours are especially sharp, since stray reflections from internal organs do not detract from the purity of the reflected colours.

Only one insect, the beetle *Serica sericea* (Scarabaeidae), produces colour by diffraction. A parallel series of minute grooves, or striae, are evenly spaced at about 0.8 $\mu$m apart. These produce an iridescence on the elytra when viewed in direct sunlight. This is actually a diffraction grating which operates by reflection. An artificial impression of the close-spaced striae, made of collodion and peeled from the elytron, will exhibit the same iridescence.

The last type of structural colour, that produced by scattering, is not iridescent and does not change with the angle of observation. Minute particles, less than 0.6 mm in diameter, reflect more of the short waves of light than long waves, and produce a blue colour. The sky is blue, for example, because dust particles and atmospheric gases reflected more blue and violet than other colours. Without the impurities in the atmosphere to reflect these colours the sky would appear black. Such a colour is called *Tyndall blue* after an early student of sky colour. Among the dragonflies, granules in the epidermal cells produce a blue colour. Granules in cells in combination with yellow pigments will give green. Waxy secretions on the surface of the cuticle in other species of drgonflies also give the Tyndall effect. Larger granules reflect all wavelengths of light, producing a structural white. Depending on the arrangement of the reflecting surfaces, a dull, pearly, or silvery white can be produced.

**Pigmentary Colours**

The second major category, pigmentary colours, includes a much wider range of colours and a diverse assortment of chemical compounds which serve as pigments. These colours are not metallic or iridescent, but otherwise very enormously. They are often excretory products which are no longer involved in biochemical pathways but which now serve an important function in the ecology of the insect. Pigments may also be absorbed from the food of the insect.

One of the most common of the pigments among animals is melanin. The colours produced by this pigment vary from black and brown to light brown and yellow. The synthesis of each colour pigment is incompletely known. Nearly all colours in this range where formerly attributed to melanin. Some are now known to be the result of other chemical compounds, such as sclerotin. Melanogenesis of the black pigment involves the oxidation of tryrosine by the enzymatic action of tyrosinase, plus several additional steps, one of which requires air by way of the tracheal system. Once formed, melanin is insoluble in ordinary solvents, but it can be bleached if a trasparent exoskeleton is desired for study. Tyrosine and tyrosinase are present in insect blood but somehow kept apart. When an insect is injured the blood that is exposed to air often blackens. Such melanization in experimental insects is usually fatal and may be prevented by excluding oxygen from the surgery or by blocking the action of tyrosinase. The hereditary lack of tyrosinase produces the familiar albino mutant.

Melanogenesis may be partly influenced by the environment. For example, insects reared in the laboratory at lower temperatures often show increased melanization. In nature, as insect with two or more generations per year often displays distinctive colours associated with the seasons during which pigments were developed. The interaction of the environment and melanogenesis is not easily unraveled. The presence of darker populations of an insect in a geographic region of lower temperatures cannot be simply explained as entirely an environmental effect. Such individuals may be genetically predisposed to greater melanin deposition, and the darker colours may have survival advantages peculiar to the geographic region, irrespective of the lower temperatures. High humidity and darker colours have also been associated in the study of geographic variation of insects, but oddly there is little evidence for a direct physiological influence. Some environmental factors have indirect effect through the nervous or endocrine systems; certain wavelengths of light received by the eyes of an insect in the pupal stage may promote melanin deposition or, as in the migratory-phase locusts, the darker gregarious phase is associated with crowding.

Yellow, orange, and red colours are produced by carotenoid pigments derived from alpha and, more commonly, beta carotene which occurs in plants. Insects are believed to be unable to synthesize these pigments and therefore must depend on plants for their supply. Some predators acquire their pigment via plant-feeding victims. The pigments are absorbed and deposited in the epidermis or fat cells or in secretions such as silk or wax. Green colours which closely resemble the chlorophyll of leaves are produced by many kinds of insects. Thus far no experiments have shown that the green plant pigment is absorbed and used directly for the colour. In several different insects, however, the green is produced by a combination of a yellow carotenoid and a blue or blue-green pigment belonging to a group known as a bilin or bile pigment. The origin of the latter is not clear, but it is probably synthesized by the insect. Curiously, a diet of fresh green leaves seems to be important for the synthesis, but it is thought that some component other than chlorophyll is the needed constituent.

# 10

# Colour

Some pigments play a vital role in metabolic processes and their production is often linked with other processes in the body. Several different classes of pigment exist and, they are responsible for many of the colours of insects. Most whites, blues and metallic colours, on the other hand, result from the physical structure of the surface of the cuticle and not from pigments. Short-team, reversible colour change resulting from the movement of pigments only occurs in a few insects, but long-term changes in the deposition of pigment commonly occur. These often result in the insect matching its background, and if the background undergoes a permanent alteration the colour of the insect may undergo a parallel evolutionary change. The colour of many insects tends to conceal them from potential predators. Other insects have markings which help to fighten off predators, or they may have a conspicuous colouration associated with distastefulness so that predators soon learn to avoid them. Different species, which may or may not themselves be distasteful, may have a similar colouration to a distasteful species so that they benefit from the learned avoidance by predators of this particular colour pattern.

Colours are also important in intraspecific recognition, and sometimes excretory products are stored in a coloured from.

**The Nature of Colour**

Colour is produced from white light when some of the wavelengths are eliminated, usually by absorption, and the remainder are reflected or transmitted. The wavelengths of the reflected or transmitted component determine the colour which is seen (Fig. 10.1).

If all wavelengths are reflected equally the reflecting surface appears white; if all are absorbed the colour is black. Differential reflection of light to produce colours occurs in one of two ways; the physical nature of the surface may be such that only certain wavelengths are reflected, or pigments may be present which, as a result of their molecular structure, absorb certain wavelengths and reflect the remainder. Colours produced by these methods are known, respectively, as physical (or structural) and pigmentary colours.

## PHYSICAL COLOURS

Surface structures are mainly responsible for the production of whites, blues and iridescent colours. Such colours may be produced by scattering, interference or diffraction.

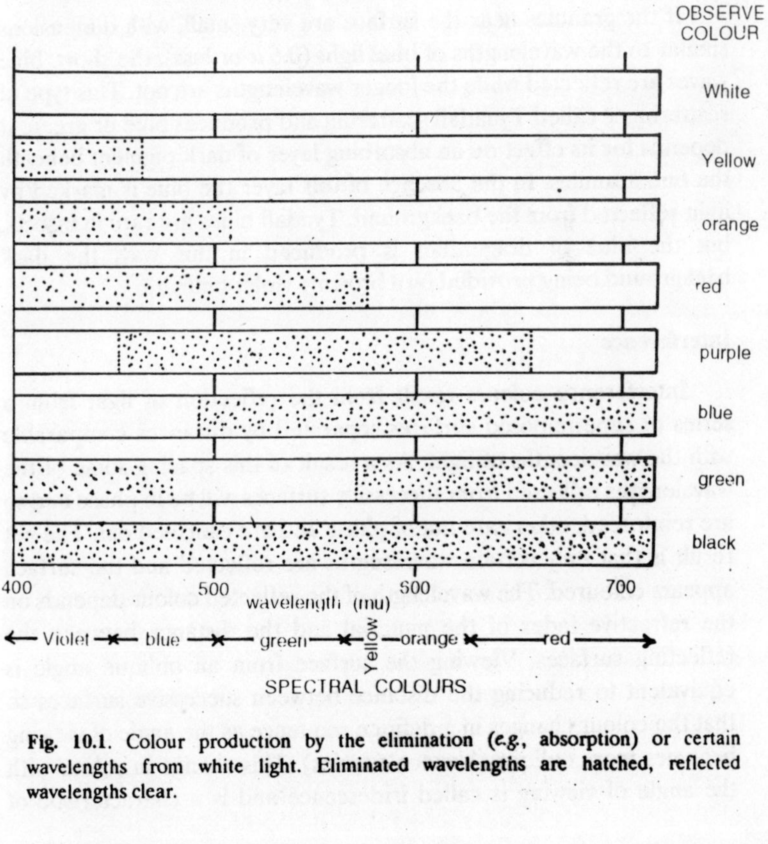


**Fig. 10.1.** Colour production by the elimination (*e.g.*, absorption) or certain wavelengths from white light. Eliminated wavelengths are hatched, reflected wavelengths clear.

### Scattering

Light may be scattered, that is, reflected in all directions, by irregularities of a surface or by granules just beneath it. If the irregularities or granules are large relative to the wavelength of light all the light is reflected and the surface appears white. Most whites in insects are produced in this way, although some white pigments also occur. Matt whites are produced by an even scattering of the light in all directions and in Lepidoptera, such as Pieridae, this results from deep longitudinal corrugations and fine, unordered striations on the surface of the scales. Pearly whites, such as occur in *Argynnis* (Lepidoptera), are produced by scattering from a number of thin, overlapping lamellae separated by airspaces. In butterflies the lamellae are the upper and lower laminae of overlapping scales.

If the granules near the surface are very small, with dimensions similar to the wavelengths of blue light (0.6 $\mu$ or less), the short, blue waves are reflected while the longer wavelengths are not. This type of scattering is called Tynadall scattering and produces blue or green. It depends for its effect on an absorbing layer of dark pigment beneath the fine granules. In the absence of this layer the blue is masked by light reflected from the background. Tyndall blues are rare in insects, but the blue of dragonflies is produced in this way, the dark background being provided by a brown – violet ommatin.

### Interference

Interference colours result from the reflection of light from a series of superimposed surfaces separated by distances comparable with the wavelengths of light. As a result of this spacing some of the wavelengths reflected from successive surfaces will be in phase and so are reinforced, others are out of phase and are canceled out. The net result is that only certain wavelengths are reflected and the surface appears coloured. The wavelength of the reflected colour depends on the refractive index of the material and the distance between the reflecting surfaces. Viewing the surface from an oblique angle is equivalent to reducing the distance between successive surfaces so that the colour changes in a definite sequence as the angle of viewing becomes more oblique (Newton's series). This change in colour with the angle of viewing is called iridescence and is a characteristic of

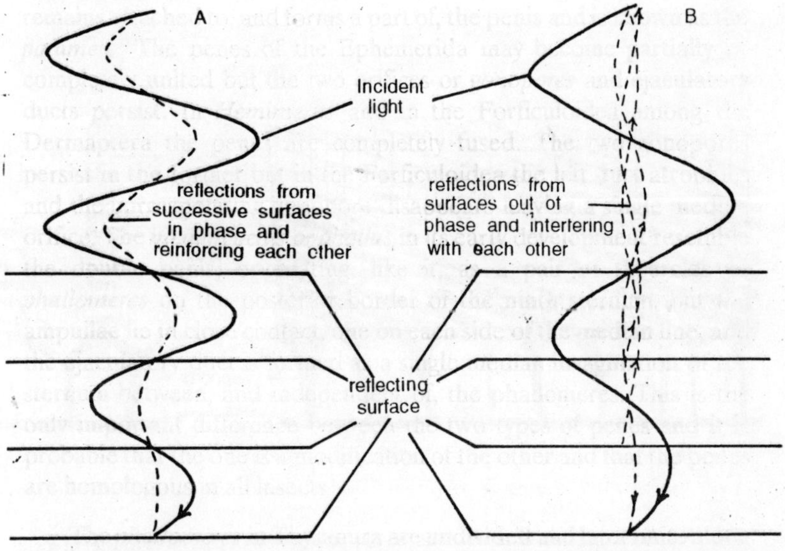


**Fig. 10.2.** Diagrammatic representation of colour production by interference. At each of the reflecting surfaces some light is reflected while the remainder is transmitted at a lower intensity (as indicated by the reduced amplitude). The diagram shows two components of incident white light, the wavelength of one of which (A) bears a simple relationship to the distance between the reflecting surfaces so that it reflections from these surfaces reinforce each other. The other component (B) has a wavelength unrelated to the distance between the surfaces so that its reflections interfere with each other.

interference colours. The brightness of the reflected colour increases with the number of reflecting surfaces.

Interference colours are common in Lepidoptera, being produced by the scales. In *Uranio* the iridescent scales are hollow with an upper lamina composed of five to ten lamellae which produce colour

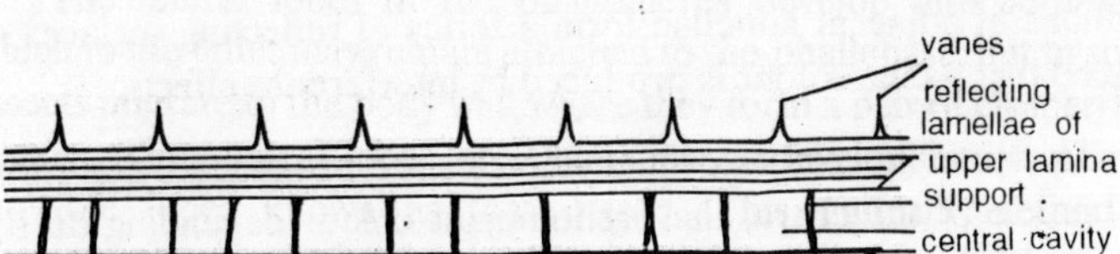


**Fig. 10.3.** Diagrammatic transverse section of a scale of *Uranio*.

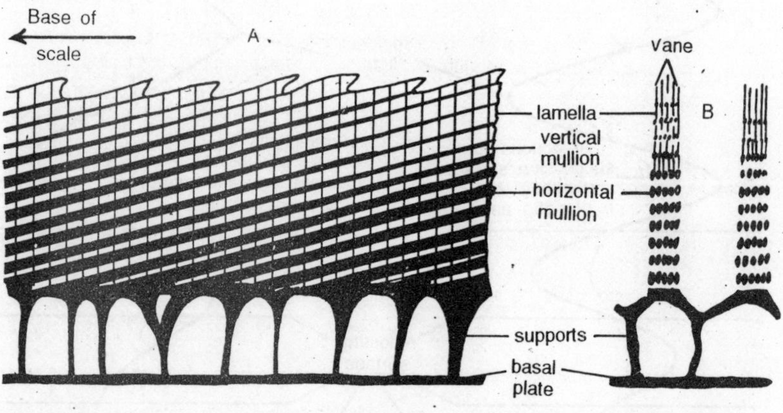


**Fig. 10.4.** A. Diagrammatic representation of a lateral view of part of an iridiscent scale of *Morpho*. B. Transverse section of part of the scale.

by interference. Viewed from immediately above, different scales appear green, blue or reddish-purple, depending on the spacing of the lamellae, changing to purple, orange and yellow-green when viewed obliquely. In *Lycaena* the iridescence is produced by the lower lamina of the scale seen through the grid-like upper surface. The blue of *Morpho* (Lepidoptera) is produced by a different type of scale which consists of a flat basal plate carrying a large number of vertical vanes running parallel to the length of the scale on Y-shaped supports. Each vane is made up of a number of delicate vertical lamellae supported by a series of vertical and obliquely horizontal mullions (thickenings). There are twelve horizontal mullions spaced about 0.19 $\mu$ apart and becoming thicker towards the base of the vane. Collectively the mullions of adjacent lamellae form a series of reflecting surfaces so spaced that a blue colour is produced by interference effects.

The metallic colours of many beetles are produced by layers in the cuticle of the elytra, the predominant colour depending on the spacing of the layers. A metallic effect is produced by an underlying layer of pigment. In the tortoise beetles, Cassidinae, the lamellae in the cuticle are separated by a fluid or a substance with a high moisture content. Normally the beetles are brassy-yellow or green, but the colours change on dehydration due to changes in the spacing of the lamellae. The occurs in life if the insect is disturbed and the colours change from gold to green to violet and finally to brown-orange, as the background pigment shows through, in less than one minute. After a short time the original colour is restored. The iridescence of membranous insect wings is due to reflection from the surfaces of a series of cuticular layers, 0.2 $\mu$ apart.

**Diffraction**

A series of fine grooves or ridges separated by spaces corresponding with the wavelength of light will split white light into its component spectral colours. This is known as diffraction. Diffraction colours are rare in insects, but the beetle *Serica* has transverse striations on the elytra at 0.8 $\mu$ intervals so that they form a diffraction grating. In diffuse light the beetle looks brown, but when viewed along its axis in a narrow beam of light the grating produces an iridescence.

## PIGMENTARY COLOURS

Pigmentary colours result from the molecular structure of certain compounds. Particularly important in the production of colour are double bons. C = C, C = O, C = N, and N = N, the number and arrangement of these being important. Particular groupings are also important. The $-NH_2$ and $-Cl$ radicals, for instance, shift the absorptive region of a particular compound so that it tends to absorb longer wavelengths. The colour-producing molecule, known as the chromophore, is often conjugated with a protein molecule, forming a chromoprotein.

**Brown and Black of Cuticle**

The black or brown colour of much insect cuticle is attributed to the pigment melanin, but, if the definition of melanin is restricted to compounds composed of polymerised indole rings, this is not entirely true. Hardening of the cuticle involves cross-linkages between the protein molecules, quinones providing the links , and some darkening does result from this tanning process. However, hardening and darkening may be independent of each other. Thus, albino *Schistocerca* have a hard, but colourless cuticle and when a normal locust moults it exhibits some dark markings before it hardens. This suggests that darkening may involve some deposition of melanin as well as the production of sclerotin.

Melanin synthesis involves tyrosine and DOPA-quinone:

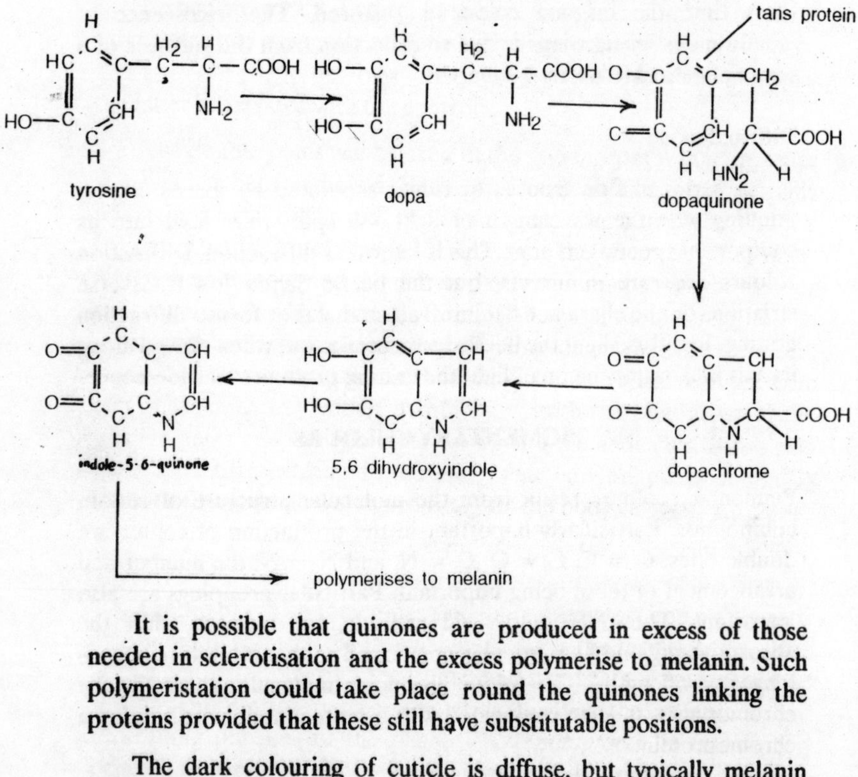


It is possible that quinones are produced in excess of those needed in sclerotisation and the excess polymerise to melanin. Such polymeristation could take place round the quinones linking the proteins provided that these still have substitutable positions.

The dark colouring of cuticle is diffuse, but typically melanin occurs in a granular from. Dark granules, which may be melanin, do exist in the epidermis of *Carausius*.

## Carotenoids

Carotenoids are a major group of pigments, soluble in fats and containing no nitrogen. They are built up from isoprene residues:

Carotenoids are of pant origin and are not synthesised by insects. There are two groups; the carotenes, and their oxidised derivatives, the xanthophylls. The latter may be obtained in the diet, but can also be produced by insects by the oxidation of carotene. Yellow, orange

$CH_2 \quad C(CH_3) \quad CH. \quad CH_2$

isoprene

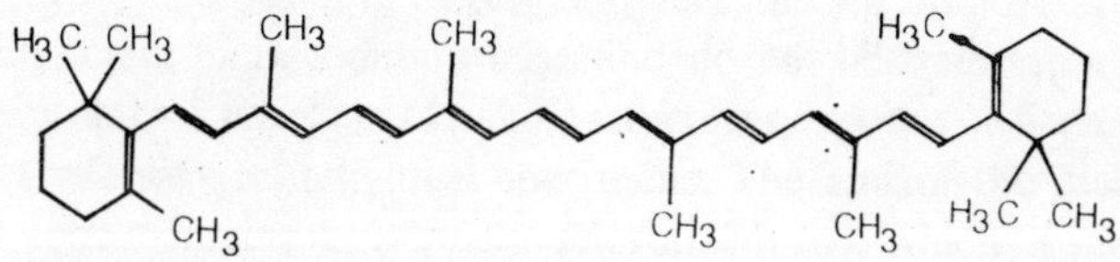


carotene

and red are commonly produced by carotenoids, the colour depending largely on the form of the terminal isoprene residues – whether or not they form a closed ring and the degree of unsaturation. The yellow of larval and mature adult *Schistocerca* is produced by $\beta$-carotenes which also occurs in internal organs, and in various insect secretions, such as the silk of *Bombyx* (Lepidoptera) and beeswax.

The red colour of *Coccinella* (Coleoptera) is due to $\alpha$- and $\beta$-carotenes together with lycopene, and the latter also produces the red of *Pyrrhocoris* (heteroptera). Astaxantin, a xanthophyll, is produced from carotene in the cuticle of *Schistocerca* and contributes to the pink colour of the immature adult, although this is mainly due to insectorubin. In grasshoppers of the genus *Oedipoda* carotenoid proteins produce the blue, red and yellow of the hind wings, of different species.

In combination with a blue pigment, usually mesobiliverdin, carotenoids produce greens (green produced in this way is sometimes known as insetovgerdin). The yellow component is $\beta$-carotene in *Carausius* (phasmida) and *Schistocerca* blood, $\beta$-carotene and astaxanthin in the integument of solitary *Schistcerca* larvae, and lutein, a xanthophyll, in the larva of *Sphinx* (Lepidoptera).

The functions of carotenoids in cell metabolism are not known, but small quantities are probably concerned in the production of the visual pigment retinene.

**Pterines (Pteridines)**

The pterines are nitrogen-containing compounds, all having the same basic structure, but differing in the radicals attached to this nucleus.

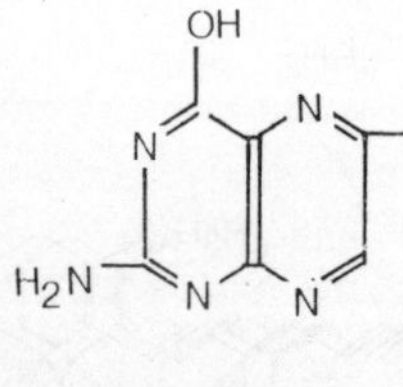

Xanthopterin

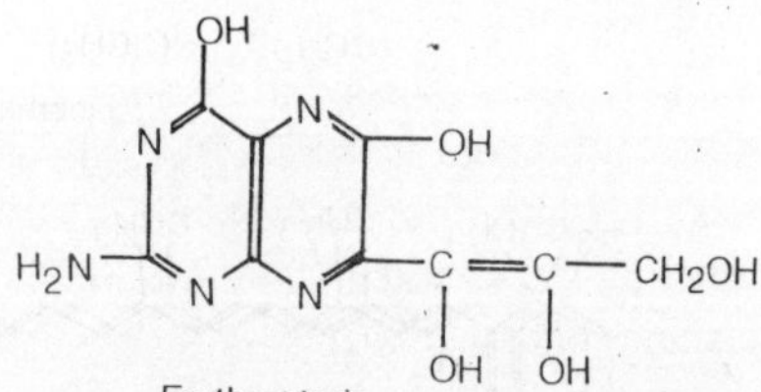

Erythropterin

Pterines may be synthesised from purines and the fall in the concentration of uric acid in the pupa of *Drosophila* when the eye pigments are synthesised supports this. Possibly flavins, also closely related compounds, are involved in pterine synthesis. White (leucopterin), yellow (xanthopterin) and red (erythropterin) are commonly produced by pterines, xanthopterin being the most widely distributed. Other pterines, such as biopterin, fluoresce in ultraviolet light, although they do not appear coloured in daylight. Pterines are important pigments in Lepidoptera. Leucopterin and xanthopterin are common in the wings of Pieridae where they supplement the structural white. The yellow of the brimstone butterfly is due to chrysopterin, the brighter colour of the male resulting from its higher concentration, while the red of the orange-tip butterfly is due to erythropterin. The yellows of Hymenoptera are produced by crystalline granules of pterine in the epidermis overlying areas of metabolically inactive tissue.

The pterines are also important eye pigments, occurring with ommochromes in the accessory pigment cells separating the ommatidia. In *Drosophila* five pterines have been isolated from the eye: two yellow compounds, another (isoxanthopterin) with a purple fluorescence, and two (including biopterin) with a blue florescence. Pterines are important metabolically as cofactors of enzymes concerned in growth and differentiation and they may act as controlling agents in these processes. Their association with ommochromes arises because they are cofactors of the enzymes involved in ommochrome synthesis. The vitamin folic acid is possibly derived from pterines.

**Ommochromes**

The ommochromes are a group of pigments derived from the amino acid tryptophan via kynure inine and 3-hydroxykynurenine.

Oxidative condensation of the 3-hydroxykynurenine gives rise to the ommochromes. This condensation is coupled with the production of DOPA quinone from DOPA, involving the same enzyme as in melanin formation. DOPA quinone acts as an electron acceptor, inducting the condensation. Ommochromes are widely distributed as masking pigments in the accessory cells of the eyes, serving to isolate the ommatidia.

Yellow, red and brown body colours are produced by ommochromes. The pink of immature adult *Schistocerca* is due to an ommochrome, innsectorubin, which is synthesised in the integument and then slowly decreases in amount as the insect gets older. Red Odonata and probably also the reds and browns of nymphalid butterflies are due to ommochromes, while in blue Odonata a dark brown ommochrome provides the background for the production of Tyndall blue. Ommochromes may be produced during the breakdown of proteins since they are found in the faeces of starved locusts, in the

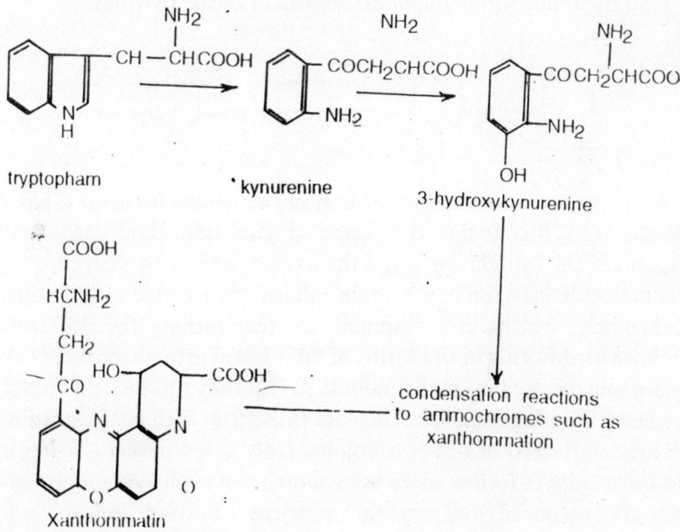

meconium of butterflies and also in the eyes of the vermilion mutant of *Drosophila* (Diptera) after starvation. Ommochromes are normally lacking from this mutant.

**Tetrapyrroles**

There are two major classes of tetrapyrroles, the porpohyrins in which the pyrroles form a ring:

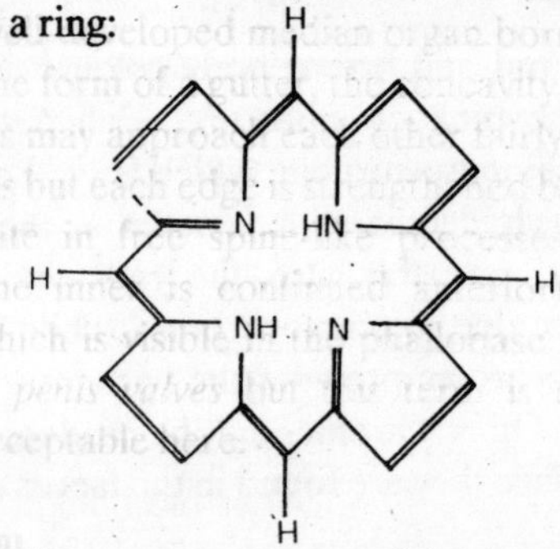


And the bilins with a linear arrangement of the pyrroles:

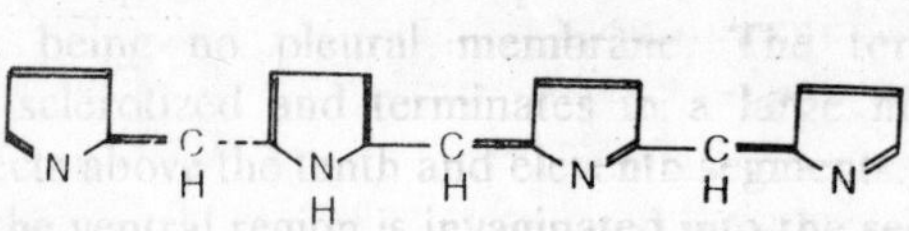


A porphyrin with an atom of iron in the centre is called a haem molecule and this forms the basis of two important classes of compounds, the cytochromes and the haemoglobins. In each case the haem molecule is linked to a protein. All insects are able to synthesise cytochromes, which are essential in respiration, the different cytochromes differing in the forms of their haem groupings. Normally they are only present in small amounts so that they produce no colour, but where as in flight muscle, they are present in high concentration, they produce a reddish-brown colour. Only a few insects, living in conditions subject to low oxygen tensions, contain haemoglobin and these are coloured red by the pigment showing through the integument. In *Chironomus* (Diptera) larva the haemoglobin is in solution in the blood, while in the larva of *Gasterophilus* it is in the fat body. Haemoglobin serves a respiratory function.

Bilins, bilirubin and biliverdin, may arise from the opening out of porophyrins as a result of oxidation. Typically they are blue or green.

In *Chironomus* Bilins from the haemoglobin of the larva accumulate in the fat of the adult and impart a green colour to the newly emerged fly. Similarly in *Rhodnius* the pericardial cells become green due to the accumulation of bilins derived from ingested haemoglobin. Mesobiliverdin is commonly present combined with protein and in association with a yellow carotenoid produces the green colour of many insects.

**Quinone Pigments**

The quinone pigments of insects fall into two categories: anthraquinones and aphins.

Anthraquinones are formed from the condensation of three benzene rings and three of these pigments, each produced by a different coccid, used to be important as commercial dyes. The best known is cochineal from *Dactylopius cacti*. The purified pigment is called carminic acid:

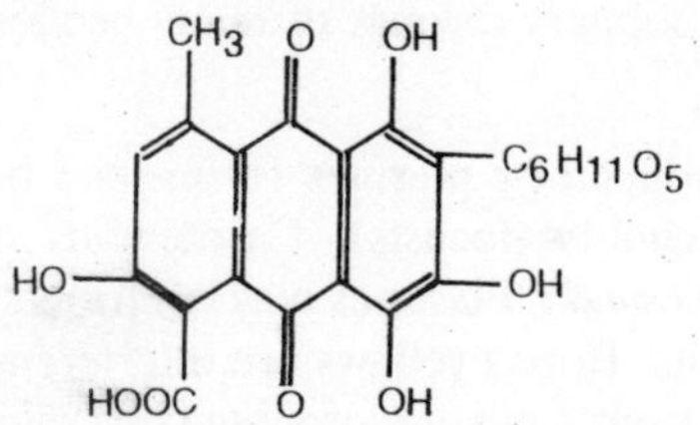


Carminic acid

The pigment, which is derived from the food-plant *Opuntia coccinellifera*, is present in globules in the eggs and fat body of the female, constituting up to 50% of the body weight. The male contains relatively little pigment. Aphins are quinone pigments with a nucleus of seven condensed benzene rings. They are found in the blood of aphids, sometimes in high concentration, and impart a purple or black colour to the whole insect. The series are known one characteristic of *Aphis*, the other of *Tuberolachnus*.

**Flavones (Anthoxanthins)**

These are plant pigments found in a few insects. They are responsible for the red colour of the bugs *Leptocoris* and *Lygaeus* and also occur in Lepidoptera. The yellow colour of the marbled white butterfly is due to a flavone obtained unaltered from the grass, *Dactylis glomerata*, on which it feeds. The related anthocyanins, responsible for the colours of many plants, have not certainly been identified in insects.

**The Colours of Insects**

The colours of insects may result from a variety of structures and pigments which it may be useful to summarise. Black and browns result from sclerotin and melanin. Aphins produce a black or very dark purple colour in aphids, while iridescent purples, as in the purple emperor butterfly, result from interference effects. Red is commonly produced by carotenoids (Coccinellidae, *Pyrrhocoris*), but may be due to pterines (orange-tip butterfly) or ommochromes (Odonata, probably the reds and browns of Nymphalidae, pink of immature *Schistocerca*). Pterines and ommochromes are also responsible for eye colours. Chironomid larvae may be red due to the presence of haemoglobin and some coccids have red quinone pigments. Orange-reds and coppery colours of many beetles are interference colours.

Yellow is produced by pterines (brimstone butterfly, Hymenoptera) and carotenoids (locusts). Carotenoids also colour some secretions (silk, beeswax). Flavones may contribute to yellow, usually with other pigments. Brassy yellows are interference colours. Green frequently results from a mixture of a blue bile pigment and a yellow carotenoid (*Carausius*, Lepidoptera larvae), but in Pieridae is produced by the juxtaposition of yellow and black scales. Bile pigments alone may produce green (adult Chironomidae) and in other (emerald moths) the pigments have not yet been isolated. Interference effects produce the metallic greens of beetles and Zygaenidae. Blues are usually produced by interference phenomena (Lycaenidae) and rarely by Tyndall scattering (Odonata). Blue pigments are uncommon, but *oedipoda caerulescens* (Orthoptera) has a blue carotenoid in the wings. White primarily results from scattering, sometimes associated with a white pigment, such as leucopterin in Lepidoptera.

## COLOUR CHANGE

Colour change may occur in two ways; by the physical movement of pigment or by the production or destruction of pigment. A change in colour produced by the movement of pigment is called a physiological change and differences in colour may be produced quickly. Colour changes involving the metabolism of pigments are called morphological changes and they take place more slowly.

## Physiological Colour Change

Physiological colour changes are unusual in insects. In *Carausius* (Phasmida) individuals which are brown in the daytime become black at night due to the movement of pigment in the epidermal cells. Similar changes due to the movement of pigment occur in the Australian grasshopper *Kosciuscola*, but in response to temperature rather than light. Above 25°C the males are blue, below 15°C. They are a dull black. Females show similar, but less marked changes. Under natural conditions the grasshoppers are black at night when the temperature is low and on a clear day they become pale two or three hours after sunrise as the temperature rises. They darken again in the late afternoon as it becomes cooler. In these changes the epidermal cells are independent effectors, responding directly to stimulation; nerves and hormones are not involved. The changes may have a thermoregulatory function since dark insects absorb more radiation than paler ones and *Kosciuscola* is a montane species with only a limited range of movement. Changes in the physical colours of Cassidinae as a result of disturbance and of Chrysididae with humidity are also known to occur.

## Morphological Colour Change

Changes in the amount of pigment may be produced in many different ways in response to external or internal factors. The colours of grasshoppers and related insects tend to have a general resemblance to the prevailing colour of the environment, a phenomenon known as homochromy, and a change in the environment leads to a change in the colour of the insect. Thus *Acrida* (orthoptera) changes from straw-yellow to green if put on to the appropriate background, some specimens changing in three days, but most taking rather longer. Such changes in the colour of the integument only occur at a moult, but this is not true of darkening which can occur in the absence of moulting. Black grasshoppers are commonly found in Africa after bush fires and some individuals of *Phorenula werneriana* change from grey to coal black in two days. This change occurs in recently emerged and old adults, but only in bright sunlight. In diffuse light the change is much less marked and in general the contrast between incident and reflected light is an important factor in this type of colour change. In *Mantis* (Dictyoptera) the green biliverdin breaks down in high light intensities forming products which are initially brown and subsequently become almost colourless. Differences due to background colour also occur in other groups. Pierid pupae, for instance, may be dark or pale according to their surroundings.

Temperature is important in pigment development. Locusts bread at 40°C have very little pigmentation and are pale yellow with a few dark markings. At lower rearing temperatures the darkening becomes progressively grater and at 26°C the larvae are largely black with some yellow pattern. Similar changes occur in Lepidoptera and probably also in other insects. Crowding influences colour in some insects. Locust larvae reared in isolation are green or fawn, while rearing in crowds produces yellow and black individuals. The colours and patterns changes as the degree of crowding alters. The larvae of some Lepidoptera, such as *Plusia,* undergo comparable changes, some of them occurring in the course of an instar, but the most marked alterations only at moulting.

Many changes occur in the course of development, the reasons for which are unknown. Thus the early larva of *Papilio demodocus* (Lepidoptera) is brown with a white band at the centre; the late larva is green with purple markings and a white lateral stripe. Some at least, of these changes are under hormonal control and just before pupation the larva of *Cerura* (Lepidoptera) turns from green to red, the change being controlled by the moulting hormone. Colour change is often associated with ageing and maturation. Male *Schistocerca* change from pink to yellow as they mature and *Mesopsis* (Orthoptera) slowly develops a black patch on the hind wings over a period of about six months, again possibly associated with maturation. The male of the dragonfly *Berachythemis leucosticta* develops dark patches on the wings as it matures.

## Seasonal Colour Change

In some insects marked differences in colour occur between successive generations correlated with seasonal changes in environment. This is particularly well shown in some African butterflies such as

*Precis octavia* which is rusty red in the wet season and violet-blue in the dry. In butterflies from temperate regions the spring and summer generations often look different and in *Arachnia levana* they are quite distinct. The form which develops depends on the length of day to which the larvae are subjected. Larvae experiencing short days give rise to diapause pupae and adults of the spring form larvae experiencing long days have no diapause in the pupa and the adults are of the summer form.

**Long-term Evolutionary Changes**

The best known evolutionary colour changes are those associated with industrial melanism. During the last 120 years there has been a marked increase in the incidence of melanism amongst moths which is well shown by *Biston betularia*. The typical form is peppered black and white, while the melanic form *carbonaria* is almost entirely black. In 1848 form *carbonaria* was a rarity in Manchester area, but by 1895 it constituted 95% of the *B. betularia* population of this region. At the same time and subsequently it has spread through Britain so that in 1958 the black form was known from all parts of Britain except the extreme west of England and the north of Scotland, often being the dominant form.

*Biston betularia* is a night-flying moth spending the day at rest on tree trunks. Originally the typical form blended well with its background of bark an lichens, but industrial pollution has blackened the bark and killed the lichens so that over the past 150 years the background against which these moths rest has been completely altered and form *carbonaria* is now much better concealed than the typical form. Field experiments show that in a polluted wood birds take more typical forms than black forms. The effects of pollution are not restricted to industrial areas because the prevailing south-westerly winds drift dust over the greater part of the country so that even in the less industrialised east of England form *carbonaria* now comprises at least 80% of the population.

As well as increasing protection from predators, the form *carbonaria* is also more viable than the typical form under prevailing conditions and the rate of spread of this form indicates a 30% advantage over the paler variety. Some 70 to 100 of the 780 macrolepidoptera found in Britain are undergoing similar changes, but only these species which spend the day on backgrounds affected by pollution and which depend on concealment are involved. Similar changes are occurring elsewhere in Europe and North America but they are not known from the tropics.

## SIGNIFICANCE OF COLOUR

The pigments of insects probably have some metabolic significance, but, it addition, the colours which they produce are of significance in the relations of the insect with other animals. Colour is frequently used as a defense against vertebrate predators and it may also be important intraspecific recognition.

Colours are interpreted in terms of human vision, but their appearance, and hence significance, may be different for other animal groups. For instance, the male brimstone butterfly, which to us appears uniform yellow, has markings on the forewing which become visible in a photograph takes on film which is sensitive to ultraviolet. Atleast some insects are sensitive to ultraviolet so that the appearance of this butterfly to them might be quite different from its appearance to humans. Of the potential predators of insects, birds and probably also lizards, are relatively less sensitive to light of short wavelength, but more sensitive to long wavelengths, while insectivorous mammals are believed to be colour blind. This, however, only means that they may be unable to discriminate between light of different wavelengths, it tells us nothing of the range of wavelengths to which they are sensitive.

### Concealment from Predators

Colour often helps to conceal insect from predators. This may result simply from a general similarity of colour between the insect and its background (homochromy), as in grasshoppers which become black on burnt ground or green in fresh grass. Given the choice these insects are able to select a background of the appropriate colour and experiments on *Biston betularia* show that this resemblance and behaviour do afford the insects some protection from predators. Often homochromy is associated with some appropriate body form and behaviour as in Phasmida and many mantids and grasshoppers. Protection may also be afforded by obliterative shading. Objects are made conspicuous by the different light intensities which they reflect

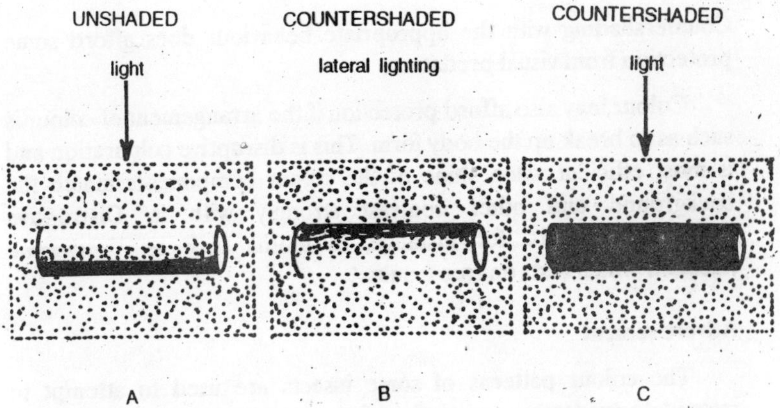


**Fig. 10.5. Diagrams to illustrate the appearance of (A) an unshaded object in overhead light, (B) a countershaded object in lateral light and (C) a countershaded object in overhead light.**

as a result of their form. Usually a solid object looks higher on the upper side and darker beneth because of the effect of shadows, but by appropriate colouring this effect can be eliminated. The object is shaded in such a way that when viewed in normal lighting conditions all parts of the body reflect the same amount of light so that it loses its solid appearance. Such countershading is well-known in caterpillars where the side towards the light is most heavily pigmented and the side normally in shadow has least pigment. To be successful this type of pigmentation must be combined with appropriate behaviour patterns since if the larva were to sit with the heavily pigmented side away from the light it would become more, not less, conspicuous.

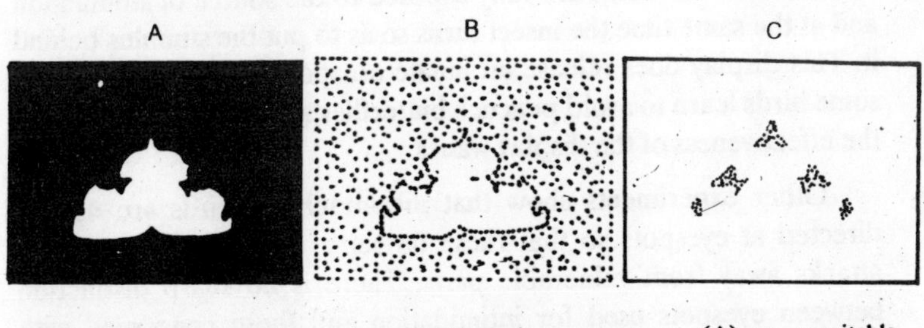


**Fig. 10.6. Diagram of a moth with distrupitve pattern on (A) an unsuitable background, (B) and (C) background blending with one of the insect colours and contrasting with the other.**

Countershading with the appropriate behaviour does afford some protection from visual predators.

Colour may also afford protection if the arrangement of colour is such as to break up the body form. This is disruptive colouration and is most efficient when some of the colour components match the background and other contrast strongly with it. Distruptive colouration occurs, for instance, in moths such as *Xanthorhöe fluctuata* which rests on tree trunks.

**Advertisement**

The colour patterns of some insects are used to attempt to intimidate predators or to deflect their attention to parts of the body which are least vulnerable. Eyespots play an important part in this behaviour and they are found in various positions on the wings of Lepidoptera and some other insects. The peacock butterfly has one eyespot on the upper surface of each wing. These eyespots are primarily black, yellow and blue, surrounded by dark red. At rest the butterfly sits with its wings held up over it back, the upper surfaces of the forewings being juxtaposed so that the eyespots are concealed. If the insect is disturbed by visual or tactile stimuli it lowers the wings so that the eyespots on the forewings are displayed and then protracts the forewings so as to expose the hindwing eyespots. At the same time the insect makes a hissing sound by rubbing the anal veins of the forewings against the costal veins of the hindwings. The forewings are then retracted and partly raised and the sequence of movements repeated, sometimes for several minutes. While displaying, the body is tilted so that the wings are fully exposed to the source of stimulation and at the same time the insect turns so as to put the stimulus behind it. This display does release an innate escape response in birds and some birds learn to avoid peacock butterflies altogether, but in others the effectiveness of the display wanes.

Other experiments show that the attacks of birds are usually directed at eyespots so that these do serve temporarily, to deflect attacks away from vulnerable parts. There is no sharp distinction between eyespots used for intimidation and those concerned with deflection. In general it may be that deflecting spots are smaller than these used in intimidation but it is possible that some may serve either function depending on the nature and experience of the predator.

Other types of colour advertisement by insects are associated with distastefulness so that predators learn to associate a particular colour pattern with distastefulness and so, subsequently, avoid taking insects displaying this pattern. Such patterns must be bold and readily recognisable and frequently involve red or yellow with black. Thus, the black and yellow of the bee and wasp is associated with a sting and the red and dark green of the burnet moth with extreme distastefulness.

In experiments, toads learned to avoid bees, some after eating only one bee, other only after repeated trials, but after week all of 33 toads had stopped eating bees. The distasteful association was remember by the toads so that when they were offered bees again two weeks after the original experiment very few of them would accept the food. Complete avoidance was soon re-established in spite of the fact that, apart from the bees, each toad had eaten only one mealworm during the whole experiment.

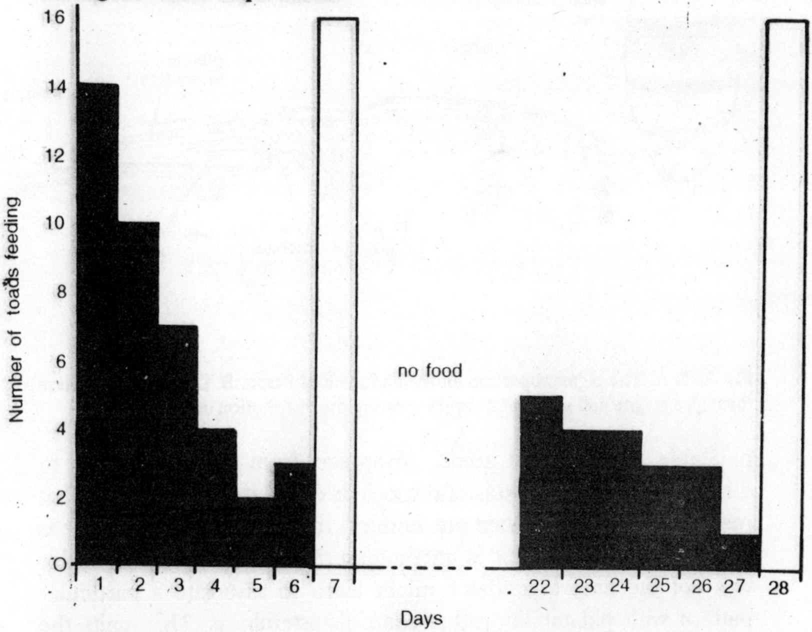


**Fig. 10.7.** The number of toads out of 18 which accepted bees (black) on successive days and the number which accepted mealworms (hatched) when they would no longer take bees.

## Mimicry

Predators learn to avoid distasteful insects with distinctive colours, but theoretically a predator must learn to avoid each individual species separately. If, however, the colour patterns of some species are similar to each other, learning to avoid one also produces an avoidance of the other so that it is an advantage for distasteful insects to look similar because in this way wastage is reduced. Thus various wasp species have the same basic black and yellow pattern, and burnet and cinnabar moths have comparable red markings on a dark green or black ground. Resemblance of one species by another is called mimicry and the principle on which it is based may be extended to include species which are not distasteful or are only mildly so. If a predator learns to avoid a particular pattern because the insect having it is distasteful it will subsequently avoid other species with the same pattern, even though they may not be distasteful. As a result of this,

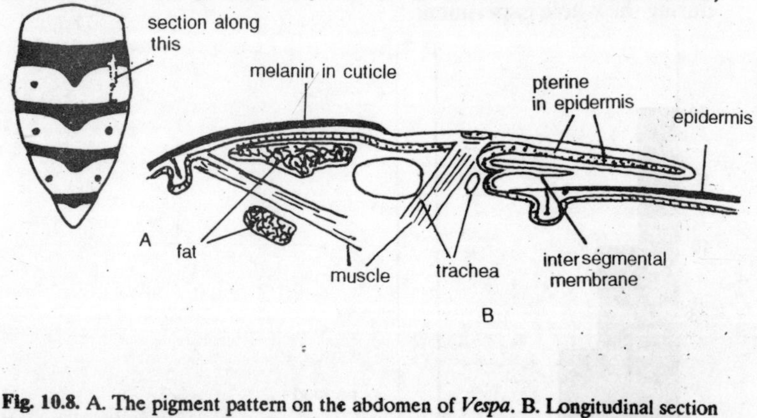


**Fig. 10.8.** A. The pigment pattern on the abdomen of *Vespa*. B. Longitudinal section through a tergite and associated tissues showing the distribution of the pigments.

palatable species gain some advantage from a resemblance to distasteful ones. The distasteful specie is called the model, those that resemble it in appearance are mimics. In this type of mimicry it is important that the mimic is uncommon relative to the model. If this was not the case a predator might learn to associate a particular pattern with palatability rather than distastefulness. This limits the numbers or distribution of a mimetic form, but such a limit may be circumvented by the mimic becoming polymorphic with each of the morphs resembling a different distasteful species. The best known example of genetic polymorphism is that of the female *paplio dardanus* (Lepidoptera) which has a large number of mimetic forms mimicking a series of quite different butterflies.

**Intraspecific Recognition**

In some insects colour is important in the recognition of one sex by the other. Thus, the male of *Hypolimnas misippus* (Lepidoptera) responds to the brown of the female wings by pursuing her, but his response is inhibited by the presence of white in the wing. Similar use of colour is known in other insects.

# 11

# Mimicry and Camouflage

Mimicry is a special kind of protective adaptation serving to ensure better survival of the species. When a perfectly harmless animal resembles in its colour and shape, with a well-protected species, the phenomena is called *mimicry*. The doctrine of mimicry was founded in 1861, in Darwin's time, by the English naturalist H.W.Bates. In a single place in the primeval forest of Brazil he saw numerous brightly coloured butterflies that, although of different families, were so similar in colour, pattern, and behaviour that they could not be distin-

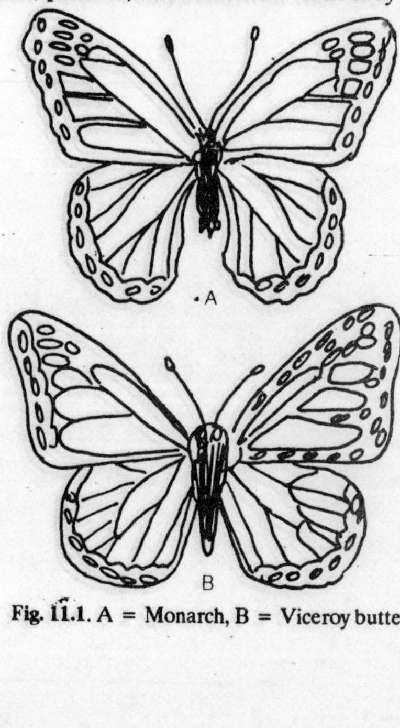


**Fig. 11.1.** A = Monarch, B = Viceroy butterflies.

guished in flight. Bates correctly assumed that one of the species in such a group, thanks to a disagreeable smell or taste, or to poisonous by fluids, was more or less shunned by birds. The several other, edible species, then, would have the advantage of being let alone because they looked like the undesirable species.

A variation in mimicry was recognized shortly thereafter by a German scientist, Fritz Müller. He also saw great groups of similar appearing butterflies. But in each of these groups two or more species were distasteful to birds. By their resemblance to one another, these distasteful species apparently increased their protection. In total individuals, the distasteful species outnumbered the intrinsically unprotected species in the company.

Later, as many as 19 species of similarly colored butterflies from different families and genera were counted in a mimicry ring. In some species of this nature only the females bear the common uniform. The males are less in need of such protection from the viewpoint of the species and also less likely to be flying in companies of this sort. They wear quite different grab, of a type characteristic of their genus and species. Butterfly mimicry occurs everywhere in the tropics. In Africa, *Papilio dardanus*, a species that ranges from the deep south as far north as Ethiopia, gives a dramatic performance of mimicry. The male occur throughout the range in the normal, long-tailed costume of the papilios. But the females have a different colouration and wing shape in each locality. In several places, the females are deceptively like other, inedible butterfly spaces. Some 24 distinct geographical races

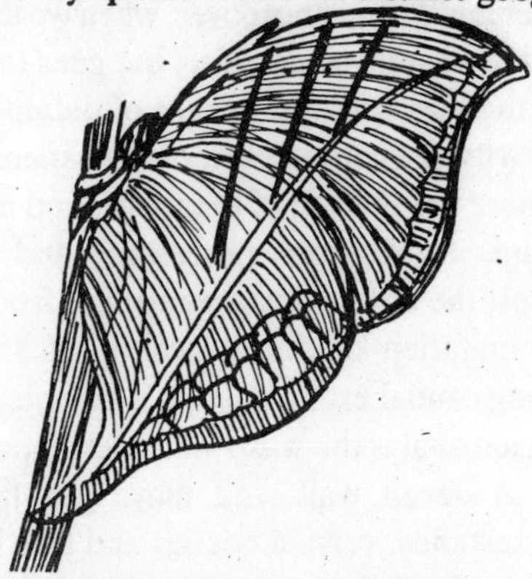

**Fig. 11.2. Kallima showing concealing mimicry.**

of the females have been identified. In rare instances, difficult to explain, sexually dissimilar patterns do not depend on geographical separation of races, but are found among the offspring of a single pair.

Many kinds of insects besides butterflies are protected by their stench or by poisons, as well as by sharp spines and various other defense measures. And many are the defenseless and unprotected insects that owe their lives to a similarity to these defensively equipped species. One must be careful, however, in designating examples of mimicry. Insects often show a remarkable resemblance to others with which they have no contact. This resemblance sometimes is incorrectly called mimicry. But true mimicry can occur only between different species that live together in the same place at the same time. How it happens that two widely separate species of insects come to resemble one another is unknown.

It is not hard to see how harmless insects might profit by a resemblance to wasps or bees. Although hunted by certain birds, wasps and bees are generally avoided because of their defensive stings. So we are not so greatly surprised that some flies and beetles, thanks to their colouring, or that certain butterflies with especially narrow, usually extended transparent wings, display such a general similarity. But we find it strange that certain of the true bugs or even grasshoppers have taken on the appearance of insects so foreign to their nature as ordinary wasps or parasitic wasps.

Our amazement grows, moreover, when we discover that mimicry is not confined to general similarities, but goes to remarkable lengths in duplicating the outward appearance of definite individual species. Thus, the similarity of a certain butterfly to a species of wasp is refined to the point where the illusion of the wasp's red mouth is given by the butterfly's palpi. (mouthpart processes) and forecoxae (foreleg segments nearest the body). Another species from the same family of butterflies, in imitation of the ovipositor of a parasitic wasp, has developed a long caudal extension that is completely alien to butterfly anatomy. Also unusual is the wasp-waist developed by certain beetles. Such models of deceit walk and move just like the insects they resemble. For instance, certain beetles and true bugs disregard their own heritage and hold their wings as the wasps do; and, in further imitation, they wave their legs and antennae, and arch their customarily straight abdomen upwards in the middle, bending the tip downward.

Among beetles, the net-winged beetles (Lycidae) are particularly unpopular as food for birds because of their toxic or at least unpleasant blood. Many net-winged beetles are resembled by beetles of other families, particularly the long-horned beetles. Some of the latter mimic certain yellowish-brown net-winged, beetles. Both have black stripes running across the broad elytra (fore wings) that, oddly enough, are flattened like the wings of moths. This mimicry circle has been joined by an entirely different order, namely moths of the family Syntomidae. Most species of this family of moths have an entirely different appearance. Moths whose legs and motions and caterpillars whose shapes are like those of bristly spiders are less widely known, but are extraordinarily striking. It is interesting to see how a moth of the family Noctuidae moves its monstrous legs, and perhaps sidles a little, as a spider does. Moths of the family Limacodidae also mimic spiders: at times during their day-long sleep they hold up their thick hairly legs in spectral stiffness. In addition, one might mention the rather rare similarity to ants displayed by certain true bugs and grasshoppers, particularly in their immature stages. There are, in addition, a few small spiders that look like ants.

**Natural Selection**

The existence of mimicry can be explained by the theory of natural selection, which holds that the fittest are selected by nature to survive. For example, the enemies of insects, mainly the birds, prefer

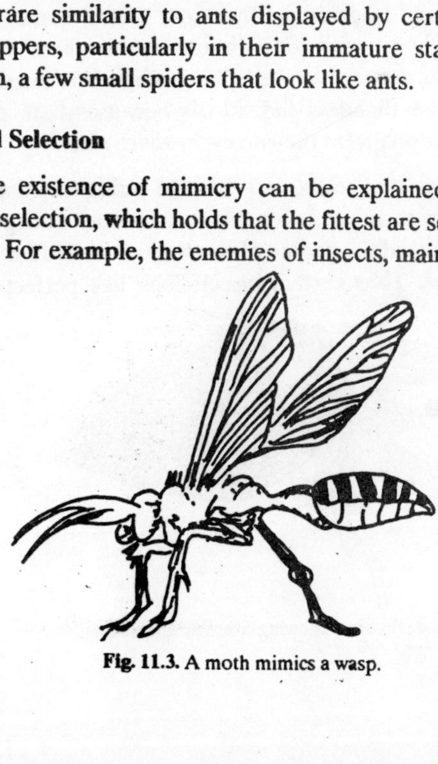

**Fig. 11.3.** A moth mimics a wasp.

certain species of insects as food. The species of insects that are the least desirable as food have the best chance of surviving and reproducing their own kind. The characteristics that enabled these species of insects to survive are passed on to succeeding generations and tend to become perpetuated. But before natural selection can begin, there must be something to be selected – in this example, the least desirable insects. A clear example of such natural selection has occurred recently in centers of heavy industry in Britain, where the light-coloured individuals of the peppered moth gradually have been replaced by dark variants, which elsewhere are present in only a small minority. In industrial areas the dark forms have a better chance of surviving long enough to reproduce, because their enemies can find them much less easily than their light-coloured relatives on the soot-covered treetrunks. This kind of selection could begin, however, only because of the presence of mutants that were dark and therefore adjusted to the changed state of affairs.

As another example, flying insects are at a disadvantage on certain islands at sea, because these islands are exposed constantly to stormy winds that tend to carry the flyers out to sea. Some of these islands now provide homes only for flightless insects, whose wings are reduced to rudiments or may even be lacking altogether. But the wind could only select the flightless individuals because these cripples occurred there occasionally, as they do everywhere else.

One can suppose that selection is able to refine an adaptation. But it can scarcely be responsible for the most complete adaptations; for often these far exceed the degree of adaptation needed for preservation of the species. Thus certain insects look like perfect leaves,

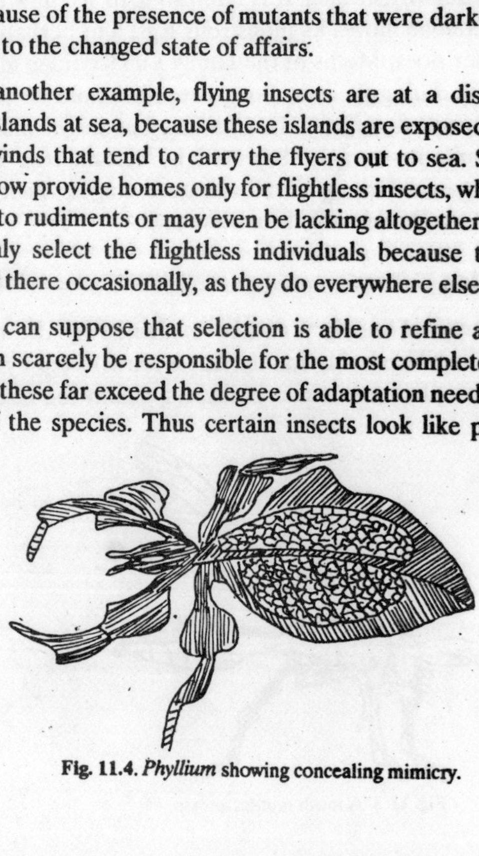

Fig. 11.4. *Phyllium* showing concealing mimicry.

whereas other insects even resemble leaves with such characteristic blemishes as moldy spots or raggedly chewed margins. But one can hardly assume that the second group is any less likely to be found and eaten by birds than the first group. Incomplete mimics may endure equally as well as complete mimics; and it may be that mimicry, as well as protective patterns and warning colouration, is responsible for the survival of only a small percentage of individuals. Even so, it may serve as a decisive safeguard for the preservation of the species.

**Protective Colouration and Form**

The commonest colours among insects are gray, yellow, and brown, in all their shades, and green; these are the colours of the ground and of plants, and thus aid in the concealment of insects. Countless insects have the simplest grab and shape, remain more or less hidden when at rest without any special expenditure on their part, and survive very well. And yet, inconspicuousness alone evidently will not suffice for all insects, for we see a wide range of protective costumes an an intriguing play between costume and posture. Some insects flatten themselves against the background, often aided by

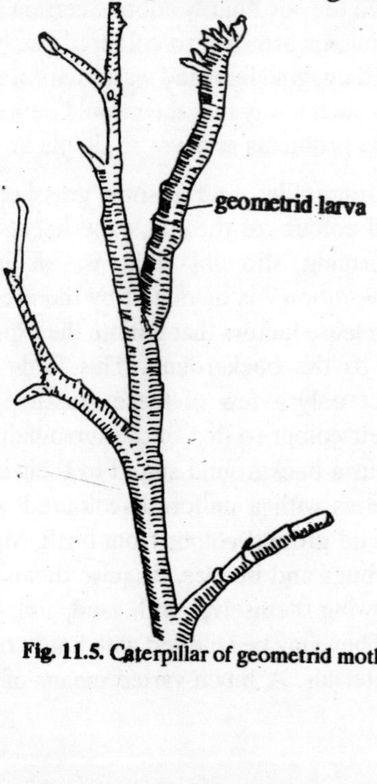


**Fig. 11.5.** Caterpillar of geometrid moth.

fringes of hairs along the lateral lines of the body or by other extensions that prevent them from casting a sharp shadow. Certain moths turn from a head-upward position on a tree trunk through an angle of 90° degrees, until their spread wings point one upward and one downward. As a result, the pattern of the moth merges precisely with the sculpturing of the bark.

Many green caterpillars have a light-coloured back and a gradually-darkening ventral surface; they hand upside-down on leaves and twigs, with their darker surface turned upward. Such a caterpillar is hardly more prominent than the flat leaves surrounding it, because its own shadow darkens its light-coloured back, white light from above illuminates its underside. Any impression of depth that the body might give is thus eliminated. The same use of this antagonistic shading is seen in certain caterpillars that cling vertically to hanging leaves; the body is light-coloured in front and grows darker and darker toward the rear.

When these insects do not take their intended stance relative to the source of light, they are more conspicuous than they would be with plain colouring. Do they ordinarily adopt a certain position because it suits their colouring, or are they so coloured because this is the way they sit? One could imagine light and warmth from above affecting the epidermal cells in such a way that they would deposit more or darker pigment. Yet these problems ar never so simple as they seem.

In nymphal dragonflies, and in some grasshoppers that may be peppered with all colours of the soil, experiments have determined that a colour-forming stimulus – that is, an imprinting of the environmental coloration – is mediated by the eyes. In other words, the optic nerves release factors that initiate the expression of a colour pattern adapted to the background. This finds fulfillment at the ensuing molt. But only a few of these grasshoppers "voluntarily" accommodate their colour to that of the surroundings. Most of them appear to seek out a background suited to their individual colouration, or, in localities with a uniformly-coloured soil, all individuals have an appropriate ground-colour from birth. Many young insects, especially young bugs and beetles, disguise themselves in a peculiar fashion by bestrewing themselves with sand, soil, or even with their own excrement. They may be studded with hooks or bristles that help to hold these materials. A much-varied means of optical deception

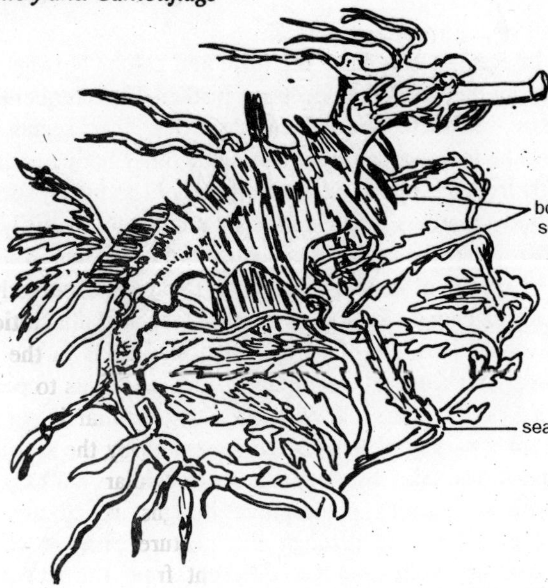


**Fig. 11.6.** *Phyllopteryx* **showing concealing mimicry.**

practiced by butterflies and other insects is the apparent dissolution of the form of the body (somatolysis) by means of strikingly sharp, visually disruptive designs. These may arrest the hunter's attention and thereby leave the true contours of the insect unrecognized. As is usual with all protective dress in insects, such effects are closely associated with the resting positions where there can be a cooperative image that extends over both the body and the wings.

Other performances aimed at invisibility go still further. False twigs, thorns, bits of wood, lichens, leaves, seed pods, and bird droppings are formed. We are peering into the curious, buffoonish world of mimesis. The ordinary shape and venation of many insect wings give them a leaflike appearance. But this is not enough. Extravagant refinements far beyond any necessity are added. The venation of the simulated leaf, for example, frequently runs directly counter to the insect's own vein-patterns. In addition, the "leaf" may appear to have translucent holes, and, as already mentioned, spots of mold, as well as irregular edges that look as though insects had gnawed on them. Even the leafstalk so sometimes included, either appearing to be hanging unattached in the air or to be connected to a twig.

Many insects not only have the form and colour of twigs, bark, and lichens, but they carry the deception further. They frequently rest in a posture, characteristic of each species, that seems quite impossible. The forepart, abdomen, or some other body region may extend stiffly from a support, or almost the whole body may stick out. Often the body contours are distorted by undulating or stiffly erect tufts of hair. A person may be acquainted with hundreds and thousands of species of moths, having seen them predominantly with symmetrically extended wings, as is usual in technical illustrations of in museum collections. But if he spends a few nights in the virgin tropical forest and observes the moths at rest, he begins to perceive how little he known, after all. What was long familiar is not to be discerned, for a variety of figures are generated by the specifically different positions take by fundamentally similar moths. These together with the moth's picutureque designs, which are given complete expression only through the posture peculiar to each species, make an impression far different from the stereotyped arrangement of a museum drawer. Many insects, such as certain butterflies, dragonflies, bees, and wasps, assemble in sleeping companies. Pressed tightly together, they often cling to twigs or straws by means of their legs or mandibles. At times they give the impression of a spray or bed of flowers. This is exemplified beautifully by little mothlike homopterous insects of the family Flatidae, whole groups of which join in sucking the juices of plant stems.

The nymphs of certain praying mantids resemble the flowers on which they sit. The resemblance results not only from their shapes and colours, but above all from their postures. Indeed, many mature mantids closely resemble either green leaves or dry leaves, an imitation whose effectiveness is enhanced by the peculiar to-and-fro movement of the insect, as though it were a leaf swaying in the wind. Particularly unusual are those few species whose prothorax bears varicoloured ventral developments that simulate a flower; when these are exposed, they attract other insects, which fly straight into a waiting pair of prehensile claws. Bright colours on the thorax and forecoxae are normally kept hidden by many mantids, only to be displayed abruptly— flashing into view and seeming to strike an enemy with terror— when the mantid attacks. These phenomena lead us to consider the complex of bright and varied colours that have irritant, startling, or warning effects.

**Flash Colours**

Concealing (cryptic) grab combined with brightly colored adornment that is hidden when the insect is at rest appears most often among the butterflies, grasshoppers, walking-stick insects, true bugs, and members of the Homoptera. The signals that shine forth so unexpectedly when the animals are disturbed are known as flash colours. Occasionally they occur on the abdomen, but almost always they are on the hind wings. They are found, as a rule, on insects that lack weapons of defense. In species that take flight hastily they usually have the form of a red, yellow, blue, or more rarely a white patch whose contrasting effect is heightened by black stripes or spots.

These contrasting colours may well cause enemies to become excited, at least initially. The pursued thus gains time for a mad, short flight, then dives into a new resting place and makes the conspicuous colours vanish suddenly by covering the bright spots on the wings or body with the nondescript fore wings. This is observed most frequently in grasshoppers, whose contrasting colouration appears on the hind wing and on the inner surface of the hind femur. But probably this coloration in grasshoppers functions primarily as an optical means for mutual attraction of the sexes.

**Eyespots**

Slower-moving or poorer-flying insects, or those that cannot take to the air fast enough when they are at rest because the first must raise their body temperature by means of respiratory or other vibrations, frequently display strikingly peculiar designs instead of simple contrasting colours. With the aid of twitching movements they create strangely nightmarish sights that are to us suggestive of evil. For example, hanging from a tree branch is an unimpressive-looking "leaf" that perhaps does not quite fool a smart bird. The bird gives it a tentative peck, and the "leaf" at once drops to the ground. There, it dances a while in a regular, clumsy rhythm, looking like a mask dominated by two now visible goggle-eyes, each of which is supplied with a pane of "glass" that flashes in the incident light. But soon it stops in a stiffly erect position. The bird hesitates,withdraws a little, moving about indecisively nearby, and then returns—to nothing. During its dance the moth was making itself ready to fly; and now a "leaf" again is hanging, but in a quite different place, where none was before.

The greatest abundance of such "startling eyes" is presented by the family of the giant silkworm moths (*Saturniidae*). Many of them do

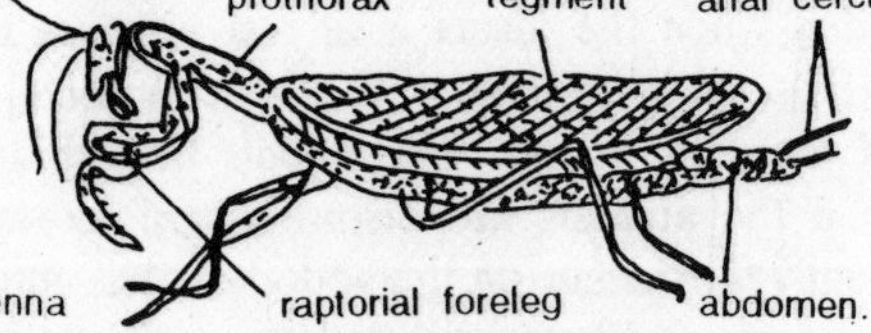


**Fig. 11.7. Praying mantis.**

not drop at once when disturbed, but are content to spread their wings suddenly and perhaps let them play in a slow rhythm of rising and falling. The central element of the saturniid pattern is an eyespot on each of the four wings. Usually it is effaced or at least reduced on the undersides.

In many species the pots on the upper surfaces of the fore wings have disappeared, and those on the hind wings are more prominent. A minority of these species has eyespots on the undersides of the fore wings, too, so that eyespots can be seen simultaneously from above or below. These species settle less frequently than others on flat surfaces that would block the underside spots from view. When there are four equally developed "eyes" on the upper wing surfaces, these are visible when the moth is at rest. The insect's active participation in the surprise effect has been eliminated.

An unusual eyespot design is displayed in the South American forests by huge butterflies of the genus *Caligo* that fly briefly at dusk. Like other related butterflies, they sit at rest in the shrubbery near the ground, with their wings clapped together over the back, excellently disguised by their wavy brown-and-light-gray colouring. But, strange to say, the disguise is broken by a huge eye on the hind wing facing the view. Where otherwise there would be practically nothing to be seen, this single eye goggles out of the darkness of the underbrush, lighting up rather devilishly whenever an errant sunbeam strikes it. Drawings of such butterflies frequently show them with extended wings, in which position they give a genuinely owl-like two-eyed effect. But in life the insects always rest with closed wings and always fly with rapid wing movements, so that the two eyespots are never visible simultaneously.

A magnificent single-eyed design is also found in certain praying mantids. It attains a special intensity of colouration through the superposition of the mantid's semitransparent wing covers, each of which bears the same design. But when the mantid assumes its defensive posture, with wings spread, suddenly two "eyes" instead of one appear. In many insects, especially beetles, the effect of eyes is produced by two spots set side by side somewhere on the body and more or less contrasting with it, or often even raised in relief.

**Movement and Design**

How great is the alarming power of eyespots and other spectacular designs? Only experiments can provide the answer, and many of thee will be difficult to carry out convincingly. Yet it is surely obvious that effectiveness is intended wherever the insect, by means of body movements, participates actively in the display of these designs. This happens in the majority of cases. For instance, a sphinx moth caterpillar pulls in its head, puffs out it thorax like a balloon, and exposes on the swollen surface the eyespots that lie on the two sides. or it lifts the front part of its body and writhes around its own axis as a means of displaying the startling designs on the lower surface of its thorax. To be sure, the general basis of such movements lies in reactions typical of certain families of these caterpillars. Even the species that lack special alarm designs raise the forepart and pull in the head. And yet a last item of behaviour is missing, such as the inflation of the body of the twisting motion, which makes sense only in combination with the appropriate patterns.

The behaviour of adult sphinx moths also illustrates an evolving series of defensive movements coordinated with colouration. Before beginning preparations for flying almost all of these moths react to disturbances with rhythmically repeated elevations of the body. The middle of the abdomen arches upward and displays, from the midst of its muted gray or brown protective garb, the strikingly contrasting colour pattern of the back. Other species, especially those that have bright display colours on the hind wings, simultaneously raise the wings as they raise the body. When the eyed sphinx moths (*Smerinthus*) follow this pattern, their blueringed eyespots, twitching in the rhythm of the body movements, are set off against the brightly shining red background of the exposed part of the hind wing.

Another peak performance is given by the death's-head moth (*Acherontia*), which takes its name from markings resembling a human skull on the forepart of its back. This satinydark, massive shadow of the night rears up in a startling, demoniac dance. The effect is produced not only by the death mask, but also by the uncanny impression given by the straddling, suggstively shaped forelegs, and a continual musical accompaniment, which sounds something like creaking leather. Over the middle of the abdomen which is raised again and again, runs a blue band; and the ochre-yellow sides, like the hind wings which now come into view, are waspishly striped in black. A truly fantastic, unique phenomenon of the insect world!

However, the movement and dances that insects use to defend themselves, that harmonize so astoundingly with shape and design, cannot be evoked at will. Rather, they quickly grow weaker with repeated annoyance, and instead and insect reacts by running away. Only after a rather long rest period can the dances be reactivated in their known as seasonal dimorphism, appears most impressively in certain butterflies both in temperate and tropical zones. The tropical butterflies have forms characteristic of the dry and rainy seasons and the temperate butterflies have forms characteristic of spring and summer. The successive generations of the same species are so unlike that they have at times been described as completely different species. The summer, and dry season forms are generally smaller and have different, brighter colours, but a less conspicuous pattern. One generation, usually the spring or wet season form, is typical of the species.

### Individual Variation

No two individuals of a species are ever alike in all particulars. There i always variation among individuals. They extent of the variation is slight for many species, but for others the extremes are so far apart that they sometimes have been described as two or more species. The nature of individual variants is such that they might all be offspring of a single pair.

### Abnormal Colouring

Occasional individuals are abnormal: they have suffered a disturbance in development or some other pathological damage. For instance the interruption of the ordinary course of colouration by heat or cold, dampness, or deficiency of light may cause a reddish pigmentation to replace the normal dark or black. This condition, called rufinism, may even become the rule for insects at high altitude, in cave, or under other extreme conditions. Again, external influences may cause the body pigments of an insect to whiten in various regions of the body and the wings. This effect is called albinism. Albinism also may occur in some insect as an inherited abonrmality.

Abnormal colouring may also be nonpathological. An example of this flavinism, the shift from a normally reddish colour toward yellow. Another is the change toward a darker form that crops up in insects again and again. This change shows itself in two ways. The dark pigments may come to predominate more and more, covering the entire surface like a veil and yet letting the underlying pattern shine through. This is melanism. Or else the black parts of the original pattern may expand to the point where they fuse, leading frequently to a complete blackening, known as nigrism. Flavinism, melanism, and nigrism, though often influenced by cold and dampness, are inherited abnormalities of insects. They are caused by mutations, those abrupt changes that originate not as the result of external influences but because of sudden alterations in the genetic material. By their very nature, mutations are of the utmost significance in investigations of the origins of species.

**Geographic Variation**

With a few exceptions, most of the widely distributed insect species have developed more or less clearly distinct local forms called geographic races, or subspecies. These races have arisen often quite abruptly through mutation. They are shape by the totality of the climatic, geologic and faunistic relationships peculiar to their range, and mingle with each other at most along the boundaries of the areas they occupy. Because there is some interbreeding here, the races seldom become completely distinct from one another. But such interbreeding at the margins cannot occur in races that occupy widely separated islands, except in the few species with extraordinary powers of flight; consequently these inhabitants of islands have gone far in their special, peculiar modifications and may even have become so different that they have attained the status of distinct species. Even on the mainland on finds sharply isolated localities, each with its own strikingly modified insect races. These localities include desertoases, mountains, isolated bodies of water, and distinctive soil formations with their special types of vegetation.

Not surprisingly in places of this sort a majority of the species, although they are from different families of insects, have certain features in common. Most desert insect display the light colour of the sand. Nordic and alpine species share the distinction of a long, thick, protective covering of hairs. And the males of certain butterflies and moths of both Australia and the Sahara Desert possess enormously enlarged antennae, with a consequent enlargement of olfactory end organs, that help them locate their females over extraordinary distances.

Somewhat less understandable is the fact that in certain countries numerous species of insects have experienced a striking coarsening of their sculptured pattern. And then there is the problem that never does every one of the species develop the adaptation common to the majority.